ONE- AND TWO-DIMENSIONAL FLUIDS
Properties of Smectic, Lamellar and Columnar Liquid Crystals

Series in Condensed Matter Physics

Other titles in the series include:

Theory of Superconductivity: From Weak to Strong Coupling
A S Alexandrov

The Magnetocaloric Effect and its Applications
A M Tishin, Y I Spichkin

Field Theories in Condensed Matter Physics
Sumathi Rao

Nonlinear Dynamics and Chaos in Semiconductors
K Aoki

Permanent Magnetism
R Skomski, J M D Coey

Modern Magnetooptics and Magnetooptical Materials
A K Zvezdin, V A Kotov

Series in Condensed Matter Physics

ONE- AND TWO-DIMENSIONAL FLUIDS

Properties of Smectic, Lamellar and Columnar Liquid Crystals

A Jákli
Kent State University, Ohio, USA

A Saupe
Kent State University, Ohio, USA

CRC Press
Taylor & Francis Group
Boca Raton London New York

CRC Press is an imprint of the
Taylor & Francis Group, an **informa** business

A TAYLOR & FRANCIS BOOK

CRC Press
Taylor & Francis Group
6000 Broken Sound Parkway NW, Suite 300
Boca Raton, FL 33487-2742

First issued in paperback 2019

ISBN-13: 978-0-7503-0969-1 (hbk)
ISBN-13: 978-0-367-39076-1 (pbk)
Library of Congress Card Number 2005035595

Library of Congress Cataloging-in-Publication Data

Jákli, Antal.
   One- and two-dimensional fluids : physical properties of smectic, lamellar, and columnar liquid crystals / Antal Jákli and Alfred Saupe.
        p. cm. -- (Series in condensed matter physics)
   Includes bibliographical references and index.
   ISBN-13: 978-0-7503-0969-1 (acid-free paper) -- ISBN-10: 0-7503-0969-5 (acid-free paper)
   1. Liquid crystals--Research. 2. Fluid dynamics--Research. I. Saupe, Alfred, 1925- II. Title. III. Series.

QC173.4.L55J35 2006
530.4'29--dc22   2005035595

Visit the Taylor & Francis Web site at
http://www.taylorandfrancis.com

and the CRC Press Web site at
http://www.crcpress.com

Preface

Self-organized anisotropic fluids such as smectic, columnar and lamellar liquid crystal phases are considered to be the next "frontier" in liquid crystal research, along with lyotopic systems for optical, biological or biomedical applications. These systems are basically the same as those of the well-known soaps, cell membranes, aqueous solutions of viruses and biopolymers, so it seems logical to treat all of them together in one book. Researchers preparing to embark on activities in this area need to have an up-to-date source of reference material to establish a solid foundation of understanding. Before this book, the only way to accomplish this goal was by extensive literature searches and reading a broad range of review articles.

Our aim is to provide both a solid foundation of fundamental concepts for students and those beginning research in the area as well as little-known facts and historical perspectives that will be of value even to those having extensive experience in this area.

The book treats mainly physical properties of liquid crystals, soaps, foams, monolayers and membranes; however, it is intended to be understandable for people with various science backgrounds, such as physics, chemistry, biology and optical engineering. To match the different backgrounds, we added four appendices that need to be read only by those who are not familiar with some basics required to comprehend the nine chapters. Appendix A briefly summarizes the basics of organic chemistry in the area of hydrocarbons and surfactants, which are components of the materials we are describing. It is mainly for those who have no chemistry background (most physicists and optical engineers). Appendix B summarizes the most important expressions and ideas in the interdisciplinary area of rheology. This is probably not needed for physicists, but would be helpful for others in understanding Chapter 4 and Chapter 9. Appendix C describes how symmetry arguments can be used to determine the relevant curvature elastic constants that are otherwise described in Chapter 4 and the piezoelectric constants that are described in Chapter 8. Appendix D describes the main concepts of dielectric spectroscopy and measurements that have importance in all areas of structured fluids.

Chapters 1 and 2 introduce the main phases and basic properties of liquid crystals and other anisotropic fluids, such as soaps, foams, mono-layers, fluid membranes and fibers. These chapters do not include difficult mathematical formulas and are probably suitable for undergraduates or for other professionals, such as K–12 teachers. Chapter 3 describes the nature of phase transitions based on the phenomenological Landau–de Gennes theories, and on the self-consistent mean-field theories that use concepts in statistical physics.

In Chapter 4 we describe, on a basic level, the continuum mechanical properties of liquid crystals, soaps and foams. Chapter 5 introduces the main tools for understanding the optical properties of anisotropic fluids, at least those that can be detected under a polarizing microscope. Due to the introductory nature of the book, we completely omit nonlinear optics and light scattering studies that would require much more advanced studies. Although after reading Chapter 5, everyone should be able to interpret colors, extinction directions, conoscopic or confocal fluorescence microscopy pictures of uniformly aligned anisotropic films; one needs to read Chapter 6 to understand the beautiful textures of anisotropic fluids with nonuniform alignment. A complete understanding of defect structures requires much more advanced analysis than included in the book; however, the most typical features of various liquid crystal phases are described. The shortest chapter of the book deals with the basics of the magnetic properties of organic anisotropic materials. They are considered to have importance only in basic research (magnetic interactions are weak and are usually easy to interpret); however, this may change in the future, especially by the developments of ferrofluids in liquid crystals. Chapter 8 describes electric interactions, which have tremendous importance in understanding various display modes, and some possible biological phenomena in connection with electric transport properties of cell membranes. Due to the myriad of current and forecasted applications, the most important examples of technological applications are described in a separate Chapter 9 at the end of the book. These examples show only the tip of the iceberg; the rest certainly will be revealed in the not-so-distant future.

This book basically covers the material of two three-credit-hour entry-level graduate courses, which were actually taught several times at the Chemical Physics Interdisciplinary Program of Kent State University. Feedback from the graduate students who took these courses was really important to help us judge how deeply we needed to go in some areas and still be able to cover the most important phenomena needed for students to do research in the field. Although the references are far from complete, we think they are detailed enough to help readers find the most important sources needed to go into detail on any subject in the area of self-organized fluids.

We are grateful to a number of our colleagues and friends (Helmut Ringsdorf, Oleg Lavrentovich, Antonio Figueiredo-Neto, Istvan Janossy, Nandor Eber, Elizabeth K. Mann, Samuel Sprunt, James T. Gleeson and Daniele Finotello) for their suggestions and comments. We also wish to thank all the graduate students in the Chemical Physics Interdisciplinary Program of Kent State University between 2001 and 2004 for their feedback and corrections.

Antal Jákli
Alfred Saupe

Contents

1

Liquid Crystal Materials

If people are asked for the different states of materials, most will only know the solid, the liquid and the gaseous states. But often in nature, the borders between the different categories are not well defined. Ordinary fluids are isotropic in nature; they appear optically, magnetically, electrically, etc., to be the same from any perspective. The liquid crystal state is a distinct state of matter observed between the crystalline (solid) and isotropic (liquid) states: they have some of the ordering properties of solids, but they flow like liquids. Liquid crystals represent a unique segment of soft matter, where the orientational order and mobility have delicate balance in determining the macroscopic properties.

We will see that, in addition to the anisometric shape, a kind of amphiphilic nature of the molecules is also needed to give rise to the unusual, fascinating and potentially technologically relevant structured fluidity. People used to distinguish the so-called "thermotropic" and "lyotropic" liquid crystals. In the thermotropic liquid crystals, the shape (rod[1], disk[2], pyramid[3], or banana[4,5]) of the molecules dictates the orientational order, and the thermal motion gives the mobility. Lyotropic liquid crystals that appear in nature in living organisms[6] acquire mobility by addition of a solvent, and their liquid crystalline properties are governed by the relative concentration of the solute. However, the distinction between the thermotropic and lyotropic liquid crystals is not complete and there are materials which exhibit both thermotropic and lyotropic liquid crystalline properties. They are called amphotropic.[7,8]

1.1 Thermotropic Liquid Crystalline Materials

Thermotropic liquid crystalline materials have been observed for over a century, but were not recognized as such until the 1880s. In 1887, Otto Lehmann used a polarizing microscope with a heated stage to investigate the phase transitions of various substances. He found that one substance would change from a clear liquid to a cloudy liquid before crystallizing, but he thought this was simply an imperfect phase transition from liquid to crystalline. In 1888, Friedrich Reinitzer (see figure)

wrote the first systematic report of the phenomena when he prepared cholesteryl benzoate (the first liquid crystal). He has consequently been given the
credit for the discovery of the liquid crystalline phase,[9] although Otto Lehman
was the first to suggest that this cloudy fluid was a new phase of matter.[10]
Until 1890 all the liquid crystalline substances that had been investigated
were naturally occurring, and it was then that the first synthetic liquid
crystal, p-azoxyanisole (PAA), was produced by Gatterman and Ritschke.
Subsequently more (over 100,000) liquid crystals were synthesized, and it is
now possible to produce liquid crystals with specific predetermined material
properties.

Thermotropic liquid crystals are composed of moderate-size (~2–5nm)
organic molecules, which are strongly anisometric: elongated and shaped
like a cigar (so-called calamitic liquid crystals), disc-shape (discotic LCs), or
bent-shape (pyramidal or banana-shape).

Although the literature is full of a variety of highly exotic shapes that
differ from the cigar shape, still a common cartoon of structure of liquid
crystals shows them as rigid, uniform rods. Generally, in isotropic phases
the molecules will be oriented arbitrarily without any long-range positional
and orientational order, as shown in Figure 1.1a. Because of their elongated
shape, under appropriate conditions the molecules can exhibit orientational
order, such that all the axes line up in a particular direction and form a so-
called nematic liquid crystal state. The molecules can still move around in
the fluid, but their orientation remains the same. The average direction of the
molecules is called the director. Depending on the type and shape of the
molecules and the conditions they find, not only orientational order can
appear, but also a positional order is possible (smectic and columnar
phases). In smectics, the orientation is fixed and the movement is limited
to layers (Figure 1.1c).

1.1.1 Nematic Phases

By decreasing the temperature from the isotropic phase, in which the molecules are randomly positioned and oriented (see Figure 1.1a), to the nematic
phase, the material gains some orientational order but no long-range positional order (see Figure 1.1b). This reordering is thought to be due to the
packing constraints of the molecules. This claim is supported by the fact that
most liquid crystal molecules tend to be long, thin molecules with a rigid
central region. Although the packing constraints (steric interactions) are
necessary, they are not sufficient conditions for the appearance of the nematic
phase. This orientational order allows us to define an average direction of
the molecules called the director and denoted by the vector $\mathbf{n}$. The material
is still a fluid, but at each point $\mathbf{r}$ of this fluid the molecules prefer to orient
along $\mathbf{n}(\mathbf{r})$. Thus the material is anisotropic.

Another important variable in nematic liquid crystals is the order parameters, which measures how the molecules are aligned with the director. It is

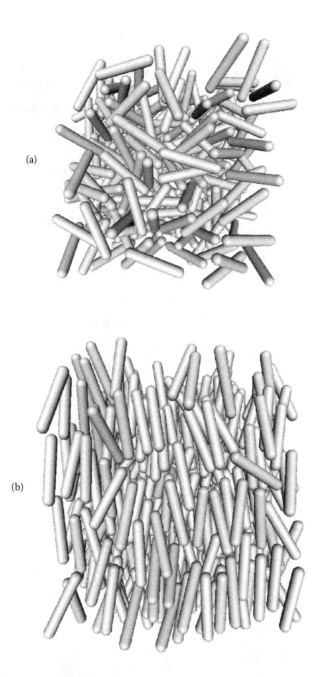

(a)

(b)

FIGURE 1.1
Usual illustration of the liquid crystals that form nematic (b) and smectic (c) phases upon cooling
from the isotropic liquid (a) and above the crystal structure (d).

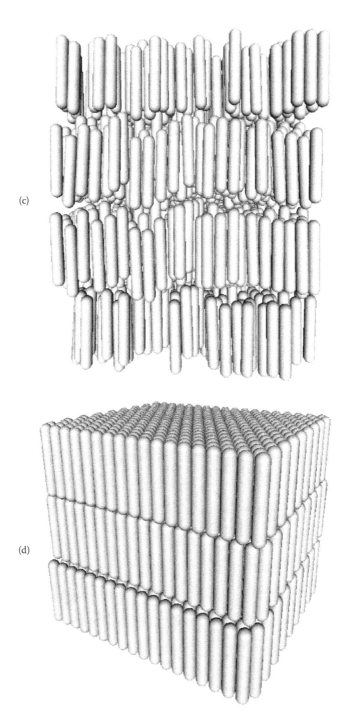

(c)

(d)

FIGURE 1.1
(Continued)

Structure	Name	Phase sequence (in °C)
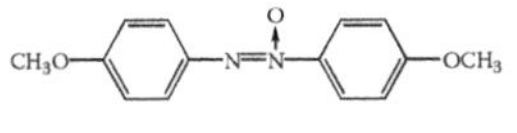 CH₃O—⟨⟩—N=N—⟨⟩—OCH₃ (with O on N)	*p,p'*-Azoxyanisole (PAA)	Cr 117 N 137 I
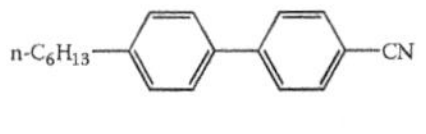 n-C₆H₁₃—⟨⟩—⟨⟩—CN	*p-n*-Hexyl-*p'*- cyanobiphenyl	Cr 14 N 28 I
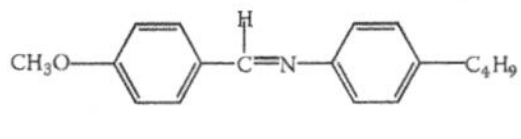 CH₃O—⟨⟩—C=N—⟨⟩—C₄H₉	*p*-methoxybenzylidene- *p-n*-butylaniline (MBBA)	Cr 21 N 45 I
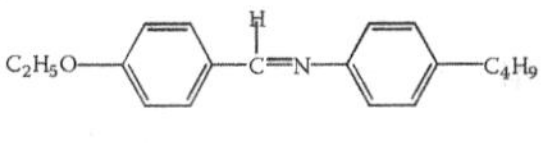 C₂H₅O—⟨⟩—C=N—⟨⟩—C₄H₉	*p*-ethoxybenzylidene-*p-n*- butylaniline (EBBA)	Cr 36 N 80 I
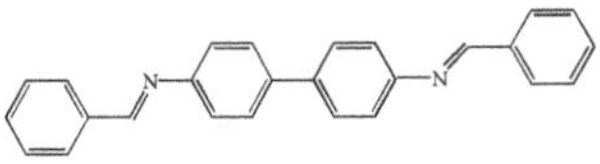 ⟨⟩—N=⟨⟩—⟨⟩—N=⟨⟩	4,4'-bis-(benzylideneamino)- biphenyl	Cr 234 N 260 I

FIGURE 1.2
A few examples of nematogens. Cr means crystal, I stands for isotropic, and N denotes the nematic phase. PAA, MBBA, and EBBA were the most studied materials until the appearance of the cyanobiphenyls, which are mostly used in displays due to their increased stability (an example is shown in row #2). The bottom row shows an example of the rigid molecules without flexible end chains. This and similar materials have nematic phase only at very high temperatures.

defined so that $S = 0$ in the disoriented isotropic fluid phase, and $S = 1$ when the orientational order is perfect. A suitable choice for the order parameter will be discussed in Chapter 3.

Molecules that form only nematic phase have shapes that are closest to the cigar cartoon. Examples for nematogens with rigid rod-shape are indeed observed at very high temperatures (see last row in Figure 1.2). However, it is important to emphasize that liquid crystalline states (especially at moderate temperatures) usually require the combination of rigid aromatic parts (typically more or less rigidly connected benzene rings), which are terminated by flexible hydrocarbon (typically alkyl or alkyloxy) chains (see Figure 1.2).

For those who are not expert in organic chemistry and nomenclature, in Appendix A we give a very brief summary about the basic definitions most important in the area of liquid crystals. As a review of over 100 years of chemistry of liquid crystalline materials, we refer the paper by Demus.[11] For those who would like to read more about the relation of the molecular structures and the liquid crystalline properties, we recommend the classical book of Gray.[12]

1.1.2 Smectic Phases

At certain temperatures, generally below the nematic or sometimes directly below the isotropic phase, the liquid crystal material may gain an amount of positional order. When this happens, the liquid crystal forms a smectic phase where the molecules, although still forming a fluid, prefer to lie, on average, in layers. Within each layer the liquid crystal is essentially a two-dimensional nematic liquid crystal. The word "smectic" is derived from the Greek word for soap. This seemingly ambiguous origin is explained by the fact that the thick, slippery substance often found at the bottom of a soap dish is actually a type of smectic liquid crystal.

There are many types of smectic phases. When the nematic-like director in each layer is parallel to the layer normal the material is smectic A (SmA). In the smectic-C (SmC) mesophase, observed first by Hermann,[13] molecules are arranged as in the SmA, except that the director tilts away from the layer normal. In the subsequent layers, the tilt can show to the same direction (syn-clinic), or it can alternate (anti-clinic) (SmC$_A$) as illustrated in Figure 1.3.

There are also smectics (typically at lower temperatures) which form layers with positional ordering *within the layers*. For instance, in smectic-B (SmB) materials, the molecules, on average, are parallel to the layer normal, just like in smectic A, but in addition, the molecules in each layer have short-range hexagonal ordering, with bond angles showing a long-range order. Similarly there are tilted smectic phases with short-range in-layer positional ordering and long-range bond order. Depending on whether the bond order is parallel with the projection of the tilt angle (c-director) or they make some

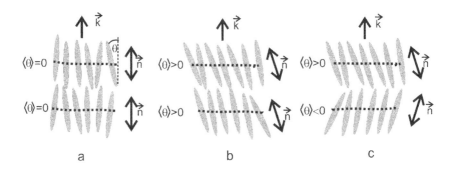

a   b   c

FIGURE 1.3
Schematic illustration of the 2D fluid smectic phases. (a) SmA; (b) syn-clinic SmC; (c) anti-clinic SmC$_A$.

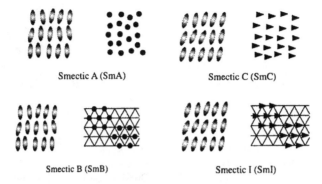

Smectic A (SmA)   Smectic C (SmC)

Smectic B (SmB)   Smectic I (SmI)

FIGURE 1.4
Types of smectic phases. Here the layer stacking (left) and in-plane ordering (right) are shown for each phase. Bond orientational order is indicated for the SmB and SmI phases, i.e., long-range order of lattice vectors. However, there is no long-range translational order within the layers in these phases.

angle with each other, we distinguish several phases and call them SmI or SmJ phases. A few examples of smectics with no in-layer or hexagonal in-layer bond ordering are shown in Figure 1.4.

On further cooling, long-range positional and bond-orientational order develop sequentially in the smectic planes, eventually leading to true 3D crystalline structure. For nontilted orientation of molecules with respect to layer normal, these phases are crystal-*B*, *CrB*, and smectic-*E*, *SmE* or *CrE*. In the tilted case the phases are smectic-*G*, -*H*, -*J*, -*K*, denoted *SmG*, *SmH*, *SmJ*, *SmK*, or frequently also *CrG*, *CrH*, *CrJ*, and *CrK*, respectively. See Table 1.1 for the distinguishing features of ordering in these phases. The hexatic phases are very special class of smectics characterized by long-range bond orientational order but short-range positional order within the layers. The hexatic-*B*, *HexB*, and smectic-*I*, -*F*, *SmI* and *SmF* or also *HexI* and *HexF*, are examples.

In the following, we will concentrate only on those smectics that are fluid along the layers, i.e., which can be considered as stacks of 2D fluids.

These pure geometrical and static pictures of the smectic orderings consider that molecules are simple rods and do not specify the position of the aromatic cores with respect to the aliphatic chains. They also neglect thermal agitation and possible diffusion of molecules through the material, especially across the layers. In other words, for rod-shape molecules, it is hard to assume a perfect lateral register. Experiments indicate that the positional ordering across the layers is often indeed small, and the density of the centers of mass of the molecules can be approximated by a simple sinusoidal distribution:

$$\rho(z) = \rho_o(1 + \psi \cos(2\pi z/d)) \qquad (1.1)$$

TABLE 1.1

Overview of the Smectic Phases with Different In-Plane Positional and Bond Orders

Phase Name	Phase Type	Orientational Order	Positional Order	Bond Orientational Order
Smectic–A Sma	Fluid	Long range ⊥ to layers	Short range in layers Quasi long ⊥ to layers	Short range in layers
Smectic–C SmC	Fluid	Long range ⊥ to layers	Short range in layers Quasi long ⊥ to layers	Short range in layers
Smectic–B SmB, HexB	Hexatic	Long range ⊥ to layers	Short range in layers Quasi long ⊥ to layers	Six fold Long range in layers
Smectic–D SmD	Plastic Crystal	None	Isotropic and cubic crystalline coexistence	None
Smectic–E SmE, CrE	Crystal	Long range ⊥ to layers	Long range in layers Long range ⊥ to layers	Six fold Long range in layers
Smectic–F SmF, HexF	Hexatic	Long range ⊥ to layers	Short range in layers Quasi long ⊥ to layers	Six fold Long range in layers
Smectic–G SmG, CrG	Crystal	Long range ⊥ to layers	Long range in layers Long range ⊥ to layers	Six fold Long range in layers
Smectic–H SmH, CrH	Crystal	Long range ⊥ to layers	Long range in layers Long range ⊥ to layers	Six fold Long range in layers
Smectic–I SmI, HexI	Hexatic	Long range ⊥ to layers	Short range in layers Quasi long ⊥ layers	Six fold Long range in layers
Smectic–J SmJ, CrJ	Crystal	Long range ⊥ to layers	Long range in layers Long range ⊥ to layers	Six fold Long range in layers
Smectic–K SmK, CrK	Crystal	Long range ⊥ to layers	Long range in layers Long range ⊥ to layers	Six fold Long range in layers
Smectic–L SmL, CrB	Crystal	Long range ⊥ to layers	Long range in layers Long range ⊥ to layers	Six fold Long range in layers

where z is the coordinate parallel to the layer normal, the average density of the fluid is ρ_o, d is the distance between layers, and ψ is the order parameter. When $|\psi| = 0$, there is no layering, and the material is in nematic phase. For $|\psi| > 0$, some amount of sinusoidal layering exists, and the material is smectic. Such a density function would result in a single Bragg peak of the X-ray scattering (see Chapter 2) and a nearly second-order nematic–smectic transition, although sometimes the phase transition is of first order, and the higher-order Bragg harmonics are also observed.[14]

Structure	Name	Phase sequence (°C)
Smectic A 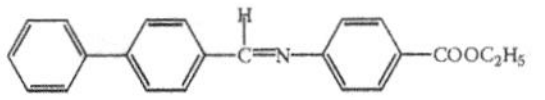 ⬡⬡-C(H)=N-⬡-COOC₂H₅	Ethyl-*p*- (*p'*-phenylbenzalamino)- benzoate	Cr 121 SmA 131 I
Smectic A 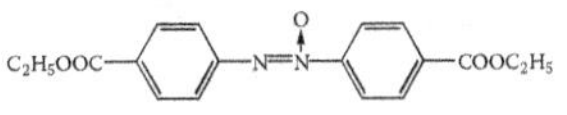 C₂H₅OOC-⬡-N=N(O)-⬡-COOC₂H₅	ethyl *p*-azoxybenzoate	Cr 113.7 SmA 122.5 I
Smectic A 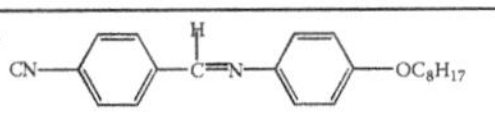 CN-⬡-C(H)=N-⬡-OC₈H₁₇	*p*-cyanobenzylidene-*p'* -n-octyloxyaniline (CBOOA)	Cr 73 SmA 82.6 N 108 I
Smectic B 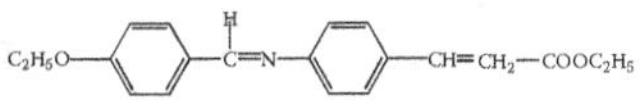 C₂H₅O-⬡-C(H)=N-⬡-CH=CH₂-COOC₂H₅	Ethyl-*p*-ethoxybenzal-*p'*- aminocinnamate	Cr 77 SmB 116 I
Smectic C 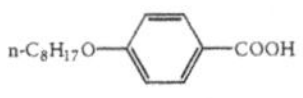 n-C₈H₁₇O-⬡-COOH	*p*-*n*-octyloxybenzoic acid	Cr 108 SmC 147 I
Smectic C 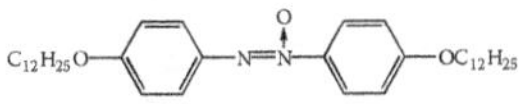 C₁₂H₂₅O-⬡-N=N(O)-⬡-OC₁₂H₂₅	*p.p'*-di-*n*- dodecyloxyazoxybenzene	Cr 81 SmC 122 I
Smectic F and G 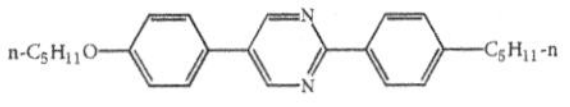 n-C₅H₁₁O-⬡-⬡(N,N pyrimidine)-⬡-C₅H₁₁-n	2-(*p*-Pentylphenyl)-5- (*p*- pentyloxyphenyl)pyrimidine	Cr 79 SmG 102.7 SmF 113.8 SmC 144 SmA 210 I

FIGURE 1.5
Examples for molecules forming smectic phases.

Smectogenic molecules, which are typically built from semiflexible rod-like cores of phenyl rings linked together by more or less flexible links such as: COO, $-CH = N-$, $-CH = CH-$, CH_2-CH_2-,[12] and one or two paraffinic chains whose length is usually 6 to 20 carbon atoms grafted in para positions. Examples for smectogens are seen in Figure 1.5.

Experiments indicate that the aromatic core and the aliphatic chains of the molecules have distinct volume behavior, which means nanophase segregation.[15] Accordingly, the 2D fluid smectic structure results from the alternate stacking of aromatic and aliphatic sublayers, separated by interfacial regions much thinner than of the layer spacing. Such a behavior is similar to amphiphiles and block copolymers, which will be discussed in the following sections.

Measurements typically show that the layer spacing d is generally less than the length of the fully extended molecules l.[16] There are basically three different models that can explain such discrepancy.[17]

The first of these models is called "diffuse cone model" and takes into account the orientational disorder ($S < 1$). The molecular tilt caused by orientational disorder first was realized by Leadbetter[18] and A. de Vries.[19] In case of orientationally disordered rigid rods with the average tilt of θ, we can write that

$$d = l\langle\cos\theta\rangle \tag{1.2}$$

This expression defines the tilt angle θ, which then would be the same both from X-ray and optical measurements. In general, however, the tilt angles defined by X-ray and optical measurements are not equal, because the molecules are not completely rigid, and because the optical signal is sensitive mainly to the rigid part of the molecules.

Expanding (10.2) to the second order in θ, we get:

$$d \approx l\left(1 - \frac{\langle\theta^2\rangle}{2}\right) \tag{1.3}$$

As the temperature decreases, $\langle\theta^2\rangle$ increases; consequently the layer spacing decreases in accordance with the X-ray observations.

As we discussed previously, the molecules cannot be considered as rigid rods, but they are the combination of terminal chains with large mobility and of the rigid aromatic core with lower mobility. Based on this notion, Diele et al.[16] proposed the "zig-zag" model. Contemplating that the rigid rods are more ordered than the tails, the molecules may exhibit a kinked conformation with their cores orthogonal, but the end tails tilted in the SmA phase.

In addition to these models, the degree of positional order may have a large impact on the SmA layer geometry. If one allows a relatively large degree of molecular interdigitation between adjacent layers, $d < l$ becomes possible. Actually, this was de Vries' first proposal to explain the d–l discrepancy, but he later rejected it, when he realized that the diffuse cone model can explain the situation in a less *ad hoc* way.[19]

At the transition to the SmC phase, the layers usually shrink just as we would see even from the simplest cartoon. However, in a few examples the layers do not shrink at all upon the transition to the SmC phase. Such transitions can be explained by assuming that the director is already tilted in the individual layers, but the tilt direction is not correlated in the subsequent layers. In these models, it is assumed that only the azimuthal distribution of the molecules becomes asymmetric, and the layer spacing may not be affected by the transition.[20]

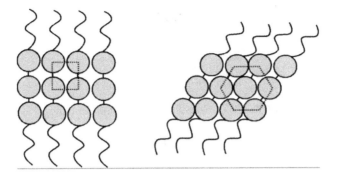

FIGURE 1.6
Schematics of the arrangement of the more realistic SmA and SmC phases. It can be seen that the tilted SmC phase has a closer packing (hexagonal) than in the orthogonal smectic A (rectangular). These illustrations, of course, assume short-range in-plane packing, which is still compatible with the macroscopic fluid nature.

Different kinds of interactions have been proposed to explain the tilt. Meyer and McMillan[21] had underlined the importance of dipolar interactions, but it appears that the molecules undergo a quasi-uniaxial rotation along their long axis, and dipolar interactions are nearly cancelled, except if the molecules are chiral.[22] It has also been argued by Barbero and Durand[23] that the gradient of the quadrupolar interactions due to the aforementioned different order of the cores and the tails induces dipoles. The depolarizing field associated with these dipoles minimizes the energy by tilting the molecules. One can also consider that steric interactions exist between the semiflexible rod-like cores. The aromatic cores can be visualized by beads of phenyl rings connected by slim flexible links. A model of such structure with several neighbor units is shown in Figure 1.6. It is seen that the closest packing of these molecules is realized for tilted structures (old honeycomb problem). The effect of steric interactions becomes important at lower temperatures, when the fluctuations are not able to wash out these details. Note that this picture could explain the occurrence of in-plane bond order and imply that each tilted smectic phase should possess some kind of bond order. This is in general true, although typically the bond order is much less pronounced than the director tilt. This indicates the importance of the dipole–dipole interactions proposed by McMillan,[24] which will be effective once the tilt occurs due to steric interaction sketched in Figure 1.6.

1.1.3 Columnar Liquid Crystals

In addition to the nematic phase (termed N_D), disc or bowl-shaped[25] molecules may stack in layers (discotic smectic), or much more often, to columns forming so-called columnar (Col) discotic or columnar pyramidic

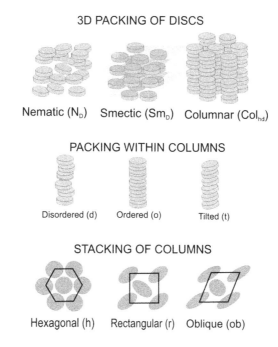

FIGURE 1.7

Schematic illustration, classification and nomenclature of discotic liquid crystal phases.[27] For the columnar phases, the subscripts are usually used in combination with each other. For example, Col_{rd} denotes a rectangular lattice of columns in which the molecules are stacked in disordered manner.

phases, respectively. A simplified picture of columnar liquid crystals[26] is that they are liquid-like along the columns and solid-like in two-dimensional array of columns. Due to this packing, some columnar liquid crystals can be considered as 1D fluids (fluid along the columns, but solid in the plane normal to the columns). Discussions of columnar ordering were initiated in 1977, when Chandrasekhar et al. first reported the occurrence of liquid crystals with flat discotic organic molecules.[2] In the last two decades, several types of columnar phases have been observed in addition to the N_D phase (see Figure 1.7).

In the Col_{hd} phase, there is a disordered stacking of discotic or bowl-shape molecules in the columns, which are packed hexagonally. Hexagonal columnar phases, where there is an ordered stacking sequence (Col_{ho}) or where the mesogens are tilted within the columns (Col_t), are also known. It should, however, be noted that the individual columns are one-dimensional stacks of molecules, and long-range positional order is not possible

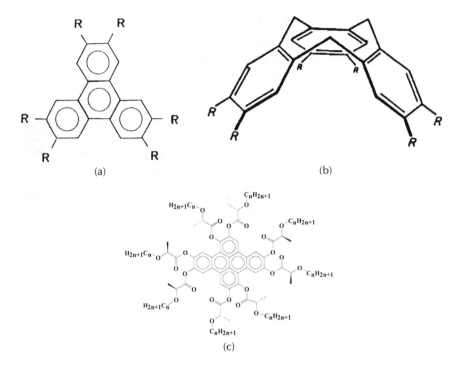

FIGURE 1.8

Discotic and bowl-shaped molecules that form columnar phases. (a) Hexa-alkanoate [R=CH$_3$-(CH$_2$)$_n$-2COO-] of triphenylene;[29] (b) hexa-noyloxy-tribenzocyclononene,[30] which forms a polar pyramidic phase; (c) 1,2,5,6,8,9,12,13-octakis-((S)-2-heptyloxy) dibenzo[e,1]pyrene, which has chiral therminal chains and forms tilted ferroelectric columnar liquid crystal.[31]

in a one-dimensional system due to thermal fluctuations.[28] Accordingly there is no sharp distinction between the ordered and disordered phases. Phases where the columns have a rectangular (Col$_r$) or oblique packing (Col$_{ob}$) were also observed.

Substances by disk or bow-shaped molecules are formed of a flat or cone-shape aromatic core surrounded by several peripheral aliphatic chains as shown in Figure 1.8. It is clear that the column is built with the same constraints as the lamellae in a smectic: the cores are parallel, a segregation between chains and cores takes place, and the paraffins are in a melted state, which insures the decoupling between adjacent columns. In comparison between the columnar and smectic phases, we note that the weight of the chains is much larger than in smectics, and the physical properties are more governed by the behavior of the chains than by the ordering of the cores inside the columns.

1.1.4 Chiral Organic Materials

So far, we have considered only nonchiral molecules, which have a plane of mirror symmetry, so the molecule is equal to its mirror image. However, there are molecules that are not symmetric when reflected, and they are called chiral. The first studies of the molecular chirality are dated back to the Ph.D. work of Louis Pasteur in 1848, when he observed the chiral separation of the crystals of tartaric and paratartaric acids in the sediments of fermenting wines. However, the definition of chiral objects was given first by Lord Kelvin only in 1893: "An object is chiral, if it cannot be superimposed on its mirror image."

Louis Pasteur

Lord Kelvin

The chirality is extremely important in living systems, because most biomolecules are chiral. For example, glucose, a sugar, exists in D conformation; the opposite enantiomer cannot be used as a food source. Biochemistry is using only one hand and not the other: on Earth the L-amino acid and the D-sugars. The obvious question is why L and not D? The reason could be the parity-breaking weak force, which may make L-amino acids slightly more stable. Finding the same handedness on different planets would therefore support the role of weak force. A case of "wrong" or different hands would teach about the evolution that discriminates one handedness over the other. In fact, a search for extra-terrestrial biology is approached as Search for Extra-Terrestrial Homochirality (SETH). Excess of L-amino acids has recently been found in the Murchison meteorite, suggesting advanced prebiotic chemistry on the parent asteroid or presolar nebula.

Over half of the organic compounds in drugs are chiral, and the different enantiomers have different effect. For example, Ritalin prevents hyperactivity in children in one enantiomer, whereas it has no effect when the opposite enantiomer is used. Another example is thalidomide: one enantiomer of thalidomide can cure morning sickness of pregnant women, whereas the other causes birth defects. Our nose is also sensitive to the chirality of the odor molecules. For example, (+) limonene has an orange smell; the opposite enantiomer smells like lemon. Similarly, (+) carvone smells like mint, whereas (−) carvone has the smell of caraway.

Among organic molecules, chirality is mainly introduced by a so-called stereo-center carbon atom that has four nonequivalent groups bonded to it. The simplest examples to illustrate the differences between the achiral and chiral molecules are propanol, which is achiral, and butanol, which is a chiral molecule. Their simplified structures are illustrated in Figure 1.9.

H
|
H - C - H
|
(a)   H - C - OH   Plane of
| symmetry
H - C - H
|
H

H
|
H - C - H
|
H - C - OH   No plane
(b) | of symmetry
H - C - H
|
H - C - H
|
H

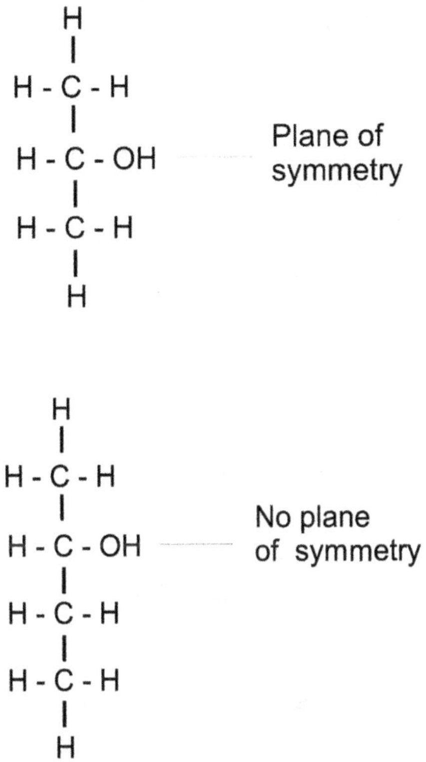

FIGURE 1.9
Illustration of the chirality of organic molecules containing stereo-center carbon. (a) 2-propanol, which is an achiral molecule, and (b) 2-butanol, which is a chiral molecule. The sketches are oversimplified; in reality a carbon atom is tetrahedrally bonded to four atoms and chemical groups.

Molecules containing one or more stereo-centers of the same sign (R or S), or with majority of one sign over the other, are called enantiomers. The sign of enantiomers are labeled interchangeably by R and S or by L (standing for *levo* = left) and D (standing for *dextro* = right). Mixtures of enantiomers in equal portion are called racemates or racemic mixtures.

In liquid crystals, the chirality from molecular level usually transfers to mesoscopic level by forming helical structures. The way in which chiral information is transferred from stereo-centers to molecular, supramolecular, macromolecular and ultimately to mesoscopic dimensions is an intriguing puzzle. For further reading of the problem, we recommend the papers by Lubensky et al.,[32] by Kubal,[33] and by Lemieux.[34]

The nematic phases of chiral molecules (N*) are often called cholesteric (Ch), because the first chiral nematic phases consisted of cholesterol

Structure	Name	Phase sequence (°C)
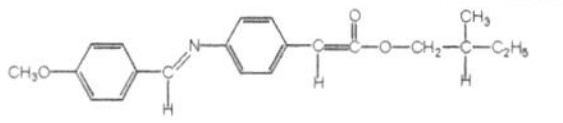	(-)-2-Methylbutyl 4-methoxybenzylidene-4'-aminocinnamate	Cr 76 N 125 I
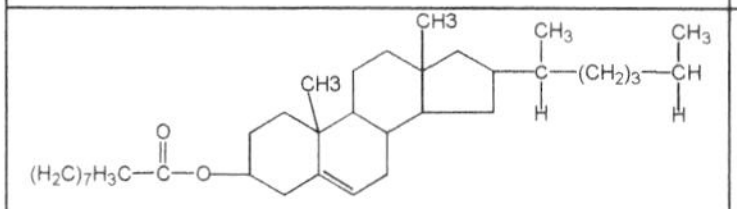	Cholesteryl benzoate	Cr 145 N 179 I

FIGURE 1.10
Typical materials forming chiral nematic (cholesteric) phases.[35]

derivatives. Typical examples of molecules forming chiral nematic (cholesteric) liquid crystals are shown in Figure 1.10.

Structure	Name	Phase sequence (°C)
	(-)-2-Methylbutyl 4-methoxybenzylidene-4'-aminocinnamate	Cr 76 N 125 I
	Cholesteryl benzoate[9]	Cr 145 N 179 I

In the N* phase, the molecules prefer to lie next to each other in a slightly skewed orientation. This induces a helical director configuration in which the director rotates through the material (Figure 1.11). A helix is much like a screw: either left- or right-handed. The mirror image of a left-hand helix is a right-hand helix, and the two cannot be superimposed to each other. This is exactly the criteria of chirality as defined by Lord Kelvin in 1893. We note that in a number of respects cholesteric liquid crystals can be regarded as periodic (2D fluid) systems, where the pitch of the helix determines the layers.

Similar difference is found between the SmC and SmC* phases. In the SmC phase, the molecules tilt in some direction, which is then maintained in the whole domain. In the SmC*, the tilt direction is slightly shifted from layer to layer, and the director is twisted to a helix (see Figure 1.11). We note here that, in case of SmC* materials, the main effect of chirality is not the induction of the helical structure, but the creation of macroscopic polarization in each layer normal to the tilt direction, i.e., the materials will be ferroelectric.[36] This chirality and tilt-induced polarity scenario and the ferroelectricity in liquid crystals will be discussed in detail in Chapter 8.

One can ask how the sign of the molecular chirality would determine the sign of the helical structure. Experimentally, the sign of the helical structure can be determined by the sense of the rotation of the polarization

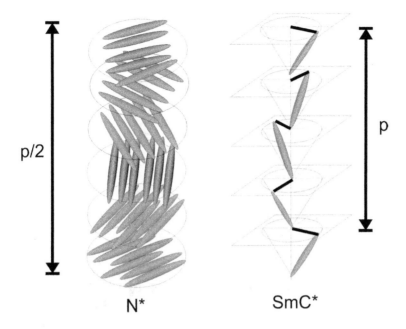

FIGURE 1.11
Cartoon of chiral nematic (cholesteric) and of the chiral tilted smectic (SmC*) structure. The pitch p is the length where the N* director or the projection of the SmC* director onto the smectic layers rotates by 360 degrees.

direction of the light traversing through the medium along the helical axis (the optical rotation will be discussed in Chapter 5). If the polarization direction rotates clockwise, we call it dextrorotatory (d), if rotates in anti-clockwise, we call it laevorotatory (l). Detailed studies of cholesteric materials with single stereo-centers of carbon atoms show that the number of atoms (n) that the chiral center is removed from the rigid central core determines the handedness of the helical structure of the chiral nematic phase. As the atom count (n) by which the chiral center is away from the core switches from odd (o) to even (e), so the handedness of the helix alternates from left to right or vice versa. Similarly, if the absolute spatial configuration of the chiral center is inverted, say from (S) to (R), so the handedness of the helix also reverses. Accordingly, the twist sense of the helix depends both on the absolute spatial configuration and on the position of the asymmetric center with respect to the rigid core.[37] Gray and McDonnell[38] suggested that the spatial configuration of the chiral center of the molecular structure is related to the screw direction of helical structures in cholesteric phases in the following way:

Sol; Red; Rod; Rel.

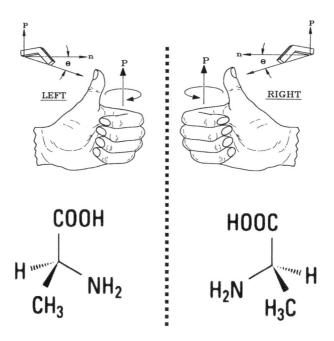

FIGURE 1.12
Illustration of the formation of chiral structures in achiral bent-core molecules if they form polar tilted smectic phase. The molecules below the hands illustrate the similarity to the typical chiral molecules s-aline and r-aline, respectively.

where (S) or (R) is the absolute spatial configuration of the chiral center, (e) or (o) is the even or odd parity for the atom count from the rigid central core, and (d) or (l) refers to the handedness of the helical structure.

A special example of chiral liquid crystals is the tilted smectic phases of bent-core molecules, which can be chiral even though the molecules do not contain any stereo-center carbons. The reason is that a bent-core tilted layer structure with polar order has a three-dimensional structure defined by the polar vector and the tilt direction, which can be illustrated either by our left or right hands of Figure 1.12. We note that the latest investigations indicate other sources of chirality of bent-core molecules, which are termed conformational chirality, due to a propeller type configuration of the two molecular legs.[39]

1.1.5 Bent-Core Liquid Crystals

Although more than 60 years ago Vorländer had already reported the synthesis of bent-shaped liquid crystals,[40] they did not attract much interest until Matsunaga et al.[41] synthesized new mesogenic compounds with "banana-shaped" molecular structures. Only recently it was realized that not only rod-shape (calamitic) or discotic molecules can form liquid crystals, but bent-core (bow-like or banana-shape) molecules can, too.[42]

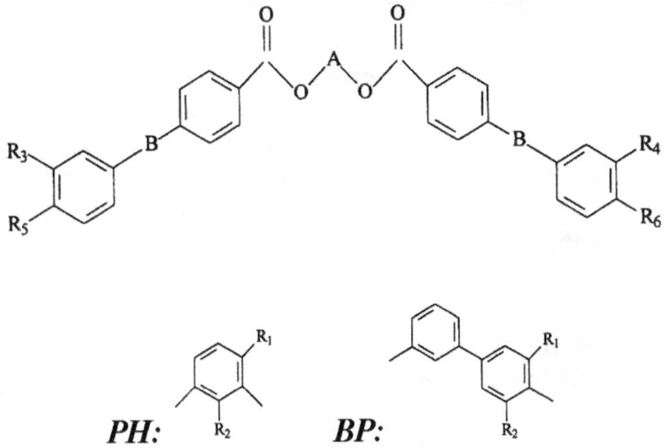

FIGURE 1.13

Structures of typical bent-shape molecules having liquid crystalline mesophases. The central part A of the molecules can be either a phenyl group (Ph) or a biphenyl group (BP). B= -N=CH (NCH) or -OOC-. R_1, R_2, R_3, R_4 are independently hydrogen or a halogen, or cyanide, and R_5 and R_6 are independently C_8–C_{16} alkyl or C_8–C_{16} alkyl-oxy.

Examples of bent-core molecules forming mesophases are shown in Figure 1.13.

In the first few years of experimental studies, seven different "banana liquid crystal" textures have been observed and labeled as B_1, ..., B_7 according to the chronological order of their observations.[43] A list of these phases, based on textural observations and X-ray measurements, is summarized in Table 1.2.

TABLE 1.2

List of Different Phases of Bent-Core Molecules

Name	Structure	Polarity	Director Tilt
B_1	Columnar	Nonpolar	Nontilted
B_2 (SmCP)	Smectic with no in-layer order	Polar	Tilted
B_3	Smectic with hexagonal in-layer order	Polar?	Tilted
B_4	Optically active solid	Polar?	Nontilted
B_5 (SmIP)	Highly viscous SmCP	Polar	Tilted
B_6	Interdigitated layers	Nonpolar	Nontilted
B_{7-I}	Columnar	Polar	Tilted
B_{7-II}	Modulated layer structure (SmCP[44] or SmC$_G$[45])	Polar	Tilted

Note: The B_2 phase was identified as polar tilted smectic (SMCP) phase.[41] Originally only one B_7 phase was proposed, but it turned out that they correspond to at least two different phases (B_{7-I} and B_{7-II}, where B_{7-I} is basically a columnar phase, and B_{7-II} has modulated smectic layer structure.

It is important to note that bent-core smectics have polar order due to the close packing of the bent-core molecules.[47] This packing usually results in a much more ordered smectic structure that manifests itself with high-order Bragg peaks,[48] and strong first-order isotropic smectic transitions.[49] It is also very interesting that the molecular planes mostly become tilted with respect to the layer normal (the reason most likely connected with the five benzene rings), which together with the polar packing lead to chiral layer structures (*layer chirality*) without the need of chiral structure in the molecular level.[50] This represents a tilt + polarity = chirality scenario, which is different from the tilt + chirality = polarity scheme found in SmC* materials.

In these phases, depending on the relative orientations of the two-fold symmetry axis and the tilt direction, the layers can be either right- or left-handed (layer chirality). Provided that the molecules are nonchiral, one expects that the left- and right-handed arrangements are equally possible, and macroscopically the layer chirality averages out.

The B_7 textures show characteristic helical filamentary growth in cooling from isotropic melt.[51,52] However they denote at least two distinct phases. X-ray measurements clarified that the original B_7 materials, which form free-standing strands,[51,53] just like columnar liquid crystals of disc-shape molecules (see Chapter 2), indeed have a columnar phase (hereafter we will label them as B_{7-I}). Other B_7-type materials (we call them B_{7-II}) were found to have modulated layer structures.[54] We note that the B_{7-II} and B_2 type banana-smectics also form strands of fibers.[55]

Since the B_i (*i = 1,...7*) nomenclature was introduced, a polar smectic-A (*SmAP*)[56] phase and nematic structures[57] were observed, too, but they are not labeled with the B_i symbols. Based on X-ray studies, B_3 and B_4 are first designated as crystalline phases; however, *AFM* and polarizing microscopic observations indicate that they are highly ordered mesophases.[43,58]

The exciting ferroelectric and electro-optical properties of bent-core liquid crystals will be also discussed in Chapter 8.

1.2   Lyotropic Liquid Crystals

Lyotropic liquid crystals[59] are found in countless everyday situations from soap–water systems to cake butters and other kitchen products. Most importantly, biological systems, such as biological membranes, DNA, RNA, tobacco mosaic virus, spider fiber during its formation, etc., display lyotropic liquid crystalline behavior. They were actually discovered long before their thermotropic counterparts were known. Around the middle of the last century, Virchow[60] and Mettenheimer[61] found that the myelin, which covers and insulates nerve fibres, formed a fluid substance when left in water and exhibited a strange behavior when viewed using polarized light. Although they did not realize that this was a different phase, they are attributed with the first observation of liquid crystals. In spite of their early observations, the research of lyotropic liquid crystals was behind that of the thermotropics up to recently when the situation seems to be reversed.

The molecules that make up lyotropic liquid crystals are amphiphilic molecules and mainly water (and sometimes oils).

Before reviewing the most important structures of lyotropic liquid crystals we need to briefly review the constituent molecules

1.2.1   Water

The water (H_2O) is the most abundant, the most important and best-known terrestrial fluid. At the same time, it is the most exceptional liquid concerning its properties: very high melting and vapor temperatures compared to its low molecular weight; large latent heat of vaporization. Water also shows anomalous temperature dependence of the density in the liquid state (density has maximum at 4°C) and the lower density in the solid (ice) state then in the liquid form.

All these peculiarities can be explained by the hydrogen bonding, i.e., that the two hydrogen atoms, which are arranged on one side of the molecule, are attracted to the oxygen atoms of other nearby water molecules (Figure 1.14).

Hydrogen atoms have single electrons which tend to spend a lot of their time "inside" the water molecule, toward the oxygen atom, leaving their outsides naked, or positively charged. The oxygen atom has eight electrons, and often a majority of them are around on the side away from the hydrogen atoms, making this face of the molecule negatively charged. Since opposite charges attract, it is not surprising that the hydrogen atoms of one water molecule like to point toward the oxygen atoms of other molecules. Of course, in the liquid state, the molecules are not locked into a fixed pattern;

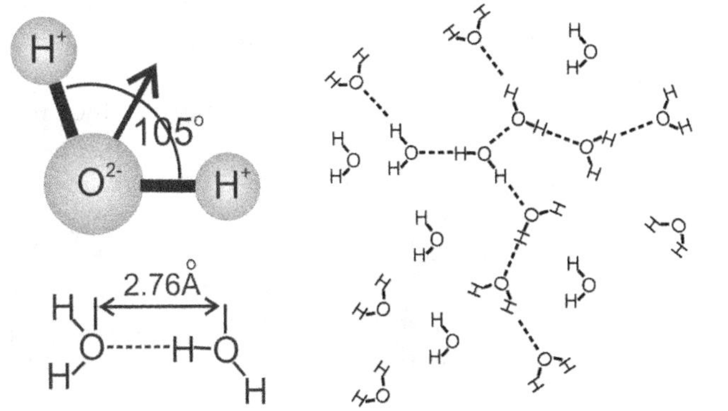

FIGURE 1.14
(a) Structure of the single water molecule and the hydrogen bond; (b) Frank-Wen flickering cluster model of liquid water.

nevertheless, the numerous temporary "hydrogen bonds" between molecules make water an extraordinarily sticky fluid.

1.2.2   Amphiphiles

The word amphiphile was coined by Paul Winsor 50 years ago. It comes from two Greek roots: first the prefix *amphi* which means "double," "from both sides," "around," as in amphitheater or amphibian; then the root *philos* which expresses friendship or affinity, as in "philanthropist" (the friend of man), "hydrophilic" (compatible with water), or "philosopher" (the friend of wisdom or science).

An amphiphilic substance exhibits a double affinity, which can be defined from the physico-chemical point of view as a polar–apolar duality. A typical amphiphilic molecule consists of two parts: on the one hand a polar group which contains heteroatoms such as O, S, P or N, included in functional groups such as alcohol, thiol, ether, ester, acid, sulfate, sulfonate, phosphate, amine or amide; on the other hand, an essentially apolar group, which is in general a hydrocarbon chain of the alkyl or alkylbenzene type, sometimes with halogen atoms and even a few nonionized oxygen atoms. The polar portion exhibits a strong affinity for polar solvents, particularly water, and it is often called hydrophilic part or hydrophile. The apolar part is called hydrophobe or lipophile, from Greek roots *phobos* (fear) and *lipos* (grease). The hydrophobic effect, i.e., extremely weak solubility of nonpolar molecules (such as alkenes) in water, can be explained by taking into consideration that nonpolar particles cannot participate in formation of hydrogen bonds. When they are isolated individually in water, they are surrounded by a "cage" of water molecules. To preserve the tetrahedral network of the hydrogen-bonded water molecules, the cage should choose only specific orientations that do not interfere with the hydrogen bonds. This additional ordering leads to a decrease of the entropy of the water molecules and costs more energy to immerse a nonpolar particle in water. This hydrophobic effect will then lead to aggregation of hydrophobic particles to reduce the loss of entropy. This results in a rule of "like dissolves like."

Because of its dual affinity, an amphiphilic molecule does not feel completely happy in any solvent, be it polar or nonpolar, since there is always one of the groups which "does not like" the solvent environment. This is why amphiphilic molecules exhibit a very strong tendency to migrate to interfaces or surfaces[†] and to orientate so that the polar group lies in water, and the apolar group is placed out of it. It is worth remarking that not all amhiphiles display such activity. It does not happen if the amphiphilic molecule is too hydrophilic or too hydrophobic, in which case it stays in one of

---

[†]In the following, the word "surface" will be used to designate the limit between a condensed phase and a gas phase, whereas the term "interface" will be used for the boundary between two condensed phases. This distinction is handy, though not necessary, and the two words are often used interchangeably, particularly in American terminology.

the phases. The amphiphiles with more or less equilibrated hydrophilic and lipophilic tendencies are likely to migrate to the surface or interface, and they are also referred to as "surfactants."

In English, the term *surfactant* (short for *surface-active-agent*) designates a substance which usually decreases the surface tension of the water or oil. Surfactants are also important to decrease the evaporation of the water and to decrease the curvature where needed. Based on their dissociation in water, the surfactants are classified as anionic, cationic, nonionic and zwitterionic surfactants. They are briefly summarized in Appendix A-2.

Amphiphiles are also often labeled according to their main use such as: *soap, detergent, wetting agent, disperssant, emulsifier, foaming agent, bactericide, corrosion inhibitor, antistatic agent,* etc.

1.2.3 Amphiphilic–Water Aggregation

The structures that amphiphiles are able to build depend on their internal structures, on their concentration and the temperature. For the sake of simplicity, in the following we assume that the amphiphilic molecules are dissolved in water, so the molecules will be arranging themselves with the polar heads in contact with the water. However, it is very straightforward to apply the arguments for nonpolar solvents.

1.2.3.1 Optimal Surface to Tail Volume Ratio

It is evident that the optimal interface between the water and amphiphilic molecules is the surface area of the hydrophilic part. The shapes in which the molecules arrange themselves depend partly on the optimal surface area, as well as partly on the fluid volume of the hydrocarbon chains and the maximum length at which they can still be considered fluid. Although many structures can fit the geometry, one is usually best from a thermodynamic perspective. Large structures create too much order, while small structures cause the surface area to be larger than optimal, so a medium-sized structure usually wins out.

The interfacial curvature is determined by the effective area of the head group, a, with respect to length l and occupied volume V of the hydrophobic tale. Their ratio determines the surfactant packing parameter as $N_s \equiv \frac{V}{al}$.

When the amphiphiles have a large optimal area a, but a small (typically single) hydrocarbon chain volume V, they often form spherical micelles in which the radius is not larger than the maximum fluid length of the hydrocarbon chain, l (see Figure 1.15). In such cases the surfactants can be approximated as cones and $V/al < 1/3$. We note that heads carrying a net charge result in a higher optimal interface area since they have a higher affinity for water.

The association number p (the number of molecules forming a micelle) for spherical micelle can be expressed by simple geometrical arguments as:

$$p = \frac{\frac{4}{3}\pi R_{mic}^3}{V} \tag{1.4}$$

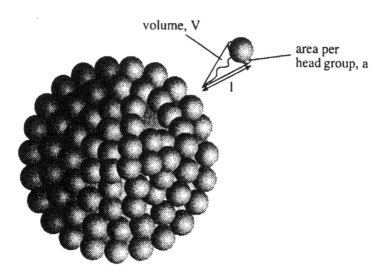

FIGURE 1.15

A spherical micelle. The packing of amphiphilic molecules is controlled by the effective cross-sectional area of the head group, *a* and the hydrophobic chain of length, *l* and volume, *V*. These quantities define the surfactant packing parameter.

In this expression, R_{mic} is the radius of the micelle. Similarly, we can express p as:

$$p = \frac{4\pi R_{mic}^2}{a} \tag{1.5}$$

These mean that

$$\frac{V}{aR_{mic}} = \frac{1}{3} \tag{1.6}$$

Since R_{mic} cannot exceed the length of a fully extended chain, *l*, we see that, for spherical micelles

$$N_s \equiv \frac{V}{al} \leq \frac{1}{3} \tag{1.7}$$

The extended length of an alkyl chain containing n_c carbon atoms in nanometers is:[62] $l = 0.154 + 0.127 \cdot n_c$. Here 0.154 nm is the length of the C–C bond, and 0.127 nm is the projection of this distance to the chain axis for all-*trans* transformations. For the volume of the hydrocarbon chain in nm³, it is

| **Surfactant** | | **Aggregate** | |
| shape | packing parameter | shape | name |

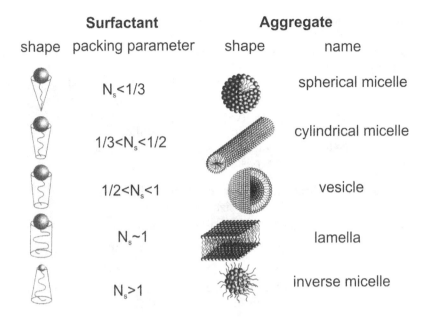

	$N_s < 1/3$		spherical micelle
	$1/3 < N_s < 1/2$		cylindrical micelle
	$1/2 < N_s < 1$		vesicle
	$N_s \sim 1$		lamella
	$N_s > 1$		inverse micelle

FIGURE 1.16
Surfactant packing parameter ranges for various surfactant aggregates.

found that $V = 0.027(n_c + n_{Me})$, where n_{Me} denotes the number of methyl groups. For single chains, $n_{Me} = 1$; for double chains, $n_{Me} = 2$.

These arguments show that spherical micelles can be considered as close-packed cones. If the molecular units are more similar to truncated cones or cylinders, the micelles are not spherical. By arguments analogous to those of spherical micelles, one can construct the structural scenario shown in Figure 1.16.

When $1/3 < N_s < 1/2$, the molecules might assemble into cylindrical micelles. Like those that make spherical micelles, molecules that form cylindrical micelles usually have single chains. Their optimal interface areas, however, are smaller. We note that amphiphiles that usually form spherical aggregates can be induced to form cylinders if a salt is added to the solution, because the ions in the salt weaken the hydrophilic reaction of the polar heads.

If $N_s \sim 1$, the amphiphiles are likely to form bilayers. These molecules often have two hydrocarbon chains. The extra chain makes them more hydrophobic and also raises the volume to length ratio. Amphiphilic bilayers are not as dynamic as micelles — molecules stay in the bilayer for a relatively long time rather than continually diffusing in and out as happens with micelles. If the head group is especially small or is anionic or is in a salt solution, or if the chains are saturated and frozen, a planar bilayer may form.

If $1/2 < N_s < 1$, the molecules usually form so-called vesicles.

Vesicles are bilayers that have folded into a three-dimensional spherical structure, sort of like a micelle with two layers of molecules. Vesicles form because they get rid of the edges of bilayers, protecting the hydrophobic chains from the water, but they still allow for relatively small layers. In order for a flat bilayer to exist without edges, it must have infinite length. Molecules that form vesicles usually have a fluid double chain and a large optimal area. Lipids found in biological membranes spontaneously form vesicles in solution.

Vesicles are usually not in thermodynamic equilibrium, but they can be kinetically stable for quite long periods. There are many methods to prepare them, resulting in different types and sizes of vesicles.

- Sonicitation (ultrasound agitation) of dilute lamellar phases results in small vesicles with broad size distribution
- Spontaneous formation by dissolving dry phospholipids in water
- Dispersion of a lamellar phase in a large excess of water
- Dispersion in organic solvents and injection of this to an excess of water
- Steady shear to a lamellar phase (an effective method to prepare multilamellar vesicles with narrow size distribution)

When $N_s > 1$, the amphiphiles can form inverse micelles, with the heads on the inside and tails on the outside. Molecules that form this structure usually have a small optimal interface area or a large chain volume to length ratio. Double chains, nonionic heads, and *cis* unsaturated chains are also common. When inverse micelles form, the solution changes from appearing as oil droplets in water to water droplets in oil.

1.2.3.2   Concentration

At low amphiphile concentrations, the solution looks like any other — particles of solute distributed randomly throughout the water. When the concentration gets high enough, however, the molecules begin to arrange themselves into *micelles* of hollow spheres, rods or disks.

The critical micelle concentration (CMC) represents a relatively sharp transition with a relatively monodisperse (20 to 30% polydispersity) micelle distribution. The major effect determining the CMC is the increase of entropy of the system of molecules in micelles compared to unassociated molecules. This is because unassociated molecules lead to an ordering of surrounding water, i.e. reduction in entropy due to the hydrophobic effect. This ordering effect is reduced when molecules associate into micelles. The gain in entropy of the water upon micelle formation outweighs the enthalpy penalty (caused by the demixing of the water and surfactant) and the loss of configurational entropy of the molecules due to the constraints imposed by the micellar structure.[63,64,65] The process leading to CMC can be simply described by the so-called *Closed Association Model*, which considers equilibrium between non-ionic surfactant molecules and monodisperse micelles characterized by the association number of p, as $pS \Leftrightarrow S_p$. Assuming that the association number is large (p > 50), and defining the CMC as the point at which it is equally likely that a molecule adds to a micelle or remains as a unimer in solution (i.e., that $\partial c_s / \partial c = 1/2$), it can be shown[63] that:

$$\ln c_{CMC} = -\frac{1}{p} \ln K \qquad (1.8)$$

where $K = \frac{c_m}{c_s^p}$ is the equilibrium constant, with c_s (c_m) being the unassociated surfactant (micelle) concentration.

The CMC and the association number depend on a number of factors, such as temperature, structure of surfactant, salt and co-solutes. These effects are described in a number of textbooks;[64] here we just briefly summarize these dependences.

1.2.3.2.1 Temperature

For most surfactants, the CMC is basically independent of temperature. An exception is the nonionic surfactants based on oxyethylene hydrophilic groups, where CMC strongly decreases with increasing temperature. For ionic surfactants, the CMC first slightly decreases, then increases with increasing temperatures. In this case, heating causes hydration of the hydrophilic group, thus favoring micellization; however, it also enhances the molecular motion, which opposes micellization.

The association number appears to decrease slightly in heating, presumably because the optimal surface area is increased. For oxyethylene-containing nonionics, heating leads to an increase of p, up to the so-called "cloud point," when phase separation occurs, because the oxyethylene becomes insoluble in water.

1.2.3.2.2 Surfactant Type

The CMC is usually much lower for nonionic surfactants than for the ionic ones (the larger is the ionic charge, the bigger is the difference) for otherwise equal hydrophobic chains. This is because electrostatic repulsions between head groups have to be overcome for ionic surfactants, but not for nonionics. The chemical nature of the hydrophobic chain is also important; for example, addition of fluorine to the chain dramatically increases the hydrophobicity, and the CMC decreases sharply.

The hydrophobicity can be increased most systematically by varying the chain length. An increase of the chain length favors micellization. This is often described by an empirical relationship of the form:

$$\log c_{CMC} = b_o - b_1 n_C \qquad (1.9)$$

where n_C is the number of carbon atoms in the chain, and typically $1.2 < b_o < 2.4$ and $0.26 < b_1 < 0.34$. Roughly, the addition of a $-CH_2$ decreases the CMC by a factor two for ionic surfactants and by factor three for nonionics. This behavior holds for chains up to about 16 carbon atoms. For longer chains, a coiling occurs, which compensates the change.

The association number in aqueous media increases with the chain length, by the simple geometrical consideration derived in Eq. (1.4).

1.2.3.2.3   *Effect of Salt and Co-Solutes*

Experimental data indicate that for ionic surfactants the addition of salt decreases the c_{CMC} and increases p. Empirically, the CMC is expressed at the function of total counterionic concentration C_i as follows:

$$\log c_{CMC} = -a \log C_i + b \qquad (1.10)$$

where a and b are constants for a given ionic head at particular temperature. The effect of the added electrolyte is to reduce the repulsion between charged head groups. It is more pronounced for longer chain surfactants, since the increase of the chain length has similar effect on the CMC. The effect of the salt also depends on the valency of the electrolyte ion, in particular on the added counter-ions. Salt has little effect on the CMC or p of the nonionic surfactant.

Addition of the co-solutes can either increase or decrease the CMC, depending on the polarity of the molecules. Highly water soluble co-solutes tend to increase of the CMC, since the solubility of surfactant molecules is enhanced. Other factors:

(i)   pH[†]: If the PH is lowered, the hydrophilic interaction of the head is reduced and the optimal interface area is lowered.

(ii)   The additional ions (that screen out the water surfaces) in solution also reduce the repulsive interactions between head groups, reducing the radius of curvature. This makes the molecules more likely to form bilayers or inverse micelles. Using these methods to reduce the interface area often has the additional effect of straightening the hydrocarbon chains.

(iii)   If the hydrocarbon chains are unsaturated or branched, their length is reduced. This increases the volume to length ratio and again makes bilayers and inverse micelles more likely.

(iv)   When two different kinds of amphiphiles are mixed, the characteristics of the solution are similar to an average of the characteristics of solutions of the individual types, provided the two types can mix freely in solution. Furthermore, carefully adding more of one kind of molecule can cause the solution to form structures of different shapes or sizes that neither molecule would form alone.

[†] The pH scale provides a convenient way to express the acidity and basicity of dilute aqueous solutions. pH=-log[H_3O^+]

The pH scale provides a convenient way to express the acidity and basicity of dilute aqueous solutions. pH = $- \log[H_3O^+]$

1.2.4   Micellar Aggregates

At low concentrations the micelles arrange in a liquid structure (no long-range translational order). It is called L_1 phase.

As the concentration increases, the micelles begin to arrange themselves into loose patterns. These patterns are the actual liquid crystal aspects of the molecular behavior. In the first liquid crystal phases, with the smallest surfactant concentrations, the micelles form a structure similar to a face-centered or body-centered cubic crystal lattice. The illustration in Figure 1.17a shows a body-centered cubic crystal structure. In these structures, the pattern of micelles is not as stable as the arrangements of atoms or ions or molecules of solid crystals. Rod-shaped micelles often form hexagonal arrays made out of six rods grouped around a central one for a total of seven, as illustrated in Figure 1.17b. Notice that the micelle surface is composed of hydrophilic heads. The hydrophobic tails are isolated inside the micelle. Hexagonal liquid crystals generally exist in solutions that are 40–70% ionic and nonionic amphiphiles.[68] They have the same structure as columnar liquid crystals of discs. Columnar mesophases have also been obtained with rod-like polymers of biological interest, such as desoxyribonucleic acids (DNA)[69] or polypeptides[70] in concentrated solution, in soaps[65] and block copolymers.[66]

At even higher concentrations ribbon type structures form that involve finite bilayers ending at cylindrical half-micelles. This ribbon phase is a precursor to the molecules from another liquid crystalline phase — the lamellar lyotropic liquid crystal (L_α) structure, where the ribbons fuse together. This structure has a double layer of molecules arranged a bit like a sandwich with polar heads taking the place of the bread and nonpolar tails as the filling (Figure 1.17). Because the sheet-like layers can slide easily past each other, this phase is less viscous than the hexagonal phase, at least in the direction of the sliding, despite its lower water content. The lamellar phase is characterized by a low viscosity along the lamellae and with elastic behavior normal to it. Its structure is analogous to the smectic phases of the thermotropic liquid crystals.

In an intermediate concentration range, where the materials cannot decide if the water or the amphiphile aggregate structure should be the continuous matrix, often so-called bicontinuous structures form. In such bicontinuous cubic phases the interfaces have saddle-splay type structures characterized by nonzero negative mean curvature and negative Gaussian curvature.[†] The most common bicontinuous cubic phase is called gyroid

---

[†] Both the mean and the Gaussian curvatures can be defined in terms of the principal curvatures $c_1 = 1/R_1$ and $c_2 = 1/R_2$, where R_1 and R_2 are the radii of curvature. The mean curvature H is defined as $H = \frac{c_1+c_2}{2}$, whereas the Gaussian curvature is defined as $K = c_1c_2$. For $R_1 = R_2$ in the inverse bicontinuous cubic phases $-1/(2R) < H < 0$ and $-1/R2 < K < 0$.

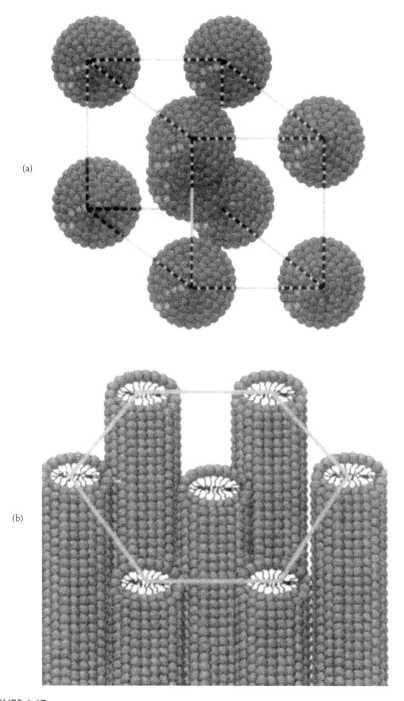

FIGURE 1.17
Normal structures and the lamellar phase. (a) Normal micellar cubic structure (I_1); (b) normal hexagonal structure (H_1).

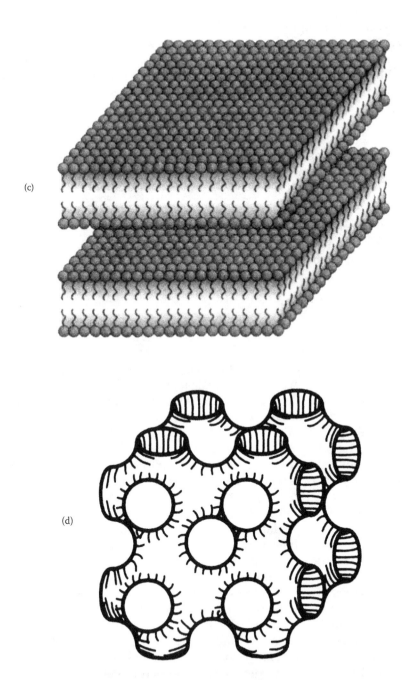

(c)

(d)

FIGURE 1.17 (CONTINUED)
(c) lamellar phase (L_α); (d) normal bicontinuous cubic structure (V_1). Here the portion of the "gyroid" structure is sketched. The amphiphilic molecules form a bilayer film separating two continuous labyrinths of water. The amphiphilic film is a network with three-fold node points, which defines the gyroid phase (also called "plumber's nightmare").

phase as shown in Figure 1.17d. A normal bicontinuous phase consists of two continuous channels of water separated by a bilayer of surfactant molecules. In the gyroid phase the connection nodes have three-fold structure. Other structures with four- and six-fold connectors are also known. Complex phases with tetragonal or rhombohedral symmetry that have been called "mesh" phases.

If the amphiphile concentration is larger than about 50 wt%, the amphiphile aggregates may revert to inverse structures (inverse hexagonal or inverse cubic structures). In these cases, the solvent becomes the minority phase. The sequence in increasing amphiphile concentrations is typically the following:

- Inverse bicontinuous phase (V_2)
- Inverse hexagonal phase (H_2) formed from rod-like water channels in an amphiphilic matrix
- Inverse micellar cubic phase (I_2)
- L_2: inverse micellar phase

Although lyotropic liquid crystals are characterized by the fact that concentration is the determining factor in their phase transitions, temperature also plays an important role. This can be seen on the phase diagram of a soap–water system, where the vertical axis is the concentration of amphiphilic molecules and the vertical axis is the temperature.[73] The concentration at which micelles form in solution, called the critical micelle concentration, is shown as a dotted line. The line at low temperatures is called Krafft temperature T_k. It separates the crystal in water part from the liquid crystalline structures. Above T_k the solutions have milky appearances, since the micelles scatter light (the larger the micelle, the milkier the solution). Below T_k the solution becomes clear, as only crystals are suspended in the solvent.

There is a strong correlation between T_k and the temperature where the hydrocarbon regions of each micelle "melt." From the schematic picture of a micelle, it is apparent that the packing of the amphiphilic molecules into micelles requires considerable bending and folding of the hydrocarbon chains, i.e., the hydrocarbon chains have to be melted. This leads to the idea that the Krafft temperature marks something akin to a solid–melt transition within the hydrocarbon region of the small micelles.[73] Below T_k the amphiphilics prefer to aggregate in configuration that involves less curvature (in this case, the packing of straight chains is less costly in energy. Accordingly, the structure below T_k is basically suspensions of crystals in water (see in the lower part of Figure 1.18); however, it is also possible that there are amphotropic liquid crystalline phases below T_k.[74]

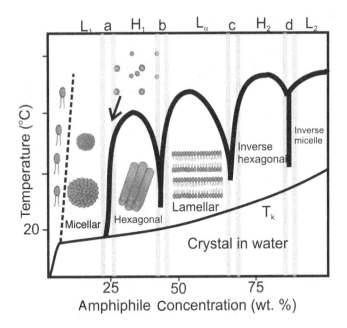

FIGURE 1.18

Phase diagram for a typical soap–water system. The nearly vertical dashed line shows the minimum concentration for micelle formation. The line separating the crystal in water part from the liquid crystalline parts is called Kraft temperature (T_K). Vertical bars and symbols at the top of the figure indicate hypothetical sequence of phases in binary amphiphile–solvent systems. Here a, b, c and d indicate intermediate phases (for example the bicontinuous cubic phase), L_2 denotes the inverse micelle solution, H_2 is the inverse hexagonal phase, L_α is the lamellar phase, H_1 is the normal hexagonal phase and L_1 is the normal micelle phase. Note, this idealized sequence has never been observed entirely and the phase boundaries are rarely vertical.

1.3   Other Lyotropic Phases

1.3.1   Lyotropic Nematic Phases

In addition to the above lamellar, columnar and optically isotropic phases, there are also lyotropic nematic phases, which usually involve mixtures of a charged amphiphilic, such as simple soap, with an alkanol (a weaker amphiphilic where the head group is an alcohol), together with water and a simple salt. They are termed nematic because, like thermotropic nematics, their optical axes are easily oriented by external magnetic fields. In contrast to thermotropic nematics, the basic units of lyotropic nematics are molecular aggregates with dimensions of about 2–10 nm. In lyotropics, the nematic phase is much less usual than in thermotropic liquid crystals. Lyotropic nematics generally form in ternary systems. The first example was found in a mixture of sodium decylsulfate (SDS), 1-n-decanol and D_2O.[75] A particularly interesting example is the potassium laurate (KL)–1-decanol–D_2O, which has two uniaxial nematic phases[76] separated by a biaxial nematic.[77,78] The uniaxial nematics are called N_L and N_C, where the indices denote calamitic and discotic, referring to the shape of the micelles. This latter one has a negative optical and positive diamagnetic anisotropy. The biaxiality of the intermediate phase N_b was proven by conoscopy and NMR measurements.[79]

Nematic states occur also in nonionic monomeric and polymeric surfactant solutions,[80] living amoeboid cells[81] and rigid or semiflexible polymeric rods, such as fd virus in water solutions.[82,83] We note that colloidal suspension of semiflexible virus particles also form nematic phases.[84,85]

1.3.2 Chromonic Liquid Crystals

What is the discotics for the thermotropic liquid crystals that is the lyotropic chromonic liquid crystals (LCLCs) for the lyotropic materials. LCLC molecules are plank-like rather than rod-like, rigid rather than flexible, aromatic rather than aliphatic.[86,87] Typical LCLC molecules and their aggregation in water are shown in Figure 1.19.

The polar hydrophilic ionic groups at the periphery of the molecules make the material water-soluble; the side groups dissociate in aqueous solutions, leaving the molecules charged. At the same time, the central core is relatively hydrophobic. Avoiding core-water contact, the molecules produce face-to-face stacking, thus forming a supramolecular aggregate. In contrast to standard lyotropic systems, where micelles are formed by a fixed number of molecules, aggregation in LCLCs does not produce a well-defined size or even a well-defined distribution of sizes.

According to Lydon,[88] aggregation of LCLC molecules is a "steady, progressive build-up of chromonic aggregates where the addition of a molecule to a stack is always associated with the same increment of free energy." On

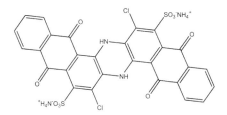
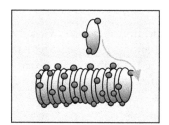

FIGURE 1.19

Molecular structure of a typical LCLC representative dye Blue 27, and the scheme of aggregation of the molecules in aqueous solutions. (Picture courtesy of O. Lavrentovich.)

the other hand, a model based on the balance of enthalpy and entropy predicts a scenario with a well-defined most probable aggregation number (or the length of the aggregate) proportional to the square root of the molar concentration of LCLC molecules.[89] The experimental data on the length distribution of rods in LCLC are absent. The same can be stated about practically all the physical properties relevant to the mesomorphic state of the LCLC materials, including the viscoelastic material coefficients, indices of refraction, surface alignment, hydrodynamic effects, etc.

Since very little is known about LCLC structures and properties, the recognition of these materials is not widespread. However, as one of the pioneers in the field, J. Lydon, writes:[90] "A single commercial application, however, could change this picture overnight. The liquid crystal literature will be as full of chromonic studies as it now is with ferroelectric chevron structures. Watch this space."

1.4 Amphotropic Liquid Crystals

The term *amphitropic* or *amphotropic* means compounds that can exhibit both types of liquid crystal formation, thermotropic as well as lyotropic. Three recent reviews have summarized the progress in this field.[91]

1.4.1 Amphiphilic Block-Copolymers

Block copolymers represent the best-understood and simplest amphotropic materials studied to date.[92] In such macromolecules, two chemically different blocks (e.g., polystyrene, PS, and polyisoprene, PI) are joined together by covalent bonds, so that these incompatible blocks cannot separate into distinct macroscopic phases. X-ray diffraction and electron-microscopy have established that the units of block copolymers segregate into microdomains (micro-phase or rather nano-phase separation), and produce well-developed liquid crystalline phases.[93]

Among these, the smectic phases (Figure 1.20a) consist of two types of sublayers, periodically and alternately superimposed, each sublayer being formed by blocks of the same nature. As the molecular dimensions of the block are usually large, the thickness of the smectic layers is usually of a few tens of nanometers. The thickness of the interfacial region is only a few nanometers, i.e., much smaller than the overall layer spacing.[95] The ability to form ordered structures and the stability of these structures depends on the degree of segregation, which is described by the term χN, where χ is the so-called Flory-Huggins interaction parameter between the different blocks (enthalpic term), and N is proportional to the molecular weight (entropic term). Segregation requires a certain minimum value of χN (>10). The degree of segregation has an influence on the sharpness of the interfaces dividing

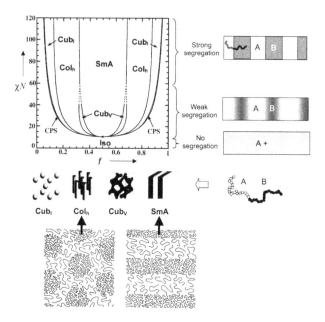

FIGURE 1.20

Illustration of phases of block co-polymers. Upper part: phase diagram for conformationally symmetric flexible AB di-block copolymer melts[94] and the illustration of the separation of the incompatible units. (CPS: disordered arrangement of spherical aggregates). Lower row: schematic representation of the smectic and the columnar structure of block copolymers. (Reprinted from C. Tschierske, *Curr. Opin. Colloidal Interface Sci.* 7, 355, 2002, with permission from Elsevier.)

the segregated domains. In strong segregating systems ($\chi N > 60$), the interfaces are sharp, whereas in weakly segregating systems the interfaces are diffuse. This means that there are not pure subphases, but instead the concentration of one component in the other changes periodically in space (see Figure 1.20). Note that, in the case of block copolymers, the variation of the ratio of the different blocks plays essentially the same role as the amphiphiles and water (or oil) in the lyotropic liquid crystals. In a sense, it is also similar to low-molecular-weight smectics, where the weak incompatibility of the rigid aromatic core and the flexible aliphatic chains lead to a nano-segregation. The composition of the block copolymer determines the mesophase morphology as shown in Figure 1.20. With increasing volume fraction of one component, the structure changes in the following order:

1. A lamellar SmA type textures;
2. Bicontinuous networks with a cubic symmetry (Cub_V);
3. A hexagonal arrangement of infinite columns (Col_h);

4. Spheroidic aggregates organized in a cubic lattice (Cub_I);

5. A disordered arrangement of spheroids (Iso).

Note that this phase sequence, can be universally observed in the thermotropic phase sequence of block copolymers and low-molecular-mass amphiphiles[96] as well as in lyotropic detergent solvent systems. In addition to the phase sequence described above, several additional intermediate phases can occur at the transitions between different mesophase morphologies. The bicontinuous cubic phases Cub are often also included in these intermediate phases.

Solvents have a strong impact on the phase behavior of block copolymers, and these solvents can be divided into three different classifications: *specific*, *nonspecific* and *amphiphilic*.

- A *specific solvent* preferentially interacts with only one part of the amphiphilics.[91,92] In this way, it increases the volume fraction of this part, which changes the composition *f* and, therefore, often leads to a change in the morphology. In addition, it modifies the strength of the intermolecular interactions, which influences the interaction parameter and, hence, changes the degree of incompatibility and, in turn, the degree of segregation. Therefore, these solvents have an influence on the mesophase stability and morphology, as well as on the sharpness of the intermaterial interfaces. The sharpness of these interfaces has a feedback effect on the morphology.

- The situation is quite different for *nonspecific solvents*, which are good solvents for both molecular segments. Here the solvent is uniformly distributed between both molecular parts, and both parts are swollen equally, i.e., the mesophase morphology is not changed.[93] This means that the amphiphiles are "diluted." This is analogous to increasing temperature, leading to a strong reduction in the order–disorder transition temperature. In this case, uniform dissolution of the molecules occurs, often leading to a complete loss of positionally ordered mesophases.

- *Amphiphilic solvents*, such as long-chain alcohols, carboxylic acids and amines, consist of at least two distinct parts and behave as co-surfactants. Hence, such systems should be regarded as mixtures of different amphiphiles rather than as truly lyotropic amphotropic systems.

1.4.2   Thermotropic Liquid Crystals in Solvents

We have seen that thermotropic liquid crystal molecules that tend to form 1D and 2D structures in the form of smectic and columnar liquid crystals have weakly amphiphilic anisometric (rod-, disc-, banana-, and bowl-shaped) molecules, where the rigid cores tend to avoid the flexible aliphatic chains. These materials tend to maintain and tune their thermotropic behavior in the presence of solvents just as, for example, block copolymers do in the presence of (specific, nonspecific and amphiphilic) solvents. In general, lyotropic mesomorphism of amphotropic mesogens requires a certain minimum size of the aromatic cores in the presence of additional attractive forces between the cores. Calamitic mesogens with short aromatic cores simply dissolve in nonpolar solvents, but disc-like mesogens with intermolecular hydrogen bonding between the cores show especially stable lyotropic Col_h and N_{Col} phases. Instead of reviewing the great number of examples, we just refer to a recent review,[8] and here we mention only three cases which may have practical importance. One of the first examples of lyotropic character of thermotropic SmA and SmC materials with organic solvents, like n-decane and n-hexane, was reported by Rieker in 1995. It was shown that the nonpolar solvents pack together, with the aliphatic chains resulting in an increase of the layer spacing and decrease of the clearing point.[97] Later examples relate to chiral disc-shaped molecules that form ferroelectric tilted columnar mesophase, which is electrically switchable, but only at high temperatures and at very high fields. It was found that, at low solvent concentration (like dodecane up to 20 wt%), the columnar mesophase of the pure compounds swells and the clearing temperatures decrease. Such swollen columnar phases have significantly smaller viscosity; therefore they can be used in fast-switching columnar mesophases with chiral tripheline derivatives such as shown in Figure 1.8c.[98] Related studies were carried out with banana-shaped mesogens, both with aromatic solvents, like p-xylene and m-xylene,[99] and with aliphatic chain, like hexadecane.[100] Similarly to the ferroelectric columnar phases, these studies may lead to significance advances toward applications.

1.4.3 Flexible Amphiphilics

Amphiphilic polyhydroxy amphiphiles[101] and carbohydrate derivatives (lipids)[102] represent the best-investigated amphotropic materials. Here we would like to concentrate only on glucolipids, which contain polar (hydrophilic) sugar head groups and apolar (hydrophobic) carbon chains. They form liquid crystalline structures in aqueous systems, depending on the concentration, as well as in their pure state, depending on temperature. Glycolipid is one of the three lipids that make up the biological cell membranes. The shorter glycolipids (with 8 to 12 carbon atoms) are mainly used as detergents,[103] whereas others glycolipids are involved in membrane fusion processes.[104,105] The more complex glycolipids (starting with three sugar head groups) are involved in cell surface recognition processes.[106] Liquid crystalline properties of lyotropic synthetic glycolipids have been rigorously studied in the last decade. An example of alkyl glucoside, the simplest class of glycolipids, with one sugar polar headgroup and an apolar carbon chain with their phase diagram in water systems is shown in Figure 1.21.

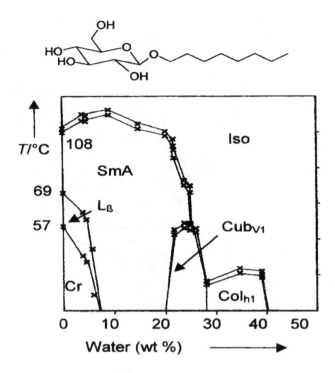

FIGURE 1.21

Structure of n-octyl-b-D-glycopiranoside and its phase sequence in water. (Reprinted from C. Tschierske, *Curr. Opin. Colloidal Interface Sci. 7*, 355, 2002, with permission from Elsevier.)

References

1. G.W. Gray, *Molecular Structures and the Properties of Liquid Crystals*, Academic Press, London (1962).
2. S. Chandrasekhar, D.K. Sadashiva, K. Suresh, *Pramana*, 9, 471 (1977).
3. H. Zimmermann, R. Poupko, Z. Luz, J. Billard, *Z. Naturforsch.*, 40A, 149 (1985).
4. D. Vorländer, A. Apel, *Ber. Dtsch. Chem. Ges.*, 1101 (1932).
5. Y. Matsunaga, S. Miyamoto, *Mol. Cryst. Liq. Cryst.*, 237, 311 (1993).
6. G. Brown, J.J. Wolken, *Liquid Crystal and Biological Structures*, Academic Press, New York (1979).
7. D. Blunk, K. Praefcke, V. Vill, Amphotropic liquid crystals, in *Handbook of Liquid Crystals*, Ed. D. Demus, J. Goodby, G.W. Gray, H.W. Soiess, V. Vill, vol. 3, p. 305, Wiley-VCH, New York (1998).
8. C. Tschierske, *Curr. Opin. Colloidal Interface Sci. 7*, 355 (2002).
9. F. Reinitzer, *Monatsch. Chem.*, 9, 421 (1888). English translation of this paper: *Liq. Cryst.*, 5, 7 (1989).
10. O. Lehmann, *Z. Phys. Chem.*, 4, 462 (1889).
11. D. Demus, *Mol. Cryst. Liq. Cryst.*, 364, 25 (2001).

12. G.W. Gray, *Molecular Structure and the Properties of Liquid Crystals*, Academic Press, London (1962).

13. K. Hermann, *Trans. Faraday Soc.*, 29, 883 (1933).

14. A.M. Levelut, J. Doucet, M. Lambert, *J. Physique*, 35, 773 (1974).

15. D. Guillon, A. Skoulios, *J. Physique*, 37, 797 (1976).

16. S. Diele, P. Brand, H. Sackman, *Mol. Cryst. Liq. Cryst.*, 16 (1–2), 105 (1972).

17. J.P.F. Lagerwall, Structures and properties of chiral smectic C liquid crystals: Ferro- and antiferroelectricity in soft matter, Ph.D. thesis, Goteborg, Sweden (2002).

18. A.J. Leadbetter, E.K. Norris, *Mol. Phys.*, 38, 669 (1979).

19. A. de Vries, A. Ekachai, N. Spielberg, *Mol. Cryst. Liq. Cryst.*, 49 (5), 143 (1979).

20. A. de Vries, *Mol. Cryst. Liq. Cryst.*, 11(4), 361 (1970); Y. Takanishi, Y. Ouchi, H. Takezoe, A. Fukuda, A. Mochizuki, M. Nakatsuka, *Mol. Cryst. Liq. Cryst.*, 199, 111 (1991).

21. R.B. Meyer, W.L. Macmillan, *Phys. Rev. A*, 9, 899 (1974).

22. R.B. Meyer, L. Liebert, L. Strzelencki, P. Keller, *J. Phys. Lett.*, 36L, 69 (1975).

23. G. Barbero, G. Durand, *Mol. Cryst. Liq. Cryst.*, 179, 57 (1990).

24. W.L. Macmillan, *Phys. Rev. A.*, 8 (4), 1921 (1973).

25. H. Zimmermann, R. Poupko, Z. Luz, J. Billard, *Z. Naturforsch.*, 41A, 1137 (1986).

26. S. Chandrasekhar, B.K. Sadishiva, K.A. Suresh, *Pramana* 9, 471 (1977).

27. A.N. Cammidege, R.J. Bushby, in *Handbook of Liquid Crystals, Vol. 2B, Low Molecular Weight Liquid Crystals* II, Ed. D. Demus et al., Wiley-VCH, New York (1998).

28. L.D. Landau, in *Collected Papers of L.D. Landau*, Ed. D. ter Haar, 209, Gordon and Breach, New York (1965).

29. C. Destrade, M.C. Mondon, J. Malthete, *J. Physique Coll.*, 40, C3 (1979).

30. H. Zimmermann, R. Poupko, Z. Luz, J. Billard, *Z. Naturforsch.*, 40A, 149 (1985).

31. H. Bock, W. Helfrich, *Liquid Crystals*, 18, 387 (1995).

32. T.C. Lubensky, A.B. Harris, R.D. Kamien, G. Yan, *Ferroelectrics*, 212, 1 (1998).

33. H.-G. Kuball, *Liquid Crystals Today*, 9, 1 (1999).

34. R.P. Lemieux, *Acc. Chem. Res.*, 34 (11), 845 (2001).

35. J.W. Goodby, Properties and structures of ferroelectric liquid crystals, in *Ferroelectric Liquid Crystals, Principles, Properties and Applications*, Ch. II, Ed. G.W. Taylor, Gordon and Breach Science Publishers, Philadelphia (1991).

36. R.B. Meyer, L. Liebert, I. Strelecki, P. Keller, *J. Phys. Lett. (Paris)*, 36, L69 (1975).

37. J.W. Goodby, Symmetry and chirality in liquid crystals, in *Handbook of Liquid Crystals*, Ch. 5, Ed. D. Demus, J. Goodby, G.W. Gray, H-W. Speiss, V. Vill, Wiley-VCH, New York (1998).

38. G.W. Gray, D.G. McDonnell, *Mol. Cryst. Liq. Cryst.*, 34, 211 (1977).

39. J. Thisayukta, Y. Nakayama, S. Kowauchi, H. Takezoe, J. Watanabe, Enhancement of twisting power in the chiral nematic phase by introducing achiral banana-shaped molecules, *J. Am. Chem. Soc.*, 122, 7441 (2000).

40. D. Vorländer, A. Apel, Die Richtung der Kohlenstoff-Valenzen in Benzolabkömmlingen (II.), *Ber. Dtsch. Chem. Ges.*, 1101 (1932).

41. Y. Matsunaga, S. Miyamoto, *Mol. Cryst. Liq. Cryst.*, 237, 311 (1993); H. Matsuzaki, Y. Matsunaga, *Liq. Cryst.*, 14, 105 (1993).

42. Y. Matsunaga, S. Miyamoto, *Mol. Cryst. Liq. Cryst.*, 237, 311 (1993); H. Matsuzaki, Y. Matsunaga, *Liq. Cryst.*, 14, 105 (1993).

43. G. Pelzl, S. Diele, W. Weissflog, *Adv. Mater.*, 11, 707 (1999).

44. D.A. Coleman, J. Fernsler, N. Chattham, M. Nakata, Y. Takanishi, E. Körblova, D.R. Link, R.-F. Shao, W.G. Jang, J.E. Maclennan, O. Mondainn-Monval, C. Boyer, W. Weissflog, G. Pelzl, L-C. Chien, J. Zasadzinski, J. Watanabe, D.M. Walba, H. Takezoe, N.A. Clark, Polarization-modulated smectic liquid crystal phases, *Science*, 301, 1204–1211 (2003).

45. A. Jákli, D. Krüerke, H. Sawade, G. Heppke, Evidence for triclinic symmetry in smectic liquid crystals of bent-shape molecules, *Phys. Rev. Lett.*, 86, (25), 5715–5718 (2001).

46. J. Matraszek, J. Mieczkowski, J. Szydlowska, E. Gorecka, Nematic phase formed by banana-shaped molecules, *Liq. Cryst.*, 27, 429 (2000); I. Wirth, S. Diele, A. Eremin, G. Pelzl, S. Grande, L. Kovalenko, N. Pancenko, W. Weissflog, New variants of polymorphism in banana-shaped mesogens with cyano-substituted central core, *J. Mater. Chem.*, 11, 1642 (2001); W. Weissflog, H. Nádasi, U. Dunemann, G. Pelzl, S. Diele, A. Eremin, H. Kresse, Influence of lateral substituents on the mesophase behaviour of banana-shaped mesogens, *J. Mater. Chem.*, 11, 2748 (2001).

47. T. Niori, T. Sekine, J. Watanabe, T. Furukawa, H. Takezoe, Distinct ferroelectric smectic liquid crystals consisting of banana-shaped achiral molecules, *J. Mater. Chem.*, 6(7), 1231 (1996); T. Sekine, T. Niori, M. Sone, J. Watanabe, S.W. Choi, Y. Takanishi, H. Takezoe, Spontaneous helix formation in smectic liquid crystals comprising achiral molecules, *Jpn. J. Appl. Phys.*, 36, 6455 (1997).

48. G. Pelzl, S. Diele, W. Weissflog, Banana-shaped compounds — A new field of liquid crystals, *Adv. Mater.*, 11(9), 707 (1999).

49. A. Jákli, P. Toledano, Unusual sequences of tilted smectic phases in liquid crystals of bent-shape molecules, *Phys. Rev. Lett.*, 89 (27), 275504-1-4 (2002).

50. D.R. Link, G. Natale, R. Shao, J.E. Maclennan, N.A. Clark, E. Körblova, D.M. Walba, Spontaneous formation of macroscopic chiral domains in a fluid smectic phase of achiral molecules, *Science* 278, 1924 (1997).

51. G. Pelzl, S. Diele, A. Jákli, C.H. Lischka, I. Wirth, W. Weissflog, *Liq. Cryst.*, 26, 135 (1999).

52. A. Jákli, C.H. Lischka, W. Weissflog, G. Pelzl, A. Saupe, *Liq. Cryst.*, 27, 11 (2000).

53. D.R. Link, N. Chattham, N.A. Clark, E. Körblova, D.M. Walba, p. 322, Abstract Booklet FLC99, Darmstadt (1999).

54. D.A. Coleman, J. Fernsler, N. Chattham, M. Nakata, Y. Takanishi, E. Körblova, D.R. Link, R.-F. Shao, W.G. Jang, J.E. Maclennan, O. Mondainn-Monval, C. Boyer, W. Weissflog, G. Pelzl, L-C. Chien, J. Zasadzinski, J. Watanabe, D.M. Walba, H. Takezoe, N.A. Clark, *Science*, 301, 1204 (2003).

55. A. Jákli, D. Krüerke, G.G. Nair, *Phys. Rev. E*, 67, 051702 (2003).

56. A. Eremin, S. Diele, G. Pelzl, H. Nadasi, W. Weissflog, J. Salfetnikova, H. Kresse, *Phys. Rev. E*, 64, 051707-1-6 (2001).

57. J. Matraszek, J. Mieczkowski, J. Szydlowska, E. Gorecka, *Liq. Cryst.*, 27, 429 (2000); I. Wirth, S. Diele, A. Eremin, G. Pelzl, S. Grande, L. Kovalenko, N. Pancenko, W. Weissflog, *J. Mater. Chem.*, 11, 1642 (2001); W. Weissflog, H. Nádasi, U. Dunemann, G. Pelzl, S. Diele, A. Eremin, H. Kresse, *J. Mater. Chem.*, 11, 2748 (2001).

58. G. Heppke, D. Krüerke, C. Löhning, J. Rabe, W. Stocker, poster of 17th ILCC, Strasburg (1998).

59. A.M. Figueredo-Neto, S.R.A. Salinas, *Liquid Crystals: Phase Transitions and Structural Properties*, Oxford University Press, Oxford (2005).

60. R. Virchow, *Virchows Arch.*, 6, 571 (1854).

61. C. Mettenheimer, Corr.-Blatt d. Verein. F. gem Arbeit zur Furderung d. wiss. Heilkunde, 24, 331 (1857).
62. I.W. Hamley, *Introduction to Soft Matter*, Wiley, Chichester (2000).
63. See for example, R.J. Hunter, *Foundation of Colloid Science, I*, Oxford University Press, Oxford (1987).
64. See for example, M.J. Rosen, *Surfactants and Interfacial Phenomena*, 2nd ed., a Wiley-Interscience Publication, New York (1989).
65. M. Costas, B. Kronberg, R. Silveston, *J. Chem. Soc. Faraday Trans. I*, 90 (11), 1513 (1994); and in *Pure Appl. Chem.*, 67(6), 897 (1995).
66. C.E. Fairhurst, S. Fuller, J. Gray, M.C. Holmes, G.J.T. Tiddy, Lyotropic surfactant liquid crystals, in *Handbook of Liquid Crystals*, Vol. 3, Ch. 7, Ed. D. Demus et. al., Wiley-VCH, Weinheim (1998).
67. J.H. Clint, *Surfactant Aggregation*, Blackie, Glasgow (1990).
68. P. Ekwall, *Adv. Liq. Cryst.*, 1, 1 (1975); A. Skoulios, *Adv. Colloid Interface Sci.*, 1, 79 (1967).
69. M. Feughelman, R. Langridge, W.E. Seeds, A.R. Stokes, H.R. Wilson, H.C.W. Hooper, M.H.F. Wilkins, R.K. Barclay, L.D. Hamilton, *Nature*, 175, 834 (1955).
70. P. Saludian, V. Luzzati, in *Poly-a-Aminoacids*, Ed. G.D. Fasman, Marcel Dekker, New York (1967).
71. A. Skoulios, V. Luzzati, *Acta. Cryst.*, 14, 278 (1961).
72. A. Skoulios, *Adv. Liq. Cryst.*, 1, 169 (1975).
73. K. Fontell, *Mol. Cryst. Liq. Cryst.*, 63, 59 (1981).
74. J.F. Nagle, H.L. Scott, *Phys. Today*, 38 (1978).
75. P.P. Pershan, *Phys. Today*, 35 (5), 34 (1982).
76. K.D. Lawson, T.J. Flautt, *J. Am. Chem. Soc.*, 89, 5489 (1967).
77. K. Radley, L.W. Reeves, *Can. J. Chem.*, 80, 174 (1975).
78. L.J. Yu, A. Saupe, *Phys. Rev. Lett.*, 45, 1000 (1980).
79. R. Bartolino, T. Chiarenze, M. Meuti, R. Compagnoni, *Phys. Rev. A*, 26, 116 (1982).
80. A. Figueredo-Neto, L. Liébert, *J. Chem. Phys.*, 87, 1851 (1987); Y. Hendrikx, J. Charvolin, M. Rawiso, *PRB*, 33, 3534 (1986).
81. B. Luehmann, H. Finkelmann, G. Rehage, *Makromol. Chem.*, 186, 1059 (1987).
82. H. Gruler, U. Dewald, M. Eberhardt, *Eur. Phys. J. B*, 11, 187 (1999).
83. K.R. Purdy et al., *Phys. Rev. E*, 67, 031708 (2003); Z. Dogic, K.R. Purdy, E. Grelet, M. Adams, S. Fraden., *Phys. Rev. E*, 69, 051702 (2004).
84. D. Lacoste, A.W.C. Lau, T.C. Lubensky, *Eur. Phys. J. E*, 8, 403 (2002).
85. Z. Dogic, S. Fraden, *Phys. Rev. Lett.*, 78, 2417 (1997); *Phil. Trans. R. Soc. Lond.*, 359 , 997 (2001).
86. X. Wen, R.B. Meyer, D.L. Caspar, *Phys. Rev. Lett.*, 63, 2760 (1989).
87. J.E. Lydon. Chromonics, in *Handbook of Liquid Crystals*, Vol. 2B, p. 981, Ed. D. Demus, J. Goodby, G.W. Gray, H.-W. Speiss, V. Vill, Wiley- VCH, New York (1998).
88. J. Lydon, Chromonic liquid crystal phases, *Curr. Opin. Colloid Interface Sci.*, 3(5), 458 (1998).
89. J.L. Barrat, J.P. Hansen, *Basic Concepts for Simple and Complex Liquids*, Cambridge University Press, Cambridge (2003).
90. C. Tschierske, Molecular self-organization of amphotropic liquid crystals. *Prog. Polym. Sci.*, 21, 775 (1996); D. Blunk, K. Praefcke, V. Vill, Amphotropic liquid crystals, in *Handbook of Liquid Crystals*, Vol. 3, Ed. D. Demus, J. Goodby, G.W. Gray, H.W. Spiess, V. Vill, p. 305, Wiley-VCH, Weinheim, (1998).

91. P. Alexandridis, U. Olsson, P. Linse, B. Lindman, Structural polymorphism of amphiphilic block copolymers in mixtures with water and oil: Comparison with solvent-free block copolymers and surfactant systems. In *Amphiphilic Block Copolymers. Self-Assembly and Applications*, Ed. P. Alexandridis, U. Olsson, B. Lindman, p. 169, Elsevier, Amsterdam (2000) (overview over the phase behavior of PEO-PPO block copolymers).

92. A. Skoulios, G. Finaz, J. Parod, in *Block and Graft Copolymers*, Syracuse University Press, Syracuse, NY (1973).

93. C.J. Tschierske, *J. Mater. Chem.*, 16, 2647 (2001); P. Fuchs, C. Tschierske, K. Raith, K. Das, S. Diele, *Angew Chem. Int. Ed.*, 41, 628 (2002); X.H. Cheng, S. Diele, C. Tschierske, *Angew Chem. Int. Ed. Engl.*, 39, 592 (2000).

94. M.W. Matsen, F.S. Bates, *Macromolecules*, 29, 1091 (1996).

95. C. Lai, W.B. Russel, R.A. Register, *Macromolecules*, 35, 841 (2002); K.J. Hanley, T.P. Lodge, C.-I. Huang, *Macromolecules*, 33, 5918 (2000).

96. W. Wang, T. Hashimoto, *Polymer*, 41, 4729 (2000).

97. T.P. Rieker, *Liq. Cryst.*, 19, 497 (1995).

98. D. Kruerke, P. Rudquist, S.T. Lagerwall, H. Sawade, G. Heppke, *Ferroelectrics*, 243, 207 (2000).

99. A. Jákli, W. Cao, Y. Huang, C.K. Lee, L.-C. Chien, *Liq. Cryst.*, 28, 1279 (2001).

100. M.Y.M. Huang, A.M. Pedreira, O.G. Martins, A.M. Figueiredo- Neto, A. Jákli, Nanophase segregation of nonpolar solvents in smectic liquid crystals of bent-shape molecules, *Phys. Rev. E*, 66, 031708 (2002).

101. X.H. Cheng, M.K. Das, S. Diele, C. Tschierske, The influence of semiperfluorinated chains on the liquid crystalline properties of amphiphilic polyols: Novel materials with thermotropic lamellar, columnar, bicontinuous cubic and micellar cubic mesophases, *Langmuir*, 12, 6521 (2002); H.M. Von Minden, K. Brandenburg, U. Seydel et al., Thermotropic and lyotropic properties of long chain alkyl glucopyranosides. Part II. Disaccharide headgroups, *Chem. Phys. Lipids*, 106, 157 (2000).

102. N.L. Nguyen, J. Dedier, H.T. Nguyen, G. Siegaud, Synthesis and characterization of thermotropic amphiphilic liquid crystals: Semiperfluoroalkyl-β-D-glucopyranosides, *Liq. Cryst.*, 27, 1451(2000); H.M. Von Minden, M. Morr, G. Milkereit, E. Heinz, V. Vill, Synthesis and mesogenic properties of glycosyl diacylglycerols, *Chem. Phys. Lipids*, 114, 55 (2002).

103. J.H. Clint, *Surfactant Aggregation*, Blackie, Glasgow (1990).

104. H. Ellens, J. Bentz., F.C. Szoka, *Biochemistry*, 25, 4141 (1986).

105. V. Vill, H.M. Von Minden, M.H.J. Koch, U. Seydel, K. Brandenburg, *Chem. Phys. Lipids*, 104, 75 (2000).

106. W. Curatolo, *Biochim. Biophys. Acta.*, 906, 111 (1987).

107. G. Platz, J. Pölike, C. Thunig, R. Hoffmann, D. Nickel, W. Rzbinski, *Langmuir*, 11, 4250 (1995); B.J. Boyd, C.J. Drummond, I. Krodkiewska, A. Weerawardena, D.N. Furlong, F. Grieser, *Langmuir*, 17, 6100 (2001).

108. H.-D. Dörfler, A. Göpfert, *J. Dispers. Sci. Technol.*, 20, 35 (1999).

2

Fluids with Reduced Dimensionality

2.1 Surfaces Effects on Anisotropic Fluids

Scientists used to say that bulk was created by God, whereas surfaces are made by the Devil. This is to refer the complexities imposed by the interfaces between two different materials or different phases of the same materials.

2.1.1 Surface Tension

The surface tension γ of an isotropic liquid interface can be thought of as a free energy per unit area of surface ($\gamma = dW/dA$), or as a force per unit length ($\gamma = dF/dx$). It results from an imbalance in intermolecular forces at the surface (liquid–vapor interaction). Since there are fewer molecules at the vapor side than on the liquid side near the surface, molecules at the surface feel a net attractive force (see Figure 2.1a), which leads to the surface tension. The concept of surface tension is simply illustrated considering a liquid film, stretched on a wire frame as shown in Figure 2.1b. The work, W needed to increase an area by dA is equal the force multiplied by dx, where $dA = l \cdot dx$ (l is the width of the frame). Accordingly,

$$W = Fdx = \gamma dA = \gamma l \cdot dx \tag{2.1}$$

Actually, a soap film has two surfaces, so both the force and the work will be twice as large as indicated in Eq. (2.1). Surface tension has the units of J/m^2, or N/m. In practice, they use units of mNm^{-1} (where m in front of N means "milli," and the m after N means "meter"). The surface tension of water at room temperature is $72\ mNm^{-1}$.

Surface tension can be measured in many ways. The most important and useful methods are illustrated in Figure 2.2. One of the most accurate methods is to measure the rise of the liquid in a capillary (see Figure 2.2a). Another method is to weigh falling drops of a liquid, called pendant drop method

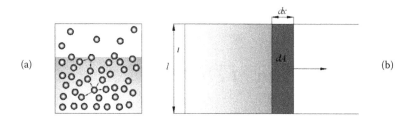

FIGURE 2.1
(a) Surface tension arises from the imbalance of forces on molecules at the liquid–gas interface.
(b) Stretching a soap film suspended on a wire frame by moving a slider through a distance dx.

(Figure 2.2b). Many commercial instruments work by either the "du Noüy" ring or Wilhelmy plate techniques (see in Figure 2.2c and Figure 2.2d).

2.1.2 Interfacial Tension

Whereas the surface tensions refer to the liquid–vapor interface, interfacial tension is defined as surface free energy per unit area between two immiscible liquids. It has the same unit as the surface tension, and it results from

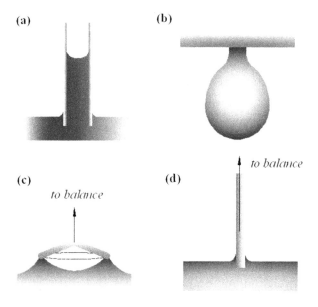

FIGURE 2.2
Methods for measuring surface tension: (a) capillary rise; (b) pendant drop; (c) du Noüy ring; (d) Wilhelmy plate.

the imbalance between two immiscible liquids.[†] Surfactants are very effective in decreasing the interfacial tension between water and organic solvents, and this is one of the mechanisms by which they act as detergents. The difference between the surface tensions of the two liquids ($\gamma_{\alpha a}$, $\gamma_{\beta a}$, where α, β denote the fluids and a stands for the "air") and the interfacial tension between them ($\gamma_{\alpha\beta}$) determines the work of adhesion as:

$$W_{\alpha\beta} = \gamma_{\alpha a} + \gamma_{\beta a} - \gamma_{\alpha\beta} \qquad (2.2)$$

For pure liquids, the work of cohesion is defined as the work required pulling apart a volume of cross-sectional area.

$$W_{\alpha\alpha} = 2\gamma_{\alpha a} \qquad (2.3)$$

Both the work of cohesion and the work of adhesion are defined for a reversible process. Surface tension can be interpreted as half of the work of cohesion. Similar to the surface tension measurements, interfacial tension can also be measured both by the du Noüy ring and the Wilhelmy plate surface tensiometers.

When a drop of liquid is placed on another with which it is immiscible, it can form a lens-shaped drop, it can spread completely into a uniform thin film, or it can form film with coexisting droplets. The actual behavior is determined by the wetting involving both short- and long-range forces[1], and is characterized by the spreading coefficient S and contact angle θ. The spreading coefficient, which is the difference between the work of adhesion between the two fluids and the work of cohesion of the top film, is related to the contact angle as:

$$S = \gamma_l(\cos\theta - 1) \qquad (2.4)$$

The condition of spreading is that $S \sim 0$. According to Eq. (2.4), a large contract angle near $180°$ means complete nonwetting, and the droplet makes a contact with a surface only in one point. In practice, the most important fluid is the water, where the wetting is directly related to the hydrophobicity of the surface; hydrophilic surfaces have small contact angles, and the water spreads over them, whereas hydrophobic surfaces make large contact angles. A smooth silane-treated hydrophobic float glass has a contact angle of about 100–$110°$. Especially interesting are the so-called super-hydrophobic surfaces, which are characterized by contact angles as large as $160°$. The best known super-hydrophobic material is the sacred lotus (Nelumbo nucifera), the symbol of purity in Asian religions, which has a contact angle of $160.4°$.[2] Even emerging from muddy waters, it unfolds its leaves unblemished and

---

[†] Note: Due to the similarity between surface and interfacial tensions, many scientists do not distinguish them, but use the term "surface tension" for both situations.

FIGURE 2.3
Illustration of the lotus effect.

untouched by pollution. When viewed microscopically, it becomes apparent that the Lotus leaves are composed of superposition of structural levels. In the micrometer scale, it contains bumps 5–10 μm tall and 10–15 μm apart (cuticles). This microstructure is coated with hydrophobic wax crystals of around 1 nm in diameter. Rough surfaces enhance hydrophobicity because of the reduced contact area (for lotus leaf, to about 2–3%) between the drop and surface. Minimization of the contact area also reduces the adhesion forces,[3] so that the slightest slope will cause the water drops not to slide but to roll off. In this way, particles such as dirt or bacteria will be picked up by the rolling water, thus cleaning the surface from contaminations. The lotus effect is illustrated in Figure 2.3.

The Lotus effect serves as a model for the development of artificial surfaces, where self-cleaning is important, for example in self-cleaning windows, roofs, paints, etc.[4] Reducing the use of cleansing agent, the Lotus effect is also beneficial for the environment. Especially interesting is the electrowetting for possible use in microfluidics. Electrowetting is essentially the phenomenon whereby an electric field can modify the wetting behavior of a droplet in contact with an insulated electrode. If an electric field is applied nonuniformly, then a surface energy gradient is created which can be used to manipulate a droplet sandwiched between two plates. Electrowetting arrays allow large numbers of droplets to be independently manipulated under direct electrical control without the use of pumps, valves, or even fixed channels.[5]

2.2 Fluid Monolayers

Benjamin Franklin carried out an experiment in 1757 on Clapham pond in which he dropped a small amount of oil onto the surface of the pond and watched how it spread out. Although he was interested only in seeing how the oil film calmed the water for life-saving purposes,[6] 40 years later his data about the volume and area of the oil film were used by Rayleigh to calculate the thickness of the film. This was the first measurement of the monomolecular thickness and represents the start of the studies of monolayer fluids. Many insoluble substances such as long fatty acids, alcohols and surfactants can be spread from a solvent onto water to form a film that is one molecule thick, called a monolayer. The hydrophilic groups (for example –COOH or –OH) point into the water, whereas the hydrophobic tails avoid it. The molecules in the monolayers can be arranged in a number of ways that reflect both the lateral forces between them and the forces between them and the supporting water. Their studies are important to learn about the nature of two-dimensional forces and structures.

2.2.1 Surface Pressure–Area Isotherms

Surface pressure is the reduction of surface tension due to the presence of the monolayer:

$$\pi = \gamma_0 - \gamma \tag{2.5}$$

i.e., it is the pressure that opposes the normal contracting tension of the bare interface.

Surface pressure isotherms are often measured for films in compressions using a so-called Langmuir trough, which is limited to nonsoluble materials (see Figure 2.4). The insoluble substance is spread from a volatile solvent to control the very small amounts of material, and the area of the monolayer is controlled through a movable barrier. The horizontal force necessary to

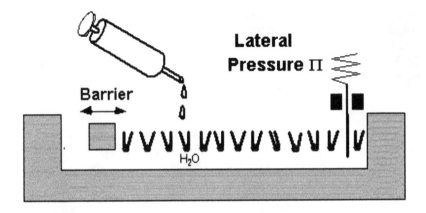

FIGURE 2.4
The schematic of the Langmuir trough.

maintain the float at a fixed position is measured using a torsion balance providing the surface tension.

A commonly used alternative to measure the surface pressure is the Wilhelmy plate. The force on this plate decreases as the surface tension decreases.

The shape of the surface pressure isotherm depends on the lateral interactions between molecules. This in turn depends on molecular packing, which is influenced by factors such as the size of head group, the presence of polar group, the number of hydrocarbon chains and their conformation (straight or bent).

Surface films also form on solutions of soluble surfactant. In that case, the surface tension isotherm is measured as a function of concentration of surfactant in the solution. The surface pressure isotherm can be basically understood by the Gibbs absorption isotherm, which in the particular case of air–water interface relates the change of the surface tension to the chemical potential of the surfactant as:[7]

$$d\gamma = -\Gamma_s \, d\mu_s \qquad (2.6)$$

where $\Gamma_s = \frac{n_s^\sigma}{A}$ is the surface excess per unit area, with n_s^σ being the number of moles in excess at the surface, and A is the surface area.

The chemical potential of the surfactant can be written as $d\mu_s = RTd \ln a$, where a is the activity of the surfactant in water, and $R = 8.315$ *Joule/K = 1.986 cal/K* is the gas constant. In the limit of an ideally dilute solution, a can be replaced by the concentration in the solution, c *(in mol/dm³)*. Thus, from Eq. (2.6), the surfactant excess per unit area at the surface can be related to the surfactant concentration in the solution through the equation:

$$\Gamma_s = -\frac{1}{RT}\left(\frac{\partial \gamma}{\partial (\ln c)}\right)_T = -\frac{c}{RT}\left(\frac{\partial \gamma}{\partial c}\right)_T \qquad (2.7)$$

This is the Gibbs adsorption equation, valid only for one surfactant.[†] Measurements of the surface tension γ versus $\ln c$ typically show the function plotted in Figure 2.5. We can see that three regimes can be distinguished.

In Regime I, the surface tension is linearly decreasing with increasing concentration, i.e.,

$$\gamma = \gamma_o - \tilde{k}c \qquad (2.8)$$

where $\tilde{k}$ is constant. In such a small concentration there is also a linear relation between the surface excess Γ_s and the surfactant concentration c, i.e.,

$$\Gamma_s = kc \qquad (2.9)$$

[†] In general, the Gibbs absorption equation is messier, since one may have more surface species, including charged surfactants, which modify the constant between 1 and 2, depending on the ratio of charged and noncharged surfactants.

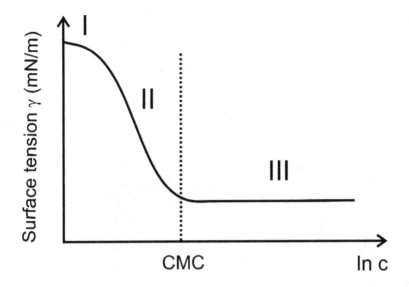

FIGURE 2.5
Typical variation of surface tension with the logarithmic of concentration for a pure aqueous surfactant solution (solid line) and a solution containing surface-active impurities such as alcohols (broken line).

where $k = \tilde{k}/RT$ has a dimension of length. For a standard surfactant, such as sodium dodecyl sulphate (S.D.S.), $k \sim 4 \ \mu m^8$. This means that, in the bulk, more than 4 μm width of volume would contain the amount of surfactant that is at the surfaces, i.e., the surface concentration is much larger than the concentration in the volume. In regime I, typically at the free surface the mean area per molecule is more than 1.5 nm^2, whereas in the bulk the mean volume per molecule is $3 \times 10^4 \ nm^3 \sim (30 \ nm)^3$. (In insoluble surfactants $k = \infty$, so that $\Gamma_s \neq 0$, while $c = 0$.)

From the definition of surface pressure of Eq. (2.5):

$$\pi \equiv \gamma_o - \gamma = \tilde{k}c \qquad (2.10)$$

Differentiating Eq. (2.8) and inserting this into the Gibbs equation, Eq. (2.7), we get that:

$$\pi = \Gamma_s RT \qquad (2.11)$$

or, from the definition of surface excess:

$$\pi A = n_s RT \qquad (2.12)$$

This shows that, in dilute solution, the surfactant behaves like an ideal gas in two dimensions, since it has the same form as of the ideal gas law in three dimensions ($pV = nRT$, Boyle-Mariott's rule). We note here that one needs extremely small surface concentrations to get this ideal-gas type behavior, and generally the surface pressure in this range is much smaller than one can actually measure.

In regime II, the decrease of γ with $\ln c$ becomes approximately linear. This occurs just below the inflection point that indicates the critical micelle concentration (CMC). This linear behavior means that surface excess Γ_s is constant (see Eq. (2.7)). It indicates that the surface excess saturates, while the bulk surfactant concentration continues to increase slightly with $\ln c$.

In regime III, above the CMC, the surface tension is nearly constant. This is due to the very weak dependence of chemical potential on concentration, and not because of the saturation of the adsorbed layer, since that already has occurred in regime II. The weak dependence of the chemical potential on the concentration above CMC is due to the fact that the newly introduced molecules add to the micelles and will not join the solution as single surfactants.[9]

To characterize the possible 2D structures the surfactant may form, one usually measures the surface pressure as the function of specific surface area occupied by one surfactant. For this, the area of the water covered by the surfactants is varied in the Langmuir trough, and the force acting on the mobile float or on the rigid barrier is measured. An experimentally measured surface pressure–area isotherm for a monolayer of n-hexadecanoic acid is shown in Figure 2.6. At large film areas, a two-dimensional gaseous (G) phase is formed, and it can be characterized by Eq. (2.12). The average area per molecule is large,

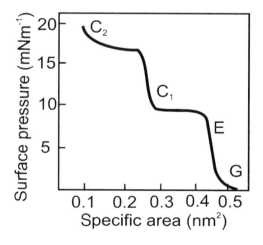

FIGURE 2.6

Schematic. Surface pressure isotherm with two condensed phases, C_1 and C_2.

although locally they tend to cluster into small islands or clumps. As the film is compressed, a transition occurs to an expanded (E) phase characterized by a plateau in the surface pressure. In this range, clusters of condensed structure form. The surface pressure does not increase in this range, because the packing in the individual clusters is unchanged; only they are brought closer to each other. The pressure will increase further only once these clusters are pressed together, i.e., when the condensed state uniformly covers the area. Some materials exhibit a range of different condensed structures showing a number of steps. At very small areas the monolayers cannot be pressed together, and multilayers form.

2.2.2 Langmuir–Blodgett (LB) Films

In the mid-twentieth century, Katharine Blodgett, a member of Irving Langmuir's laboratory, investigated the transfer of monolayers from the trough to a substrate, such as a glass slide dipped through the surface of the liquid with surfactant monolayer. She found that it is possible to build up multilayers by successive deposition of monolayer films onto a solid substrate using the Langmuir–Blodgett technique (see Figure 2.7).

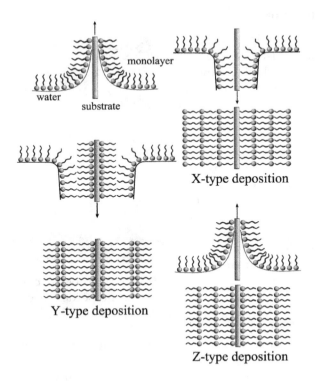

FIGURE 2.7
Deposition techniques of monolayers onto substrates. Y- and Z-type deposition require hydrophilic surfaces, like glass, whereas X-type deposition requires hydrophobic substrates.

To be deposited successfully on a solid substrate in an LB film, a monolayer should be in the condensed phase. The most common deposition mode is termed Y-type. Here the monolayer is picked up on a hydrophilic substrate (often glass) as it is pulled from the monolayer. The hydrophilic head groups attach on the substrate, leaving a surface that is now covered by hydrophobic chains. The substrate plus monolayer is then dipped back into the water, picking up the monolayer in the reversed orientation. The resulting LB film is therefore a stack of bilayers.

In some cases (with hydrophilic substrate and only weakly amphiphilic materials) the monolayers are deposited only in the upstroke with the same orientation, leading to a Z-type film. In case of hydrophobic substrates the monolayers are only deposited on the downstroke, and the result is the so-called X-type film. The combination of different deposition types is also possible. We note that similar methods were used by traditional candle makers.

2.3   Soap Films

Soap films have intrigued children as well as many historical scientific figures such as Hooke, Newton and Gibbs. In the late seventeenth century, some of the first recorded observations using soap films were reported by Hooke and Newton during their efforts to understand the reflections, refractions and colors of light. This laid the groundwork for the modern theory of light.[10,11] Specific properties of soap films, such as their stability, elasticity, their topology when stressed on various frames, have

been studied for a long time. Plateau,[12] Gibbs[13] and Boys[14] described several remarkable experiments demonstrating the properties of capillary forces. Recent excellent overviews are available by Isenberg,[15] Mysels et al.,[16] and Rusanov and Krotov.[17]

The standard view of a soap membrane is that it is a micrometer-thick sheet of water covered on either side by surfactant (soap) molecules. Without the surfactant, the liquid sheet would be unstable and would break up into droplets due to the high surface tension of the pure water. Surfactants lower the surface tension and also enable variations of the surface tension across the curvature of a film. This allows the film (bubble) to adjust to external forces that would otherwise destroy it. In addition, surfactant molecules slow down the evaporation of the water molecules; thus the lifetime of the film becomes longer. Basically, the surfactants endow the film with elasticity, or restoring force against any local thinning that could lead to rupture. Common soaps are made of a mixture of several surfactants, and the water used is normally not pure. The soap solutions typically used is some dishwashing detergent (for example, Procter & Gamble's Dawn is a good choice) mixed with water at a concentration of 1 to 2%.[18] (To increase the lifetime, usually 0.2–0.4% of glycerine is also added.)

According to the composition of the surfactant solution from which they are produced, the films were classified in three different categories by Mysels et al.:[16]

- Rigid films, such as those made from sodium dodecyl sulphate (S.D.S.), in the presence of dodecanol. The surfactant monolayers that border the liquid film are usually compact and rigid.[19] Such rigid films can last very long (even for a year) if the evaporation is halted, e.g., in a closed container saturated with water vapor.

- Simple mobile films, where rapid turbulent motion is visible close to the borders, while the thickness of the center is uniform.

- Irregular mobile films, obtained with concentrated surfactant solutions, where various different thicknesses are observed simultaneously even in the film center. In such films, so-called stratification of the films was observed a long time ago[20] and is still the subject of present studies.[21] Stratification is connected with the layered (lamellar) ordering of the molecules or micelles inside the films. These layers produce a stepwise thinning, which proceeds irregularly thorough the film surface. Similar effects will be described later for the free-standing smectic membranes.

2.3.1 Equilibrium Shapes

The scientific study of liquid surface, which has led to our present knowledge of soap films and soap bubbles, is thought to date from the time of Leonardo da Vinci, a man of science and art. Since the fifteenth century, researchers have carried out investigations in two distinct camps. In one camp there are the physical, chemical and biological scientists who have studied the macroscopic and molecular properties of surfaces with mutual benefit. The other camp contains mathematicians who have been concerned with problems that require the minimization of the surface area contained by a fixed boundary and related problems. These studies were initiated in the nineteenth century by a Belgian physicist, Joseph Plateau, who showed that dipping wire frameworks into a bath of soap solution could produce analogue solutions to the minimization problems. After withdrawing a framework from the bath, a soap film is formed in the frame, bounded by the edges of the framework, with a minimum surface area. All the minimum surfaces were found to have some common geometrical properties. This work has inspired mathematicians to look for new analytic methods to enable them to prove the existence of the geometric properties associated with minimum area surfaces and to solve the minimum area problems. However, it is only relatively recently that important steps have been made in this direction.

2.3.2   The Motorway Problem

One of the most interesting minimization problems for which an analytic solution is yet to be found is the problem of linking n points on the same plane in the shortest possible path. For example, let us consider the problem of linking four towns A, B, C and D by roads if they are situated at the corners of the square of unit length.

What would you think is the shortest path between the four towns? Some possible ways are shown in Figure 2.8.

Utilizing Plateau's observation that the soap film would like to form the smallest area, we easily can design an experiment where the formation of soap would show us the solution. In order to reach an analogue solution which will take advantage of the soap film's unique feature of reducing surface space to a minimum, we may construct two parallel clear Plexiglas plates joined by four pins, perpendicular to the plates. The pins arranged at the corner of a square will represent the four towns. When this arrangement is immersed in a bath of soap solution and withdrawn, a soap film will form between the two plates. It will reach the equilibrium configuration in which the area, and hence the length, of the film will be a minimum. Doing this experiment we find interconnected strips of soap film, which pattern looks like sketched in Figure 2.8e. The film consists of five straight strips that

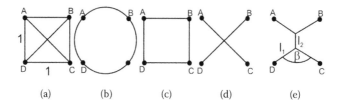

| (a)   (b)   (c)   (d)   (e) |

FIGURE 2.8

Possible ways connecting corner points of a unit length square A, B, C and D together. (a) The network that will enable us to travel from any town to any other town by the shortest possible rout. Its total length is $4 + 2\sqrt{2} = 6.83$. (b) A circular road passing through A, B, C and D has a length of $2\pi(\frac{1}{2}\sqrt{2}) = 4.44$. (c) Each circular arc can be replaced by a straight line producing the square roadway configuration with length of 4.00. (d) One might think that the shortest road is the crossroad system with a length of $2\sqrt{2} = 2.82$. (e) The line structure connecting four points at the corner of a square, corresponding to the minimum total length $1 + \sqrt{3} = 2.73$, when $\beta = 120°$.

intersect in two lines at 120°. Its total length is $1+\sqrt{3}=2.73$, which is about 4% shorter than the crossroad system.

To show that in this special four-pin system, indeed, the 120° intersection angle corresponds to the minimum, we have to find the minimum of the total length $L = 4l_1 + l_2$ with respect to the angle β (see Figure 2.8e). To simplify the calculation, we rather minimize the energy with respect to $r = sin(\beta/2)$. We see that:

$$L = \frac{2}{r} + 1 - \frac{\sqrt{1-r^2}}{r} \qquad (2.13)$$

The minimum condition requires that $dL/dr = 0$, which gives that:

$$-2 + \sqrt{1-r^2} + \frac{r^2}{\sqrt{1-r^2}} = 0 \qquad (2.14)$$

The solution of (2.14) is $r = \sqrt{3}/2$, i.e., $\beta = 120°$ angle at the junctions.

For any number of points, it is generally true that the minimum routes consist of straight lines forming a number of intersections. In the case of n points, the number of intersections will be in the range from zero to a maximum of $(n - 2)$.

A soap film contained by a fixed boundary of any three-dimensional framework will also acquire a minimum area. Consequently, soap films can be used to solve mathematical problems requiring the minimization of a surface area contained by a boundary. In order to obtain analogue solutions, we require a frame to form the boundary of a surface. When the frame is withdrawn from a bath of soap solution, a soap film will form which will attain its minimum area configuration on reaching equilibrium. Based on experimental observations, Joseph Plateau postulated that soap films contained by a framework always satisfy three geometrical conditions:

1. Three smooth surfaces of a soap film intersect along a line.
2. The angle between any two tangent planes to the intersecting surfaces, at any point along the line of intersection of three surfaces, is 120°.
3. Four of the lines, each formed by the intersection of three surfaces, meet at a point and the angle between any pair of adjacent lines is 109°28′.

These rules can be relatively easily proven for frameworks with high symmetry, such as Platonic polyhedrons with regular faces, all congruent, with equal face angle at every vertex and where all the angles between adjacent faces are equal. The images in Figure 2.9 show the minimal surface of the following forms: the tetrahedron, the cube, the octahedron and the dodecahedron. However, in the most general cases, the proof is very complicated,

FIGURE 2.9
Minimal surfaces for tetrahedron, the cube, the octahedron and the dodecahedron.

and it is only recently that Frederick J. Almgren Jr. and Jean E. Taylor[22,23] have shown that these conditions follow, in general, from the mathematical analysis of minimum surfaces and surfaces containing bubbles of air or gas at different pressures.

The tendency to form minimal surface is manifested itself in the formation of bubbles, which have the smallest possible surface when no other constraints exist. When one bubble meets with another, the resulting union is always one of total sharing and compromise. Since bubbles always try to minimize surface area, two bubbles will merge to share a common wall. If the bubbles have the same sizes, this wall will be flat. If the bubbles have different sizes, the smaller bubble, which always has a higher internal pressure, will bulge into the larger bubble. The larger is the difference in size, the faster is the bulging. The interaction of bubbles with different sizes somehow reminds us of the interaction of human beings. (When equally powerful, we can do compromises and share resources, but more powerful persons will take the resources from others, and the majority of resources will be distributed only among a small percent of people.) Even the dynamics of the redistribution of the resources seems to be analogous to the interaction of soap bubbles.

2.3.3 Stability and Elasticity of Soap Films

The relative stability of soap films is partially due to their elasticity. To see this, we consider a film with an equilibrium state represented by the surface excess Γ_s and the dissolved surfactant concentration c. A local stretching disturbs these parameters. If the disturbance occurs on a short time scale, the soap molecules do not have time to diffuse out from the inner fluid to the surface, so the same number of soap molecules remains both inside and at the surface. As the area increases, Γ_s decreases, which results in an increase of the surface tension according to Eq. (2.11), thus opposing the stretching. This process is called the Marangoni elasticity[8] and explains the film stability to rapid disturbances. If the time scale of the stretching is longer, the molecules of the interstitial fluid have time to diffuse out of the surface. At each time there is a thermodynamic equilibrium between Γ_s and c, as described by the Gibbs equation, and the increase of the elasticity now is called Gibbs elasticity.

In principle, the limit between the Marangoni and the Gibbs elasticity is given by the time scale $\tau_D = h^2/D$, which characterizes the time needed to diffuse away by the thickness h. Since the diffusion coefficient is $D \sim 4 \times 10^{-10} \ m^2/s$, for films of 1 μm thick τ_D is in the order of 0.01 s. However, it is argued[17] that impurities make the equilibrium more difficult to reach, so that Marangoni elasticity can be observed up to the order of a second.

The elastic modulus of the film can be defined as:

$$E = -2A\frac{d\pi}{dA} = -2h\frac{d\gamma}{dh} \tag{2.15}$$

For small concentrations $\pi = \Gamma_s RT$ (see (2.11)), i.e., the elastic modulus simplifies to:

$$E = 2RT\Gamma_s \tag{2.16}$$

where $\Gamma_s = kc$ (see (2.9)).

In thick films, there is no direct interaction between the two surface layers, so that the film thickness can have any value. Therefore, we can assume that the concentration of the soap is uniform and can be described as:

$$c_o = c + 2\frac{\Gamma_s}{h} \tag{2.17}$$

where c_o is the total surfactant concentration. When c and Γ_s have no time to change during the deformation, (2.17) and (2.11) give that:

$$\Gamma_s = c_o\frac{hk}{h+2k} \tag{2.18}$$

resulting in the Marangoni elasticity as:

$$E_M = 2RTc_o\frac{hk}{h+2k} \tag{2.19}$$

To calculate the Gibbs elastic constant E_G, we allow the variation of c and Γ_s, i.e., from (2.15) we write:

$$E_G = -2h\frac{d\gamma}{dh} = -2h\frac{d\gamma}{dc}\cdot\frac{dc}{dh} \tag{2.20}$$

From (2.8) and (2.17), (2.20) gives the expression for the Gibbs elastic constant:

$$E_G = 4RTc\frac{hk^2}{(h+2k)^2} \tag{2.21}$$

It is related to E_M by:

$$E_G = \frac{2E_M k}{h+2k} \tag{2.22}$$

We see, therefore, that in the limit of thin films ($h \ll 2k$), $E_G \sim E_M$, because the interstitial fluid is too thin to provide soap molecules to the surface.

The usual soap bubbles observed in white light show bright interference colors. Their thickness ranges from 0.1 to 10 μm, and they are called *"thick" films.* When such thick films placed in vertical frames, their thickness varies. This variation can be easily calculated for insoluble surfactants, where $\Gamma_s \neq 0$ and $c = 0$. In this case $\pi = \Gamma_s RT$(see (2.11)) with (2.17) gives that:

$$d\pi = RT\frac{c_o}{2} dh \tag{2.23}$$

In equilibrium, the gravity is balanced by the gradient of the surface pressure of the two surfaces, i.e.,

$$-2\frac{d\pi}{dz} = g\rho h(z) \tag{2.24}$$

where ρ is the density for unit length. Using (2.23), (2.24) can be rearranged to give:

$$-\int_{h_o}^{h}\frac{dh}{h} = \int_0^z \frac{\rho g}{RTc_o} dz \tag{2.25}$$

which gives:

$$\ln(h/h_o) = -\frac{\rho g}{RTc} z \tag{2.26}$$

i.e.,

$$h = h_o e^{-\frac{\rho g}{RTc}z} \tag{2.27}$$

This shows that, in gravity, the thickness of vertical thick film profile follows the density profile of an isothermal atmosphere. The only difference is that,

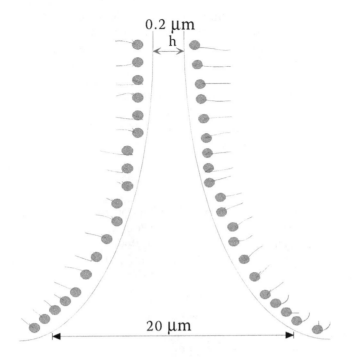

FIGURE 2.10
Thickness profile of a vertical soap film.

in the atmosphere, the gas carries its own weight only, whereas soap molecules carry the interstitial fluid as well. As a result, at room temperature, the vertical variation of the density is much larger, and a 30-cm-high soap film simulates an atmosphere a few kilometers thick. In a 30-cm-high frame, the thickness varies typically from 0.2 µm at the top to 10 µm at the bottom (see Figure 2.10).

2.3.4 Thinning of Soap Films

As we have seen in the previous example of the vertical soap film (Figure 2.10), thick film surfaces can be curved. However, as the films lose water (drainage), they progressively flatten due to interactions between the surfaces.

After drainage some films may become very thin, and they hardly reflect light. They are called *black films* and can exist in two states: the common black film has a thickness of the order of 30 nm; the second (Newton's black film) is only 4.5 nm thick. In thin films the two surface layers are also separated by some interstitial fluid, but the strong interaction between the two surface layers determines the film thickness, which will remain uniform even if placed vertically. The black films are in metastable states, they can burst, but cannot undergo further drainage, and they also cannot become thicker later. To describe the thinning process leading to the black films, a number of models have been developed.[24]

When the thickness of the film becomes comparable to the range of inter-action forces between the film surfaces, the applied pressure must be sup-plemented by the so-called "disjoining pressure" introduced by Deryaguin,[25] which results from the overlap of molecular interactions between the film surfaces. These interactions can be quantified as an excess pressure versus the separation distance (i.e., the film thickness, h), called the disjoining pressure isotherm, $\Pi(h)$. Note that the term disjoining is somewhat mislead-ing, since it also contains attractive forces that are "conjoining" by nature. The disjoining pressure contains a number of terms:

$$\Pi(h) = \Pi_{dl} + \Pi_{van} + \Pi_{st} \tag{2.28}$$

where the subscripts indicate the following contributions: dl = electrostatic double layer forces; van = London–van der Waals dispersion forces, st = steric and short-range structural forces. In the following we only briefly review each of the components of Eq. (2.28). For a comprehensive review several textbooks and monograms are available.[24,25,26,27]

2.3.4.1 Electrostatic Double-Layer Forces

Charged surfaces separated by perfect insulators attract or repel each other due to electrostatic forces. In case of fluid films, there are always free charges that can move to the surfaces and screen out the electrostatic forces. From the classical Debye–Hückel theory, it can be seen that the ions form a diffuse atmosphere characterized by the Debye screening length, λ.[28] It is found that the decay length is inversely proportional to the square root of the ion concentration in the fluid. If the thickness of the fluid h is much larger than 2λ, $\Pi_{dl} \sim 0$; however, when the ionic clouds start to overlap ($h < 2\lambda$), the ions cannot completely screen out the electrostatic forces between the two charged surfaces, and $\Pi_{dl} \neq 0$. There is an extensive literature concerning the calculation of the exact electrostatic repulsion forces,[29] and we do not describe them here.

2.3.4.2 The London–van der Waals Forces

Because of the constant motion of the electrons, an atom or molecule can develop a temporary (instantaneous) dipole when its electrons are distrib-uted asymmetrically about the nucleus. A second atom or molecule, in turn, can be distorted by the appearance of the dipole in the first atom or molecule (because electrons repel one another), which leads to an electrostatic attrac-tion between the two atoms or molecules (see Figure 2.11). Π_{van} depends on the polarizability of the interacting molecules and is inversely proportional to the sixth power of separation.

The London–van der Waals dispersion forces have long been recognized as being important in thin liquid films. These forces have been calculated for different surfaces by pairwise summation of the individual dispersion

FIGURE 2.11
Illustration of the dispersion (induced dipole–induced dipole) forces. Large dots represent nucleus; smaller dots are the electrons; and $\delta+$ and $\delta-$ indicate the net local charges due to the asymmetric distortions of the atoms.

interactions between molecules by Hamaker.[30] Casimir and Polder later[31] perfected this approach by including the corrections for electromagnetic retardation. For example, for two- plane parallel surfaces separated by a vacuum gap, the pressure reads:[27,30]

$$\Pi_{van} = -\frac{A_{12}}{6\pi h^3} \tag{2.29}$$

where A_{12} is known as the Hamaker constant. When the interaction between two different bodies is mediated, e.g., by an aqueous film, the Hamaker constant changes. Elaborate models have been developed to handle different geometries and more complex systems.[32]

2.3.4.3 Steric (Entropic) Forces
Entropic confinement forces occur at ultrathin (<5 nm) surfactant films and between bilayers in solution. They are mainly responsible for the stability observed in so-called Newton black soap films. It arises from the steric repulsion occurring when adsorbed layers overlap. These forces operate by various modes, like undulation, peristaltic fluctuations, or by head-group overlap.[33]

The schematic of a soap film disjoining pressure isotherm is shown in Figure 2.12. It is important to note that thermodynamically metastable films can exist only in regions with negative slopes. Hence, the region with positive slope separates metastable regions between the common black films (~50 nm) and the Newton black films (~4 nm). Basically, common black films' stability is due to the electric double-layer forces, while the Newton black films should be the results of entropic forces.

2.3.5 Bursting Soap Films

The death of a bubble (popping) is called bursting in science. It has been studied by many authors for more than a century.[34] A burst always starts with hole that grows due to the surface tension that wants to minimize the surface area. A simple approach for predicting the velocity of the growth of the hole is given by Culik.[35] It is based on assuming that the water from the

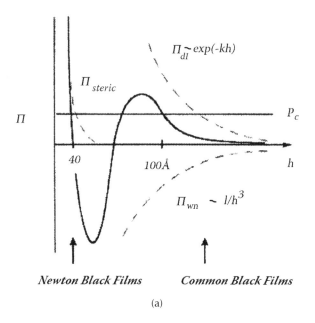

(a)

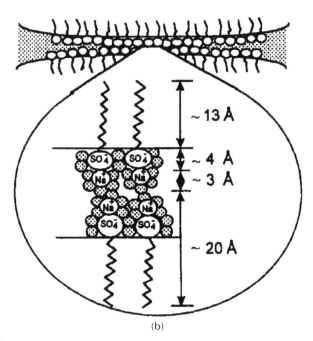

(b)

FIGURE 2.12

(a) Schematic representation of a disjoining-pressure isotherm that includes contributions from π_{dl}, π_{van} and π_{st}. (b) Schematic structure of Newton black films.

central part of the film was collected into a rim, and that there is no motion ahead of the rim. With these, the Newton's law can be written in the form:

$$\frac{dP}{dt} = \frac{d}{dt}(mv) = F = 2\gamma \tag{2.30}$$

where P is the momentum per unit length, m is the mass per unit length, and F is the pulling force related to the surface tension γ. It was experimentally observed that the velocity of the rim of the hole v is constant,[34-37] so the force is coming from the change of the mass:

$$\frac{dm}{dt} = \rho h v \tag{2.31}$$

where ρ is the mass density, and h is the initial thickness. Combining (2.30) and (2.31) we get:

$$v = \sqrt{\frac{2\gamma}{h\rho}} \tag{2.32}$$

All this seems to be simple, but the strange thing is that the work done by the capillary forces is:

$$\int Fv dt = mv^2 \tag{2.33}$$

i.e., exactly twice of the standard kinetic energy: $1/2mv^2$. This means that the dissipation is exactly equal to the kinetic energy. This situation is similar to the case of a conveyor belt, where sand is poured on the belt with constant rate, and thereafter it is transported with constant speed v. In this case the force is $F = vdm/dt$, just as observed for bursting soaps. Considering that the collision of one grain with the belt surface can be described as in the reference frame of the belt, the grain hits the belt at constant horizontal velocity, v, and then stops. This means that the collision is inelastic, and the dissipation is exactly equal to $1/2mv^2$. This analogy intrigued a number of scientists starting from Dupre[36] and Lord Rayleigh,[37] followed by Mysels[16] and most recently by de Gennes.[38]

2.4 Smectic Membranes

As early as the 1920s, Georges Friedel recognized[39] that one could prepare free-standing films (just like soap films) from smectic liquid crystals. (That could be the reason for calling them smectic, which means soap in Greek). A simple technique was developed, and systematic studies were started at

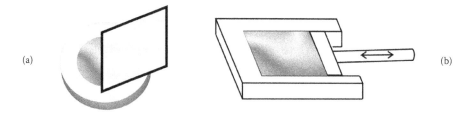

FIGURE 2.13
Freely suspended smectic film geometry. (a) Schematic of the film holder based on fixed frame and a wiper. (b) A frame with variable size.

Harvard in the late 1970s.[40] According to their method, one wets the edges of a hole in a glass cover slide with a smectic liquid crystal. A second glass slide is then drawn across the hole. By varying the amount of material and the rate of drawing, it is possible to form stable films from two to a few hundreds of molecular layers thick. The schematic of the drawing device, together with the smectic film structure, is shown in Figure 2.13a. With this technique, it is very convenient to produce highly perfect structures of smectic membranes. However, it is not easy to control the number of layers in the film.

For a better control, the fixed frame and wiper system is replaced by a rectangular frame of variable surface made of two mobile chariots supported by and gliding between two fixed parallel rails.[41] In this device, one can control the surface very precisely if one of the chariots is coupled to a precision translator stage. In this design, the production starts by depositing a small amount of liquid crystal in the almost completely closed chariot. At this stage, the liquid crystal behaves like an ordinary, but very viscous liquid; it fills the frame and forms a thick layer limited by two curved menisci. When the film is slowly opened, near the center a flat circular membrane appears. The number of layers in the film depends on the speed of the opening. For a constant membrane thickness, the surface can change by the transfer of molecules between the meniscus and membrane. In order to remove one smectic layer, one exerts a short rapid traction on the membrane by a stepwise increase of the surface area. As a consequence, a pore appears in one or more layers. Eventually, the pore increases and reaches the frame, leaving a membrane with a smaller number of layers. Conversely, the number of layers can be increased by a stepwise decrease of the surface, which results in a formation of an island or elementary step that can eventually cover the whole membrane. At the critical radius, r_c, an island is in equilibrium (does not shrink or increase). It is found that $r_c \sim 50 \ \mu m$.[41]

The number of layers, N, contained in the film can be determined by measuring its optical reflectivity, which, for thin film, is proportional to N^2. The thickness of a fresh film usually varies from a few layers in the middle to a thickness comparable to optical wavelengths in the edges, where it shows bright interference colors. By keeping the temperature near the higher end of the smectic phase (annealing), eventually the thin uniform range extends in cm size area. Once the film forms, it is stable and even can be cooled into the tilted smectic C phase, or to a layered crystalline smectic B phase. We note that free-standing films can be drawn directly in the smectic C phase, too.

Thermotropic smectic membranes are similar to the Newton black soap films in respect to their structure, thickness, and because both are spanned on frames and can exchange molecules with the meniscus. However, the smectic membranes are more complex in the respect that the number N of monolayers is variable, whereas the Newton black films are usually bilayers. Smectic membranes are also similar to vesicles by means of their layered structures; however, the vesicles are rather isolated unframed systems in the sense that the number of molecules is conserved, but the surface is free to evolve.[42]

Besides the variable thickness Nl (l is the thickness of one monolayer) of a smectic membrane, another obvious characteristic is that the membrane stays nearly flat in spite of its weight. This indicates that the membrane must be subject of tension, just like the membrane of a drum. When this tension is released, the thickness of the membrane increases by the addition of new layers. The tension of the membrane can be measured by different methods.[41,43,44] The most popular (and perhaps the most precise) method is to measure the eigenmodes of the mechanical vibrations of the membrane induced, for example, by applying a potential difference:

$$V = V_{dc} + V_{ac} \sin(\omega t) \tag{2.34}$$

between a metallic frame and a number of electrodes situated below the membrane.

According to the theory of Rayleigh,[45] the frequency of eigenmodes of a circular membrane of radius R in vacuum is:

$$\omega_{n,s} = (u_{n,s}/R)(\gamma/\rho)^{1/2} \tag{2.35}$$

where ρ is the two-dimensional mass density, and $u_{n,s}$ is the *sth* root of the Bessel function $J_n(r)$.[46]

Measurements provide that the tension is typically in the order of *50 dyn/ cm = 0.05 N/m*. It was also observed[41] that the tension is decreasing during stretching, i.e. increases with thickness. Experimentally the tension was found[41] to be a linear function of the thickness, i.e.,

$$\tau = \tau_0 + \alpha N \tag{2.36}$$

where α (~10^{-6} N/m) is the tension difference between films with thickness differing by one layer. This observation and other behavior of the islands and holes can be explained by a simple thermodynamic model worked out by the Orsay group.[47,41] The membranes can be considered as a two-dimensional system. The third dimension appears only through global quantities as the thickness. Accordingly, a two-dimensional thermodynamic model should successfully describe the behavior of the membranes. To describe membranes with coexisting areas of different numbers of layers, any uniform areas with a constant number of flat layers are called "fields." Each field can be considered as a different two-dimensional phase separated by interfaces — the steps.

We can assume that heat, as well as molecules, can be exchanged between the different layers, different fields and between the membrane and the meniscus. Therefore in equilibrium, not only the temperature, but also the chemical potential μ of molecules in all layers, fields and in the meniscus must be identical. Let $f(N,a,T)$ be the average free energy per molecule in the field of thickness N, where a is the average surface area per molecule. Since smectic membranes with any number larger than $N = 2$ can be stable in the limit of weak constraints, for each N (at constant temperatures T) there should be an equilibrium a_N of the surface per molecule, and $f(N,a,T)$ are well defined near a_N. Hence we can expand $f(N,a)$ as:

$$f(N,a) = f_o(N) + \frac{1}{2} B_N |a - a_N|^2 + \cdots \tag{2.37}$$

We assume that the distance between the molecules in a layer only slightly depends on the number of layers in a field. We also can assume that:

$$a_N \approx a_1/N \tag{2.38}$$

where a_1 is the typical surface per molecule in a monomolecular layer. Since the variation of the free energy per molecule can only depend on the relative change of the surface per molecule $(a - a_N)/a_N$, from (2.38) we see that B_N should be proportional to N^2.

The contribution of the surface layers is different from that of internal layers, i.e.:

$$f_o(N) \approx f_i + 2(f_s - f_i)/N \tag{2.39}$$

With these considerations, we get

$$f_o(N) \approx f_i + 2(f_s - f_i)\frac{a_N}{a_1} = f_i + \tau_o a_N \tag{2.40}$$

This means that the minima $f_o(N)$ are distributed on a line of slope τ_o.

The equilibrium between the two fields separated by steps means that both the chemical potentials and the tensions (since no other forces act on the steps) are equal. When a step is submitted to a force (for example due

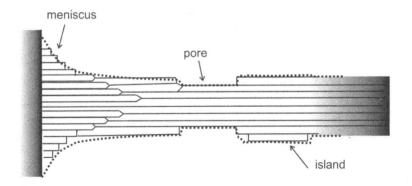

FIGURE 2.14
Simplified model of a meniscus consisting of a series of straight elementary steps.

to interactions with other steps), the tension on the two sides must be different, and equal to the force, $F \neq 0$.

$$\tau_N - \tau_{N-1} = F \quad (2.41)$$

We note that, in the case of curved steps, a force $F = \gamma/r$ appears, due to its own tension γ of the step. Obviously such a step can be canceled by the difference of tension on its own side.

The meniscus connecting the membrane to the frame may have quite a complicated structure, depending on the quality of the frame surface and on profile of the frame. However, the simplest case is when the membrane material wets the plane surface normal to the membrane, and when the volume of the meniscus is small enough, it can be considered as a collection of steps[48] (see Figure 2.14), so the considerations described above seem to be valid even when dealing with the equilibrium between the film and the meniscus.

In practice, the volume of the meniscus is much larger than that of the membrane, and it can be considered as a reservoir characterized by a chemical potential μ_{men}.

It is important to mention that for the approximations of (2.37) and (2.40), the model provides that in the limit of large N:

$$\tau_N \approx \tau_o + \frac{\Delta\mu}{\alpha_1} N \quad (2.42)$$

where $\Delta\mu = \mu_i - \mu_{men}$. This result is in agreement with the experimentally observed increase of the membrane tension with the thickness.

When smectic membranes are exposed to external pressure from one side, they bend until a balance between pressure difference and surface tension is established.[49] In this way, self-supporting spherical bubbles can be produced.[50]

The bubbles consist of highly ordered membranes of nanometer to micrometer thickness and are analogous to soap bubbles. They can reach radii of several centimeters.

2.4.1   Importance of Free-Standing Smectic Membranes

There are a number of reasons for the interest in free-standing smectic membranes. First, thin smectic membranes are models of two-dimensional fluids (SmA and SmC) and crystals (SmB). By varying the film thickness, one therefore can study the crossover from three- to two-dimensional behavior as well as the influence of surfaces on the morphology and the phase behavior.

One striking difference between three and two dimensions is the destruction of any rigorous long-range order in two-dimensional systems by long-wavelength thermal fluctuations. Although these fluctuations are present in systems of any dimensions, their effect turns out to be increasingly significant as the dimension of the system is reduced. This was first recognized in the 1930s[51] and is known as Landau–Peierls instability, although rigorous proofs were worked out only in the 1960s.[52]

Accordingly, for 3D and 2D crystals, the amplitude of the fluctuations remains finite even for infinite sample sizes, whereas in smectic liquid crystals the thermodynamic fluctuations logarithmically diverge with the linear size of the sample, L.

Numerical estimates (see Chapter 4) show that, for even as large as 1 km of smectic material, the fluctuation is only slightly larger than 1 nm. We could therefore say that these fluctuations practically do not matter in the films and systems we study. However, Michael Kosterlitz and David Thouless suggested,[53] provided that short-wavelength fluctuations did not first disrupt the order, that long-wavelength fluctuations might control the phase behavior at a phase transition. These fluctuations, in turn, are controlled by the behavior of defects in the system.

Study of free-standing membranes is also extremely important to map the polar nature of ferroelectric, ferrielectric and antiferroelectric liquid crystals of chiral[54] and bent-core achiral molecules.[55] An excellent overview of smectic membranes with detailed results and list of literature has been published recently by de Jeu et al.[56]

2.5   Fluid Foams

Foams are cellular materials made up of soft bubbles in which gas is dispersed in a continuous liquid or solid, when the volume fraction of the gas is larger than of the liquid or solid. Foam structures occur, or are conjectured to do so, on every length scale, from the Planck scale (10^{-35} meters) to that of the large-scale structure of the universe. On the everyday length scale of your favorite pub, foam is what we get when bubbles rise out of a liquid, such as beer, so it is in the millimeter scales. Examples for solid foams include cork, bread, sponge, pumice of volcanoes, "Damascus sword," polyurethane foams. Pumice is formed as gas, trapped in hot liquid lava, expands into bubbles that become fixed as the lava cools and solidifies. Some foams of pumice contain so many gas-filled cavities that they are less dense than water and can float. The sponges have vast internal storage capacity, which is another feature of foamy geometry. Corks (the barks of the cork oak tree) have been mostly used as stoppers in wine jars for over 2000 years, because of their excellent sealing property. Cork has a closed cell structure, with cells filled with air and separated by walls made of a waxy substance. This makes cork also resilient, as you force a cork into a wine bottle, it compresses and then expands as much as possible, thus completely sealing the neck of the bottle. In baking, the first step is to change starch, the main component of wheat into sugar, which is then fermented by yeast, a single-cell living fungus, which combines with sugar to make alcohol and carbon dioxide. The alcohol evaporates during baking, but the carbon dioxide is trapped by gluten, which forms a network of proteins (this linking tendency makes gluten useful as an adhesive). The final result is a solid material punctuated with numerous cells, arranged in a random pattern. More examples of solid foams and their applications can be found in numerous books, e.g., the one by S. Perkovitz.[57]

Semiliquid (cells, viruses) and liquid foams (shaving cream, whipped cream, champagne, beer, soap foam), all contain surfactants to minimize the surface energy under the volume constraint. Understanding the liquid foams is not only a deep mathematical challenge, but also provides insights into the beautiful physics of equilibrium, metastability, phase transitions, and rheology, as well as industrial applications. Like sand, liquid foam can be solid or liquid

Lord Kelvin

depending on circumstances. When you shave, it is a solid while it sticks to your skin, but it is easily removed because it is a fluid when subject to sufficient force. One may drink Guinness without much effort, yet a skilled barman can sketch a shamrock in its stiff foamy surface. This dual nature must have appealed to Sir William Thomson (later Lord Kelvin) — he corresponded a lot with his lifelong friend Gabriel Stokes on the search for what we would today call a complex fluid, as a model for the ether, the mysterious substance that was supposed to carry light waves as its vibrations. The trouble with the ether was that it seemed to have inconsistent properties: you could move through it, so it had to be a fluid. But it carried light as transverse vibrations, so it had to be a solid. For this reason Kelvin thought that the ether of space must be a foam. He applied himself to it with characteristic energy. In due course, even his wife was pressed into service to make a pin-cushion that would illustrate the ideal structure that he had conceived for the ether. Kelvin supposed that all bubbles were of equal volume in his ideal ether foam, and formulated the following question, as his starting point.

What foam structure will minimize energy, which is just the total surface area of all of the films? This is the Kelvin's problem. The solution of the problem in 2D was conjectured by him to be the honeybee's comb structure. This conjecture was proven recently by Thomas Hales[58] for infinite structure or for finite structures with periodic boundary conditions. Besides this, only the N = 2 case (the double-bubble problem) has been solved in 2D and 3D. Cases for N larger or equal to 3 in 3D have been studied only partially.[59] Concerning 3D infinite structures, Kelvin came up with the body-centered cubic structure, which he called tetrakaidecahedron. However, recently an alternative structure with a lower energy was computed by Weaire and Phalen.[60] This has a more complicated structure with two different kinds of cells (see Figure 2.15).

Concerning experimental observations it is clear that the real foams have much more irregular structure than the theoretical ones. In 2D structures one sees not only hexagons but also squares, pentagons, hexagons and heptagons are observed. This is called "topological disorder." It is proposed that there is an analogy between the 2D electrostatics and the 2D foam-statics so that the potential V corresponds to the pressure P, the electric field $-\vec{\nabla}V$ to the curvature $-\vec{\nabla}P$, and the electric charge e to the topological charge q.[61] The "topological charges" quantify the deviation from the hexagonal lattice. An n-sided bubble has a charge:

$$q = (6-n)\frac{\pi}{3} \tag{2.46}$$

Charge is additive; the charge of a collection of bubbles is the sum of their individual charges.[62] A foam with periodic boundary conditions has zero total charge and, hence, an average of six sides per bubble.[61]

The topological disorder is true for foams in 3D, too. Similarly, the sizes of the cells are different. These variations result in the strong whitish light scattering of foam, as noted first by Boyle after doing some foam experiments with egg white, and other liquids, and he gave a very correct account of it in terms of light scattering.[63]

The vertices of the polyhedra are called *plateau borders* after the blind

Joseph Plateau

Belgian physicist Joseph Plateau (see picture). At equilibrium the foam obeys the Plateau rules:[63,64] bubble edges are circular arcs that meet in triplets at $2\pi/3$ angles.[65] According to Laplace's law, their algebraic curvatures ($\kappa_{ij} = -\kappa_{ji} > 0$

(a)

(b)

FIGURE 2.15
(a) Illustration of the Kelvin cells and their joining. (b) An extended view of the Weaire–Phelan structure.

when bubble i is convex compared with bubble j) are related to the 2D pressure P_i inside bubble i as

$$\kappa_{ij} = \frac{P_i - P_j}{\gamma} \qquad (2.47)$$

Thus the algebraic curvatures of the three edges that meet at the same vertex must add to zero:[66]

$$\kappa_{ij} + \kappa_{jk} + \kappa_{ki} = 0 \qquad (2.48)$$

This equation holds for any closed contour crossing more edges. It is also valid in 3D, with κ being the mean curvature of a bubble face.

Just as we discussed in connection with the stability of the soap films, foams are not thermodynamically stable, due to their large interfacial area and, thus, surface free energy. However, some foam, particularly those formed by addition of small amounts of foaming agent such as soaps and surfactants, can be metastable, and the bubbles may keep their stability due to Marangoni and Gibbs elasticity.

In spite of this metastability, the bubble changes its shape from time to time in sudden topological rearrangements, and it grows or shrinks as gas diffuses between it and its neighbors.

One mechanism that leads to the gradual change of the foam is the drainage, although it is somewhat retarded due to the surfactant. Drainage leads to thinning of the liquid, and rupture — due to random disturbances, such as impurities, evaporation, strains, temperature changes — which is the death of the foam. Foam that initially contains bubbles can develop into foam containing polyhedral cells as a result of drainage. The tendency for regions of high gas content to rise to the top of the foam, leaving denser regions close to the bulk liquid, is called creaming. Each individual bubble therefore has an eventful life. Born at a small pit in the glass, it rises and joins its colleagues in wet foam at the surface of the liquid. Continuing to rise with its neighbors, the bubble is surrounded by less and less liquid, as the pull of gravity drains that liquid away.

In addition to the rearrangements and the gravity-driven drainage, coarsening can take place also due to the so-called *Ostwald ripening,* which is the selective growth of large bubbles at the expense of small ones.

Very recently foams of thermotropic smectic liquid crystal 8CB (4-n-octylcyanobiphenyl)[67] were reported. The foam was made in the nematic phase by bubbling nitrogen through the pure liquid crystal. The coarsening behavior was investigated both in the nematic and smectic phase. The mean bubble radius $<R>$ has been measured as a function of time and it was found that $<R> \sim t^{\lambda}$. In classical wet soap foams, the growth exponent is typically $\lambda \sim 0.33$, whereas in the liquid crystal, $\lambda \sim 0.2$ was observed both in the nematic and smectic phases. This may be explained by the presence of defects at the surface of the bubbles, thus slowing down the coarsening. Such statement is corroborated with the observation that, in the isotropic phase, the foam rapidly ruptures.

2.6   Fluid Fibers

Fibers are very important objects in our life. Examples include natural silk spun by silkworms and spiders, synthetic fabrics such as nylon, polyester, or optical fibers for communication networks. Fibers can be drawn only from viscous fluids that harden during the pulling process. The hardening can be achieved either by cooling, such as in glass fibers or by loosing water, e.g.,

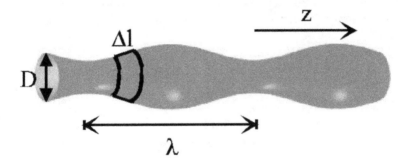

FIGURE 2.16
Description of a fluid jet under periodic fluctuation.

in spinning spider silks. As surface tension causes fluids to have as little surface as possible for a given volume, Newtonian fluids (described by a constant viscosity) cannot form fibers, but only short "bridges" with lengths smaller than their circumference (Plateau–Rayleigh instability).[68] Following Plateau's (1873) arguments for Newtonian fluids in gravity-free environment, we consider a small periodic fluctuation as:

$$R = R_o(1 + \varepsilon \cos kz) + \varepsilon^2 R_1 \qquad (2.49)$$

where R_o is the equilibrium diameter of the cylindrical fiber drawn in z direction, and k is the wave number of periodic distortion.

The volume under such perturbation over a wavelength $\lambda = 2\pi/k$ can be written as:

$$V = \pi R_o^2 \lambda + 2R_o \varepsilon^2 \pi \lambda \left[R_1 + \frac{R_o}{4} \right] + O(\varepsilon^3) \qquad (2.50)$$

The volume remains constant if $R_1 = -R_o/4$. In this case, the surface area, S can be expressed as:

$$S = 2\pi \int R(z)dl = 2\pi \int R(z)\sqrt{1 + \left(\frac{dR(z)}{dz} \right)^2} \cdot dz$$

$$= 2\pi R_o \lambda + \frac{\varepsilon^2 \pi R_o \lambda}{2} \left[\left(\frac{2\pi R_o}{\lambda} \right)^2 - 1 \right] + O(\varepsilon^3) \qquad (2.51)$$

We see that the surface area decreases if:

$$\lambda > 2\pi R_o \qquad (2.52)$$

For a Newtonian fluid cylinder, there is only a surface contribution to the potential energy, W, for which — in case of the distortion discussed by Plateau — we get:

$$W = (k^2 R_o^2 - 1)\gamma \pi \varepsilon^2 R_o / 2 \qquad (2.53)$$

where γ is the surface tension, and V is the excess surface potential energy per unit length. The Newtonian fluid cylinder is stable against the perturbation (i.e., the perturbation dies out) if the surface area (thus the free energy) is increasing, and the perturbation is growing (the cylinder is unstable) if the surface area is decreasing, described by the condition of (2.52). This is happening first for the largest wavelength of perturbation, which (considering the requirement that the diameter at the plates is constant) is the length of the cylinder. This therefore means that a stable fluid bridge has to be shorter than its periphery. This is known as Plateau–Rayleigh instability (Rayleigh considered a dynamical model to get the growth rates of the instability[69]).

In the case of non-Newtonian materials, where the viscosity depends on the strain (see Appendix B), long slender columns of liquids can be stabilized during the pulling for high strain rates characterized by sufficiently large Deborah number:[70] $De > 0.5$ ($De = \dot{\varepsilon} \cdot \tau$, where $\dot{\varepsilon}$ is the strain rate, and τ is the relaxation time of a deformation). In such case a strain hardening occurs (see Appendix B), leading to a homogeneous extensional deformation and a uniform column in the mid-region.

Liquid crystals are complex fluids with wide ranges of viscous and elastic properties (details will be described in Chapter 4). Liquid crystalline polymers are viscoelastic and can easily form fibers, just as conventional isotropic polymers. The stability of nematic liquid crystalline polymers was extensively studied theoretically.[71,72] Experimentally, it was found that liquid crystalline polymer fibers or tubes have exciting features, such as very high oxygen and water vapor permeability. These make them increasingly important in packaging as "super barrier" materials.[73] Recently it was shown[74] that even the spider silks have liquid crystalline structures in the early duct portion of the silk-producing gland.

Columnar liquid crystalline phases of disc-shaped molecules are one-dimensional fluids, and they can form free-standing fibers, as it was realized in the early 1980s.[75]

It was observed that as long as $L = 0.2$ *mm* and as narrow as $D = 1.5$ *μm*-diameter stable filaments (slenderness ratios $S \equiv {}^L/_D \sim 10^2$) can be drawn in the lower-temperature tilted-columnar phase, which then remain stable even in the higher-temperature orthogonal-columnar state. Photomicrograph and the sketch of the corresponding strands are shown in Figure 2.17.

In columnar liquid crystals, there is an additional bulk energy contribution arising from the compression of the columnar structure.[76] This compression energy is $2\pi\varepsilon^2 B$, where B is the compression modulus associated with a

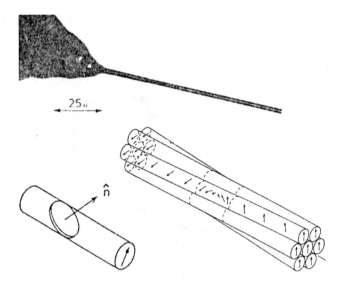

FIGURE 2.17
Transmission photomicrograph of freely suspended strands of triphenylene hexa-n-dodeconate in the tilted columnar phase, whose structure is illustrated below the photomicrograph. (Picture and illustrations are reproduced from Reference 75 with the permission of American Physical Society Publishing.)

change in the column density in the plane normal to the strands. With this W can be written as:

$$W = (k^2 \gamma R_o^2 + 2BR_o - \gamma)\pi \varepsilon^2 R_o / 2 \qquad (2.54)$$

The strand is stable against all fluctuations if $W > 0$, and instability occurs for cylinder diameters smaller than $R_c = \gamma/2B$, if $k < k_c = \frac{1}{R_o\sqrt{1-R_o/R_c}}$. Accordingly, filaments of subcritical diameters are stable only if they are short enough to suppress long-wavelength fluctuations ($L < 1/k_c$). With $\gamma = 0.03\ J/m^2$ and $B = 10^5\ J/m^3$, we obtain that $D \sim 0.15\ \mu m$, which is an order of magnitude smaller than where rupture was observed experimentally. This discrepancy was devoted to the strain-induced breaking of columns that lead to a plastic flow similar to solid strings that also experiencing plastic flow at extensive stresses.

It was also pointed out that this situation is essentially different from that of freely suspended planar smectic films, which are stable against all long-wavelength thickness fluctuations, since bulk and surface energies are positive. Experimentally, generally it is observed that smectic (2-D fluid) and nematic (3-D fluid) liquid crystals do not form fibers, but only bridges that collapse at slenderness ratios of $S \approx \pi$ (corresponding to the Plateau–Rayleigh instability) and at $S = 4.2$, respectively.[77] Rather, as we have seen in the previous paragraphs, smectic liquid crystals of rod-shape molecules generally tend to form thin membranes.[40] Based on these observations, it is tempting to conclude that the stable free-standing macroscopic objects of the materials mimic their microscopic order: 3-D fluids (isotropic liquids or nematic liquid crystals) can form only 3-D objects; 2-D fluids, like smectics can form 2-D fluid objects, like membranes; and 1-D fluids, like columnar liquid crystals, can form 1-D fluids, such as strings. After 20 years of unsuccessful attempts to produce stable smectic filaments, it was a great surprise to find recently that bent-shape molecules ("banana-smectics") may form stable fibers instead of films.[78,79] As we already have mentioned in Chapter 1, liquid crystals of bent-core molecules attracted considerable interest only when it was observed that they are ferro-electric,[80] and their structure is chiral without possessing chirality on molecular level.[81] The first banana materials that were found to form fibers also formed spectacular helical superstructures such as helical filaments[78,82] under cooling from the isotropic melt. Based on their characteristic texture formation, they are tentatively called B_7 materials (B stands for "bent-core" and 7 refers to the time order of their observation). It is known now that materials with similar B_7 textures can have different phases with different structures. Some (for example, the one where the first fiber formation was observed) have columnar structures, so their fiber formation is not surprising, although in that bent-core material the slenderness ratio is about 5000, which is an order of magnitude larger than of the columnar strands consisting of disc-shape molecules. Other B_7 materials have modulated smectic layer structure,[83] possibly due to the presence of an out-of-layer electric polarization[84] component. Whatever is their exact structure, they surely are not columnar; still they form stable and slender fibers with slenderness ratios in the order[85] of 1000. In addition, the antiferro-electric, electrically switchable fluid tilted smectic (so-called B_2) materials also form stable strands of fibers with slenderness ratios larger than 100.[85]

For small strains, the materials behave similar to Newtonian fluids, i.e., as the distance between the supports increases, they develop concave shapes with decreasing curvature radius. However, before the diameter of the mid-point reaches zero, typically at D ~ 15–20 µm, a homogeneous elongation deformation occurs, and the diameter of the filament remains constant, provided that the fluid reservoirs at the end plates have enough materials. On further increasing the length, the diameter is decreasing in about 1.5 to 2-µm discrete steps. Fibers with diameters larger than 6–8 µm show no extinction when rotating between crossed polarizers. They are not single filaments, but bundles of twisted coiling strings. The B_7 fibers remained stable until the diameter decreased to 1.5–3 µm. In this case, the filament appears to be uniform and uniaxial with extinction direction parallel to the filament axis. The B_2 fibers remain stable only in bundles at diameters of about 10 µm. Frequently, spinning of the fibers can be observed directly by optical observations. Typical fibers at their meniscus in the B_2 and in B_7 textures are shown in Figure 2.18a.

Similar to smectic membranes, periodic electric fields (although in this case the field is parallel to the fiber axis) can induce mechanical vibration of the strings. Studying the eigenfrequencies of the vibrations of the strings under

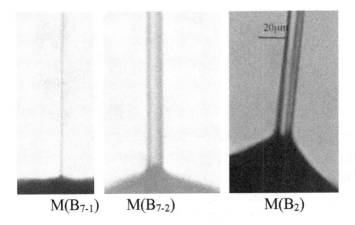

$$M(B_{7\text{-}1}) \qquad M(B_{7\text{-}2}) \qquad M(B_2)$$

FIGURE 2.18
Typical fibers at the meniscus of some bent-core materials, $M(B_{7\text{-}1})$[78,82] has a columnar B_7 phase; $M(B_{7\text{-}2})$,[84,86] has a modulated B_7 smectic structure at higher temperatures and a nonmodulated B_2 mesophase at lower temperatures; $M(B_2)$[87] has a tilted smectic B_2 "banana liquid crystal" phase with fluid in-layer structure.

electric fields, one can estimate the elastic constant of the fiber. The first harmonic of the natural oscillation of an elastic spring with Young modulus E, diameter D and length L fixed at the two ends can be expressed as:

$$f_{\min} = \frac{1}{2\pi} \cdot \frac{22.4}{L^2} \sqrt{\frac{E \cdot \theta}{\rho \cdot S}} \tag{2.55}$$

Here $S = R^2 \pi$ is the area of the cross section, and $\theta = \frac{\pi R^4}{4}$ is the inertial moment with respect to any axis in the plane of the cross section. It is observed that $f_{\min} \sim 10$ Hz, from which Eq. (2.55) gives $E \sim 3 \cdot 10^4 N/m^2$. Assuming that the Young modulus E is due to the surface tension $\gamma_\parallel$ parallel to the fiber axis, we get that $E = 2\gamma_\parallel / R$. This gives $\gamma_\parallel \sim 2 \cdot 10^{-2} N/m$, which is in good agreement with a more recent and more precise results of $\gamma = 26$ mN/m obtained by Eremin et al.[89] in a measurement wherein they have pulled a B7 filament with the cantilever of an Atomic Force Microscope and measured the force.

Such an internally fluid structure can be achieved by rolling the smectic layers into concentric cylinders (so-called "jelly roll" structure) (see Figure 2.19).

In explaining the fiber formation of "banana smectics," we have to recognize the analogy between the cross section of the banana smectic and the columnar fibers. This means that the compression of the smectic layers will have the same effect as the compression of the columns in columnar materials. Accordingly, we can use the same arguments and equation that was used to describe the stability of columnar fibers.[75] Even the typical surface tension and compression modulus data are similar ($\gamma = 2 \times 10^{-2} J/m^2$ and $B \sim 10^5 J/m^3$) so we get $D_c \sim 0.15 \mu m$. Again, similar to columnar fibers,[75] this is an order of magnitude smaller than the experimentally observed smallest

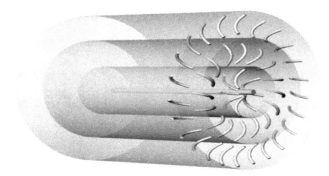

FIGURE 2.19
Proposed "jelly-roll" structure of the banana smectic fibers.

diameters. It can indicate plastic flow effects due to defects as was suggested for columnars.[75]

Presently it is not clear why smectic phases of rod-shape molecules do not form fibers, but banana smectics do. It is believed that the fiber forming ability of the smectics is related to the asymmetry of the surface tension.

Recently Stannarius et al.[90] found that the geometrical structures of two phases of the same material can be practically identical, whereas filaments are stable only in the higher temperature mesophase. They also conclude that probably the spontaneous curvature of the layers, which might be caused by the out of plane polarization component of the polarization, favors the formation of the cylindrical fibers rather than the free-standing films.

References

1. P.G. de Gennes, *Rev. Mod. Phys.*, 57, 827 (1985).
2. W. Barthlott, C. Neinhuis, Purity of the sacred lotus, or escape from contamination in biological surfaces, *Planta*, 202, 1 (1997).
3. H. Keller, Nanostructures with the Lotus effect: Building blocks for superhydrophobic coatings, *Journalists and Scientists in Dialogue Nanotechnology in Chemistry — Experience Meets Vision*, Mannheim, October (2002).
4. N.A. Patankar, Mimicking the Lotus effect: Influence on double roughness structures and slender pillars, *Langmuir*, 20, 8209 (2004).
5. P. Paik, V.K. Pamula, M.G. Pollack, R.B. Fair, Electrowetting-based droplet mixers for microfluidic systems, *Lab on a Chip*, 3, pp. 28 (2003); V. Srinivasan, V. Pamula, M. Pollack, R. Fair, A digital microfluidic biosensor for multianalyte detection, *Proceedings of the IEEE 16th Annual International Conference on Micro Electro Mechanical Systems*, p. 327 (2003).
6. C. Tanford, *Ben Franklin Stilled the Waves: An Informal History of Pouring Oil on Water with Reflections on the Ups and Downs of Scientific Life in General*, Duke University Press, Durham and London (1989).

7. See for example, J. Lyklema, *Fundamentals of Interface and Colloid Science, Volume I: Fundamentals*, Academic Press, London (1991).

8. Y. Couder, J.M. Chomaz, M. Rabaud, *Physica D*, 37, 384 (1989).

9. A nice summary of this subject can be found in I.W. Hamley, *Introduction to Soft Matter*, John Wiley & Sons, LTD, Chichester (2000).

10. R. Hooke, Communicated to the Royal Society, March 1672; *The History of Royal Society of London*, Vol. 3, Ed. Birch (London), p. 29.

11. I. Newton, *Optics, Book II*, Obs. 17-19, Dover, New York, p. 214 (1952), based on the 4th ed., London (1730).

12. J. Plateau, *Statique Experimentale at Theoretique des Liquides Soumis aux Seules Forces Moleculaires*, Gauthier-Villars, Paris (1873).

13. J.W. Gibbs, *The Collected Works*, Longmans Green, New York (1931).

14. C.V. Boys, *Soap Bubbles and Forces Which Mould Them*, Society for Promoting Christian Knowledge, London (1890) and Anchor Books, New York (1959).

15. C. Isenberg, *The Science of Soap Films and Soap Bubbles*, Dover, New York (1992).

16. K.J. Mysels, K. Shinoda, S. Frankel, *Soap Films, Studies of their Thinning*, Pergamon, New York (1959).

17. A.I. Rusanov, V.V. Krotov, *Prog. Surf. Membrane. Sci.*, 13, 415 (1979).

18. W.I. Goldburg, A. Belmonte, X.L. Wu, I. Zusman, *Physica A*, 254, 231 (1998).

19. D.O. Shaha, N.F. Djabbarah, D.T. Wasan, *Colloid Polym. Sci.*, 256, 1002 (1978).

20. E.S. Johnnott, *Philos. Mag.*, 11, 746 (1906); J. Perrin, *Ann. Phys.*, 10, 160 (1918).

21. A.A. Sonin, D. Langevin, *Europhys. Lett.*, 22, 271 (1993).

22. Frederick J. Almgren, Jr., Jean E. Taylor, The geometry of soap films and soap bubbles, *Scientific American*, July, 82 (1976).

23. Jean T. Taylor, The structure of singularities in soap-bubble-like and soap-film-like minimal surfaces, *Ann. Math.*, 103, 489 (1976).

24. I. Ivanov (Ed.), *Thin Liquid Films*, Vol. 29, Marcel Dekker, New York (1988).

25. B.V. Derjaguin, *Theory and stability of colloids and thin films*, Transl. R.K. Johnston, Consultants Bureau, New York (1989).

26. B.V. Derjaguin, N.V. Churaev, V.M. Muller, *Surface forces*, Ed. J.A. Kichener, Consultants Bureau, New York (1987).

27. J.N. Israelachvili, *Intermolecular and Surface Forces with Applications to Colloid and Biological Systems*, Academic, Orlando, FL (1985).

28. B. Chu, *Molecular Forces*, based on the Baker Lectures of Peter J.W. Debye, Wiley-Interscience, New York (1967).

29. S. Usui, S. Hachisu, *Interaction of Electrical Double Layers and Colloid Stability. Electrical Phenomena at Interfaces: Fundamentals, Measurements, and Applications* (Surfactant Science series, 15), Ed. A. Kitahara, A. Watanabe, Dekker, New York (1984).

30. H.C. Hamaker, *Physica*, 4, 1058 (1937).

31. H.B. Casmir, D. Polder, *Phys. Rev.*, 73, 270 (1948).

32. B. Vincent, *J. Colloid Interface Sci.*, 42, 27 (1973).

33. J.N. Israelachvili, Wennerström, *J. Phys. Chem.*, 96, 520 (1992).

34. W. Ranz, *J. Appl. Phys.*, 30 (1950); W. Mc Entee, K. Mysels, *J. Phys. Chem.*, 73, 3018 (1969).

35. F.E. Culik, *J. Appl. Phys.*, 31, 1128 (1960).

36. A. Dupre, *Ann. Chim. Phys.*, 11, 194 (1867).

37. Lord Rayleigh, *Proc. R. Inst*, 13, 261 (1891).

38. P.G. de Gennes, *Faraday Discuss.*, 104, 1 (1996).

39. G. Friedel, *Ann. Phys. (Paris)*, 18, 273 (1922).

40. C.Y. Young, R. Pindak, N.A. Clark, R.B. Meyer, *Phys. Rev. Lett.*, 40, 773 (1978); R. Pindak, C.Y. Young, R.B. Meyer, N.A. Clark, *Phys. Rev. Lett.*, 21, 140 (1980).

41. P. Pieranski, L. Beliard, J.- Ph. Tournellec, X. Leoncini, C. Furtlehner, H. Dumoulin, E. Riou, B. Jouvin, J.-P. Fenerol, Ph. Palaric, J. Heuving, B. Cartier, I. Kraus, *Physica A*, 194, 364 (1993).

42. F. Brochard, P.G. de Gennes, P. Pfeuty, *J. Phys. (Paris)*, 37, 1099 (1976).

43. E.B. Sirota, P.S. Pershan, L.B. Sorensen, J. Collet, *Phys. Rev. A*, 36, 2891 (1987).

44. K. Miyano, *Phys. Rev. A*, 26, 1820 (1982).

45. J.W.S. Rayleigh, *The Theory of the Sound*, Dover, New York (1945).

46. See for example, http://en.wikipedia.org/wiki/Bessel_function.

47. C. Furthlehner, X. Leoncini, Structures dans les membranes, Stage du magistere, Universite Paris Sud, Laboratorie de Physique des Solides, Orsay (1991).

48. J.C. Geminard, R. Holyst, P. Oswald, *Phys. Rev. Lett.*, 78, 1924 (1997).

49. P. Oswald, *J. Physique*, 48 (1987).

50. R. Stannarius, Ch. Cramer, *Europhys. Lett.*, 42, 43 (1998); R. Stannarius, Ch. Cramer, *Liq. Cryst.*, 23, 371 (1997).

51. F. Bloch, *Z. Phys.*, 61, 206 (1930); R.E. Peierls, *Ann. Inst. Henri Pincare*, 5, 177 (1935); L.D. Landau, *Phys. Z. Swjetunion II*, 26 (1937).

52. P.C. Hohenberg, *Phys. Rev.*, 158, 383 (1967), N.D. Mermin, *Phys. Rev.*, 176, 250 (1968).

53. J.M. Costerlitz, D. J. Thouless, *J. Phys. C*, 6, 1181 (1973).

54. C. Bahr, D. Fliegner, *Phys. Rev. Lett.*, 70, 1842 (1993).

55. D.R. Link, G. Natale, N.A. Clark, J.E. Maclennan, M. Walsh, S.S. Keast, M.E. Neubert, *Phys. Rev. Lett.*, 82, 2508 (1999); D.R. Link, N.A. Clark, B.I. Osrovskii, E.A. Soto Bustamante, *Phys. Rev. E*, 61, R37 (2000).

56. W.H. de Jeu, B.I. Ostrovskii, A.N. Shalaginov, *Rev. Modern Phys.*, 75, 181 (2003).

57. S. Perkovitz, *Universal Foam: From Cappuccino to Cosmos*, Walker & Company, New York (2000).

58. T.C. Hales, *Sci. News, Washington, DC*, 156, 60 (1999).

59. J. Sullivan, in *Foams and Emulsions*, Vol. E354 of Nato Advanced Study Institute, Series E: Applied Science, Ed. J.F. Sadoc, N. Rivier, Kluwer, Dordrecht (1999).

60. D. Weaire, *Proc. Am. Phil. Soc.*, 145, 564 (2001).

61. F. Graner, Y. Jiang, E. Janiaud, C. Flament, *Phys. Rev. E*, 63, 011402 (2000).

62. T. Aste, D. Boose, N. Rivier, *Phys. Rev. E*, 53, 6181 (1996).

63. D. Weaire, S. Hutzler, *The Physics of Foams*, Oxford University Press, Oxford (1999).

64. N. Rivier, in *Disorder and Granular Media*, Ed. D. Bideau, A. Hansen, p. 55, Elsevier, Amsterdam (1993).

65. J.E. Taylor, *Ann. Math.*, 103, 489 (1976).

66. J. Foisy, M. Alfaro, J. Brock, N. Hodges, J. Zimba, *Pac. J. Math.*, 159, 47 (1993).

67. M. Buchanan, arXiv: cond-mat/0206477v1 25 Jun (2002).

68. S. Chandrasekhar, *Hydrodynamic and Hydromagnetic Stability*, p. 515, Dover, New York (1981).

69. J.W. Strutt (Lord Rayleigh), *Proc. Lond. Math. Soc.*, 10, 4 (1879).

70. M. Reine, *Phys. Today*, 62, January (1964).

71. A-G. Cheong, A.D. Rey, P.T. Mather, *Phys. Rev. E*, 64, 041701 (2001).

72. M.G. Forest, Q. Wang, *Physica D*, 123, 161 (1998).

73. R.W. Lusignea, *Packag. Technol.*, October (1997).

74. P.J. Willcox, S.P. Gido, W. Muller, D. Kaplan, *Macromolecules*, 29, 5106 (1996); D.P. Knight, F. Vollrath, *Proc. R. Soc. Lond. B*, 266, 519 (1999); F. Vollrath, D.P. Knight, *Nature*, 410, 541 (2001).

75. D.H. Van Winkle, N.A. Clark, *Phys. Rev. Lett.*, 48, 1407 (1982).
76. J. Prost, N.A. Clark, in *Proceedings of the International Conference on Liquid Crystals*, Ed. S. Chandrasekhar, p. 53, Heyden, Philadelphia (1980).
77. M.P. Mahajan, M. Tsige, P.L. Taylor, C. Rosenblatt, *Liq. Cryst.*, 26, 443 (1996).
78. G. Pelzl, S. Diele, A. Jákli, C.H. Lischka, I. Wirth, W. Weissflog, *Liq. Cryst.*, 26, 135 (1999).
79. D.R. Link, N. Chattham, N.A. Clark, E. Körblova, D.M. Walba, p. 322, Abstract Booklet FLC99, Darmstadt (1999).
80. T. Niori, T. Sekine, J. Watanabe, T. Furukawa, H. Takezoe, *J. Mater. Chem.*, 6(7), 1231 (1996); T. Sekine, T. Niori, M. Sone, J. Watanabe, S.W. Choi, Y. Takanishi, H. Takezoe, *Jpn. J. Appl. Phys.*, 36, 6455 (1997).
81. D.R. Link, G. Natale, R. Shao, J.E. Maclennan, N.A. Clark, E. Körblova, D.M. Walba, *Science*, 278, 1924 (1997).
82. A. Jákli, C.H. Lischka, W. Weissflog, G. Pelzl, A. Saupe, *Liq. Cryst.*, 27, 11 (2000).
83. D. A. Coleman et al., *Science*, 301, 1204 (2003).
84. A. Jákli, D. Krüerke, H. Sawade, G. Heppke, *Phys. Rev. Lett.*, 86 (25), 5715 (2001).
85. A. Jákli, D. Krüerke, G.G. Nair, Liquid crystal fibers of bent-core molecules, *Phys. Rev. E*, 67, 051702 (2003).
86. G. Heppke, D.D. Parghi, H. Sawade; *Ferroelectrics*, 243, 269 (2000).
87. G. Pelzl, S. Diele, W. Weissflog, *Adv. Mater*, 11(9), 707 (1999).
88. L.D. Landau, E.M. Lifsic, *Theoretical Physics*, VII, Nauka, Moscow, (1965).
89. A. Eremin, A. Nemes, R. Stannarius, M. Schulz, H. Nádasi, W. Weissflog, *Phys. Rev. E.*, 71, 031705 (2005).
90. R. Stannarius, A. Nemes, A. Eremin, *Phys. Rev. E*, 72, 020702 (R) (2005).

3

Phase Transitions

3.1 Transition between the Minimal Surfaces of Soap Films

Study of transitions between minimal surfaces of soap films discussed in Chapter 2 is an excellent exercise to understand the basics of phase transitions in much more complicated systems. Consider, for example, the four-pin arrangement between two parallel plates shown in Figure 3.1.

Obviously there are two equivalent solutions: in one the bridge is vertical (a); in the other it is horizontal (b). We can easily make transition from one configuration to the other by changing the distances AC and BD. Let us start with the pins A,C and B,D being very close together (x is small). There are four arms, each of length $x/\sqrt{3}$., and the central bridge of $(a - x/\sqrt{3})$. This gives the total length for the film:

$$L = a + \sqrt{3} \tag{3.1}$$

It is observed that, in increasing x, the 120° rule is maintained, and the film remains in this configuration until the central bridge vanishes and the two vertices touch. This occurs at $x = \sqrt{3}a$., when the total length of the film is 4a. At this point the film jumps into a configuration with four arms of lengths, $a/\sqrt{3}$., and of the central bridge of length $x - a/\sqrt{3} = 2a/\sqrt{3}$. Further increasing x, the total length is described by:

$$L = x + \sqrt{3}a \tag{3.2}$$

After that, if we start to reduce x, the configuration jumps back to the vertical bridge arrangement, when $x = a/\sqrt{3}$. This corresponds to the total length of $4a/\sqrt{3}$. Right after the transition, the length jumps back to 2a. Putting this data together, we can plot dL/dx versus x/a (right hand side of Figure 3.1). It can be seen that there is a portion of the graph where the gradient is equal to $\sqrt{3}$. This corresponds to the case when $L = a + \sqrt{3}$. There is another part of the graph with a gradient 1, which is described by (3.2). The change in gradient occurs when the configuration changes. This is the phenomenon of hysteresis, which has numerous examples in condensed matter physics, such as switching electric polarization in ferroelectric materials.

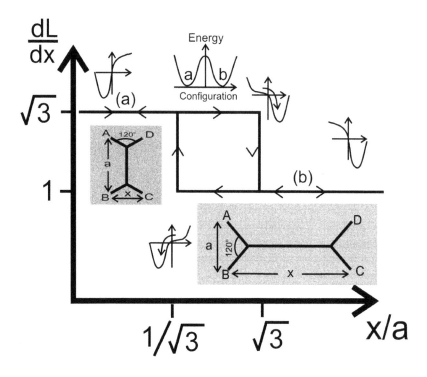

FIGURE 3.1

Variation dL/dx with x for the first order phase transition model and the equilibrium soap film configurations within the four-pin model, with the illustration of the corresponding energies with respect the configurations a and b. (a) Pair of pins are close to each other; (b) pairs of pins are widely separated. Note that the lines in reality are two-dimensional, but their sizes do not change in the other dimension.

3.2 Landau Description of Phase Transitions

Basically the same considerations apply for temperature-driven phase transitions in condensed matters with the difference that instead of the energy, we deal with the free energy, which depends on the so-called order parameter. The order parameter S is a quantity which is nonvanishing in the ordered state and zero in the disordered state. The change from the disordered state to the ordered one can be brought about experimentally by varying a suitable thermodynamic variable, like the temperature T or pressure P. In a first-order transition, the change in the order parameter is discontinuous at the transition temperature T_t, while in second-order transition, it is continuous at $T = T_c$. In the Landau theory, one tries to phenomenologically understand the transition by examining the behavior of the free energy F close to the transition temperature.

The order parameter can be a very complicated quantity (as we will see later). First we suppose that the order parameter is a scalar and the free energy is insensitive to the sign of S. In this case, near to T_c the free energy can be expressed as:

$$F(S) = F_o + \alpha S^2 + bS^4. \tag{3.3}$$

where F_o is a reference level, and β is a positive constant . The coefficient α is smooth function of T in the vicinity of T_c with a leading behavior:

$$\alpha \simeq a(T - T_c), \ a > 0 \tag{3.4}$$

The thermodynamically stable state S_o of the system is found by minimizing F with respect to S. The stability conditions are:

$$\left(\frac{\partial F}{\partial S}\right)_{S_o} = 0; \quad \left(\frac{\partial^2 F}{\partial S^2}\right)_{S_o} > 0 . \tag{3.5}$$

A simple calculation then shows that for $T > T_c$, $S_o = 0$ is the stable state, whereas for $T < T_c$, $S_o > 0$ values are possible, with the temperature dependence of S_o given as:

$$S_o = \sqrt{\frac{a}{2b}} \sqrt{T_c - T} . \tag{3.6}$$

This dependence, together with the free energy functions, is illustrated qualitatively in Figure 3.2.

Just as the example of the soap film geometries showed, not all transitions are of second order, but the majorities are of first order. To describe phenomenologically a first-order transition, we need to incorporate a third-order term in the Landau free energy.

$$F = F_o + \frac{1}{2}a(T - T_c)S^2 + \frac{1}{3}bS^3 + \frac{1}{4}cS^4 . \tag{3.7}$$

The minimization of the free energy gives:

$$\frac{\partial F}{\partial S} = 0 = a(T - T_c)S + bS^2 + cS^3 . \tag{3.8}$$

which means:

$$S = 0 \ or \ S = \frac{-b \pm \sqrt{b^2 - 4ac(T - T_c)}}{2c}. \tag{3.9}$$

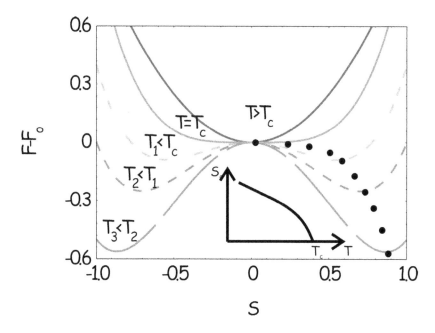

FIGURE 3.2

Plot of the free energy at the function of order parameter S corresponding to (3.3) at different temperatures with respect to the I–N phase transition T_c. The positions of the minima determine the equilibrium order parameter S_o. The temperature dependence of S_o is plotted in the insert.

Stability analysis shows that the negative sign of the nonzero expression of S is not stable. The phase transition temperature T_t is determined by the condition that at T_t the free energies per unit volume of the ordered and disordered phases are the same. Because at $S_o = 0$, $F - F_o = 0$, we get the condition that:

$$\frac{1}{2}a(T - T_c) + \frac{1}{3}bS + \frac{1}{4}cS^2 = 0 .$$ (3.10)

Combining this requirement with (3.8), we get that:

$$T_t = T_c + \frac{2b^2}{9ac} .$$ (3.11)

This means that at T_t the order parameter discontinuously jumps from $S_t = 0$ to

$$S_t = -\frac{2b}{3c} .$$ (3.12)

as shown in Figure 3.3

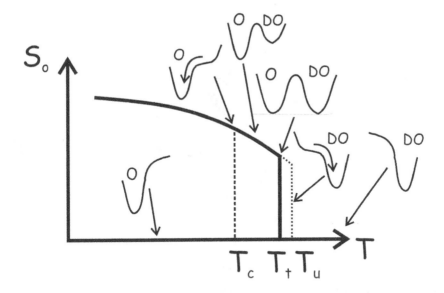

FIGURE 3.3
Schematic plot for the temperature dependence of the equilibrium order parameter S_o, with illustration of the free energies near the local and global minima. O denotes the ordered state; DO stands for the disordered state. T_c corresponds to the supercooling limit, and T_u is the superheating limit.

This result has two consequences: first, the expansion of (3.7) will be valid only if the jump is small (weakly first order transition); second, b must be negative to get a positive order parameter. Note that the second derivative of the free energy is zero at $S = 0$ when $T = T_c$. This indicates that $S = 0$ is no longer a local minimum for temperatures below T_c. Thus T_c represents a temperature below the transition temperature T_t, where the isotropic phase becomes thermodynamically unstable (see Figure 3.3). In other words, it can be interpreted as the lower limit of the supercooling of the isotropic phase. Similarly, we can give the upper limit of the superheating T_u of the ordered phase; setting both the first and second derivatives equal to zero yields at T_u, $\frac{\partial^2 F}{\partial S^2}|_{S \neq 0} = 0$ and

$$S = -\frac{2a(T_u - T_c)}{b} \quad and \quad a(T_u - T_c) = \frac{b^2}{4c}. \qquad (3.13)$$

Solving these for T_u gives

$$T_u = T_c + \frac{b^2}{4ac} = T_t + \frac{b^2}{36ac}. \qquad (3.14)$$

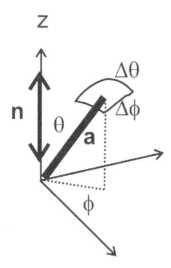

FIGURE 3.4
Illustration of an instantaneous position of an individual molecule with respect to the average molecular direction, the director.

This shows that the upper limit of superheating is eight times closer to the transition than the lower limit of the supercooling. In other words, the ability of supercooling and superheating means a hysteresis similar to that shown for the four-pin soap film system in Figure 3.1c. Concerning the temperature dependence of the order parameter, one finds that generally the Landau function is more linear than the measured behaviors, in spite of the four adjustable parameters (*a,b,c and T_c*).

3.2.1   The I–N Transition

The phase transition from the isotropic liquid to the nematic liquid crystal in frame of the Landau theory was first described by de Gennes, and is usually referred as Landu–de Gennes theory. The first step of this theory is to find the right order parameter.

In case of simple rods with uniaxial shape, the axis of a rod is defined by unit vector *a*. This is making an angle θ with the director **n,** which is chosen parallel to z (see Figure 3.4).

Accordingly,

$$a_x = \sin\theta\cos\varphi, \quad a_y = \sin\theta\sin\varphi, \quad a_z = \cos\theta. \tag{3.15}$$

The state of alignment is described by the distribution function $f(\theta,\varphi)d\Omega$. This gives the probability of finding rods in a small solid angle $d\Omega = \sin\theta d\theta d\varphi$ around the direction given by the angles θ and ϕ.

We assume cylindrical symmetry about $\vec{n}$., i.e., that $f(\theta,\phi)$ is independent of ϕ and utilize the head–tail symmetry, i.e., $f(\theta) = f(\pi - \theta)$. The simplest possibility to define the order parameter would be to take $\langle \cos\theta \rangle = \langle \vec{a} \cdot \vec{n} \rangle = \int f(\theta)\cos\theta d\Omega$., but it vanishes because of the head–tail symmetry. On the other hand, the second Legendre polynomial,

$$S = \langle P_2 \rangle = \frac{1}{2}\langle 3\cos^2\theta - 1 \rangle = \int f(\theta)\frac{1}{2}(3\cos^2\theta - 1)d\Omega . \qquad (3.16)$$

is a good choice, because this gives $S = 0$ for random orientation ($\langle \cos^2\theta \rangle = 1/3$), and $S = 1$ if $f(\theta)$ peaked around 0 and π. Another advantage of using <P_2> as an order parameter is that nuclear magnetic resonance (NMR) measurements detect exactly this.[†]

Since S is a scalar order parameter, the prior arguments leading to Eqs. (3.3)–(3.11) are valid. The isotropic–nematic phase transition is characterized by a small latent heat (0.5 kJ mol[-1]), and small a (~0.06 J cm[-3] K[-1]), which is nearly two orders of magnitude smaller than that at melting of a solid either to liquid crystal or isotropic liquid,[1] indicating the delicate nature of these transitions. Due to the small latent heat, there are pretransitional effects on either side of the transition.

Actually there are a few micellar nematic liquid crystals[5] where the I–N transition is exactly of second order ($T_t = Tc$). In these materials one generally finds a point is in the I–N transition line, where the two second-order transition lines end, and the three nematic phases and the isotropic phase become identical. This point is called Landau point.[2] In the case of the potassium laurate (KL)-1-decanol-D_2O system, this point is shown in Figure 3.5.[3]

Although the $I \rightarrow N$ transition is the most studied one in liquid crystals, there are still fascinating problems that are not completely resolved. Landau argued that for purely geometrical reasons the transition should be first order,[4] but after a flurry of experimental measurements in the 1960s and 1970s, the challenge became to explain why it was observed to be so *weakly* first order in calamitic liquid crystals. A key parameter of the transition is the value of $T_t - T_c$, i.e., the difference between the actual transition temperature T_t and the extrapolated or "virtual" second-order transition temperature T_c. This difference can be measured by magnetically induced optical birefringence or light scattering, and one finds a value of about *1–2K* in typical calamitics.[5] However, the simplest theories do not come close to this value. The Landau mean field theory[6] including Gaussian fluctuations yields $T_t - T_c = 24K$. Including higher-order effects via a renormalization group calculation[7] produced $T_t - T_c = 12.8K$. Further refinements[8] gave a somewhat improved result of $T_t - T_c = 7.46K$. However, these results remained far from the experimental values.

To do better, one needs to extend the theory to include secondary order parameters, such as the mass density ρ. The physical idea is simple: when a collection of rod-shaped molecules go from a state of complete orientational

---

[†]Generally the order parameter is a tensor is: $Q_{ij} = S_{(2)}[\bar{n}_i(nh_j(n - \frac{1}{3}f_{ij})]$.

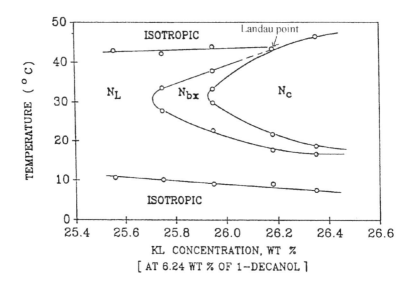

FIGURE 3.5
Phase diagram of the potassium laurate (KL)-1-decanol-D_2O system, with the Landau point.[8] N_C and N_L are uniaxial nematic phases with cylindrical and lamellar micelle structures, and N_{bx} denotes the biaxial nematic phase. (Figure reproduced from Reference [8] with permission of American Physical Society.)

disorder to a more ordered state of all pointing on average in the same direction, one might anticipate a tighter packing of the molecules and, thus, a slight increase in ρ. Mukherjee[10] carried out a similar analysis within the Landau–de Gennes framework, producing a so-called "extended" mean-field theory and, after including fluctuations, obtained similarly favorable results. The $T_t - T_c$ puzzle was declared solved, but unfortunately that is not the end of the story. It has not yet explained why the discontinuity in the nematic order parameter at T_t is as small as it is observed to be.[11] Due to these unsolved problems, study of the I–N transition is still an active research area.[12,13,14]

3.2.2 SmA–N Transition

The Landau–de Gennes theory for the nematic–isotropic transition can be also extended for the SmA–N transition. At the transition to SmA, the translational symmetry along one of the three dimensions becomes broken. Choosing the coordinate system so that the density modulation will be in a direction z normal to the layers, we can write the density function as:

$$\rho(r) = \rho(z) = \rho_o + \psi(q)\cos(qz - \Phi) . \tag{3.17}$$

where $q = 2p/d$ is the wave number of the periodic modulation, with d being the layer spacing, and Φ is an arbitrary phase. In the nematic phase $\psi = 0$, so a natural choice for the *SmA* order parameter is ψ. Experimentally, one usually finds that the N–SmA transition (except some special cases) is of second order, so in the vicinity of the N–SmA transition, the free energy can be expanded in powers of ψ as:

$$F(|\psi|,T) = F_N + \frac{1}{2}a(T - T_c)|\psi|^2 + \frac{1}{4}\beta|\psi|^4 + \frac{1}{6}\gamma|\psi|^6 . \quad (3.18)$$

Here F_N is the free energy per unit volume of the nematic phase above the SmA, and the absence of the cubic term allows the second-order transition.

To include the fact that there is also an orientational order in the smectic phase, a term is added to the free energy, expressing that a variation in the orientational order at the phase transition will change the free energy per unit volume. In addition, there is a positive coupling between ψ and S (larger S enhances the density modulation, and larger density modulation leads to an increase of the orientational order, too).

With these considerations we can write that:

$$F(|\psi|,T) = F_N + \frac{1}{2}a(T - T_c)|\psi|^2 + \frac{1}{4}\beta|\psi|^4 + \frac{1}{6}\gamma|\psi|^6 + \frac{1}{2}\frac{(\delta S)^2}{\chi} - \mu(\delta S)|\psi|^2 \quad (3.19)$$

where $\delta S = S - S_t$ (S_t is the order parameter in the nematic phase at the N–SmA transition); χ and μ are positive constants. χ depends on the width of the nematic phase. If the nematic phase is wide, S is large at the N–SmA transition, and χ is small, indicating that it becomes more and more difficult to modify the nematic order parameter by the layering. If the width of the nematic phase is small, then χ is large. Since F must be a minimum with respect to δS, the partial derivative of F with respect to δS must be zero. This enables us to express δS in terms of $|\psi|^2$.

$$\delta S = \chi\mu|\psi|^2 . \quad (3.20)$$

With this, the free energy expression yields:

$$F(|\psi|,T) = F_N + \frac{1}{2}a(T - T_c)|\psi|^2 + \frac{1}{4}\beta'|\psi|^4 + \frac{1}{6}\gamma|\psi|^6 . \quad (3.21)$$

where

$$\beta' = \beta - 2\chi\mu^2 . \quad (3.22)$$

We see that β' can change sign, depending on the magnitude of χ (i.e., on the width of the nematic range). If the nematic range is narrow (χ is large), β' is negative, and the transition is discontinuous. If the nematic range is wide (χ is small), the transition is continuous, and ψ increases smoothly at the transition to the SmA.

P.G. de Gennes pointed out[15] that there is an analogy between the N–SmA and normal–superconductor transition, taking into account that the density modulation (that gives the order of the layering) and the phase Φ (which account for layer spacing fluctuation) can be combined to a complex order parameter: $\psi^*(r) = \psi(r)e^{i\Phi(r)}$. Such a complex order parameter is also used in describing superfluid helium (Bose condensate), where the macroscopic numbers in one quantum state are characterized by a complex number. Recently similar analogy between the smectic A and the polar smectic C phase of the bent-core materials, and of the anisotropic d-wave pairing in a two-dimensional superconductor which stabilizes three distinct superconducting phases,[16] has been put forward by Lorman and Mettout.[17] Transitions between other phases, such as SmA-SmC, N-SmC, I-SmA and I-SmC have also been investigated, and complicated properties have been observed. For example, SmA-SmC and SmA*-SmC* transitions can be both of second and first order, whereas I-SmC transitions are always of first order.

The Landau–de Gennes theories of other phase transitions (SmA–SmC, N–SmC, I–SmA and I–SmC) have also been developed,[16] but we will not discuss them here. For a recent review of the transitions between various liquid crystal phases, we recommend the review of P. Collings and references therein.[18]

3.3 Molecular Approaches

The main assumption of the molecular theories is that the molecules are embedded in a "sea" of many other molecules, and every molecule feels the same average potential. Accordingly, each molecule experiences the same forces on average as any other molecule. Such a model in thermodynamics is called mean-field theory. The first attempt to explain the occurrence of the anisotropic phase was given by M. Born,[19] who adapted Langevin's[20] and Weiss's[21] molecular field theory of magnetism (known today as the Curie–Weiss theory) to liquid crystals. His basic assumption was that the liquid crystal molecules interact via their dipole moments (dipole–dipole interactions). However, this assumption is wrong, because there are nonpolar molecules which also exhibit liquid crystal phase.[22] In addition, it predicts that nematic liquid crystals would be ferroelectric (macroscopically polar), which so far has not been observed even for strongly polar molecules. Without knowing Born's theory (it was during the First World War), a year later Grandjein[23] also suggested a molecular field theory for liquid crystals based on the Curie–Weiss theory. Without requiring strong molecular dipoles, he only assumed a coefficient playing the same role as the magnetic moment in ferromagnets. However, neither the nature of the coefficient analogous to the magnetic moment nor the temperature dependence of the order parameter was discussed, and Grandjein's theory remained unknown until recently.[24] Today there are basically two types of prevailing theories that can describe the orientational order and the phase transition between orientationally disordered and ordered phases, such as the isotropic nematic transition. The first assumes that orientational order results from short-range steric interactions; the second considers long-range attractive dispersion interactions. The best-known example for the first type is the Onsager model,[25] in which the excluded volume for rod-like particles is calculated as a function of their volume fraction. The second theory was developed in the 1950s by Maier and Saupe in a series of papers,[26,27] and it can successfully describe all thermotropic nematic liquid crystals.

3.3.1 The Onsager Theory

This theory has been used to describe nematic ordering in solutions of rod-like macromolecules, such as tobacco mosaic virus (TMV) or poly(benzyl-L-glutamate). The orientational distribution is calculated from the volume excluded to one hard cylinder by another. It assumes that the rods cannot penetrate to each other. Denoting the length of the rods by L and the diameter by D, the volume fraction of the rods is expressed by $\Phi = \frac{1}{4}c\pi LD^2$. (c is the concentration of rods).

Assuming that $\Phi \ll 1$, and that the rods are very long (L >> D), it is found that the nematic phase exists above a volume fraction $\Phi_c = 4.5D/L$. This theory predicts jumps in density and in the order parameter <P_2> on cooling from the isotropic phase. However, these changes are much larger than observed for thermotropic liquid crystals. It is basically an athermal model, so the quantities like the transition density are independent of temperature. For this reason, it is mainly used for lyotropic liquid crystals composed of large rods (TMW, DNA, etc.). An Onsager theory of nematic–isotropic phase equilibria of length polydisperse hard rods was developed recently.[28]

3.3.2 Maier–Saupe Theory

This theory in the first approximations[27] assumes:

- Only pure dispersion forces (second-order perturbation terms of pure Coulomb interactions). Thus, it is assumed that the interactions between permanent electrical dipole (and higher harmonics) moments, as intermolecular interactions, are important only for the arrangements of the centers of gravity of the molecules and for the energy content of the isotropic distribution along the axes; however, they are not particularly important for the orientational order.

- Because it was observed that nematic phase can occur in apolar molecules (in contrast to Bom's model[20]), it is assumed that the dominant force between molecules is an interaction between induced dipoles. A momentary dipole moment of one molecule induces a momentary moment on the neighboring molecule, resulting in attractive dispersion force. Such a force varies with the distance to the minus sixth power, or with the minus second power of the volume occupied by an average molecule.

- It is assumed that the molecules are cylindrically symmetric about their long axes. Accordingly, the potential energy between two molecules can only depend on the angle between their long axes, with an angular dependence proportional to the second Legendre polynomial of this angle (which was already considered in choosing the order parameter).

- Finally, it is assumed that the degree of orientational order of the molecules enters into the mean-field potential in a linear way (the larger is the orientational order, the larger is the effective potential).

With these assumptions, we give the mean-field potential in the following form:

$$U_i(\theta_i) = -\frac{A}{V^2} S\left(\frac{3}{2}\cos^2\theta_i - \frac{1}{2}\right). \tag{3.23}$$

where θ_i is the angle between the long axis of the molecule and the director, V is the volume occupied by the molecules, and A is a temperature independent constant. The temperature dependence of the factor $A/V^2 = A\rho^2/M^2$, therefore, is given by the temperature dependence of the density ρ (M is the molar mass).

The fact that the system must be in thermodynamic equilibrium demands that the probability of a molecule being oriented at an angle θ_i from the director is given by the Boltzmann factor,

$$P_i(\theta_i) = \frac{1}{Z}\exp\left(-\frac{U_i(\theta_i)}{k_B T}\right). \tag{3.24}$$

where the partition function Z is given with azimuthal angle, ϕ_i, as:

$$Z = \int_0^\pi \exp\left(-\frac{U_i(\theta_i)}{k_B T}\right)\sin\theta_i d\theta_i \int_0^{2\pi} d\phi_i . \tag{3.25}$$

The order parameter can be expressed with the help of this probability function as:

$$S = \left\langle \frac{3}{2}\cos^2\theta_i - \frac{1}{2}\right\rangle = \int_0^\pi \left(\frac{3}{2}\cos^2\theta_i - \frac{1}{2}\right)P_i(\theta_i)\sin\theta_i d\theta_i \int_0^{2\pi} d\phi_i$$

$$= \frac{1}{Z}\int_0^\pi \left(\frac{3}{2}\cos^2\theta_i - \frac{1}{2}\right)\exp\left(-\frac{U_i(\theta_i)}{k_B T}\right)\sin\theta_i d\theta_1 \int_0^{2\pi} d\phi_i \tag{3.26}$$

Since U_i itself contains S, we have a self-consistent equation involving S, T and V. With the help of this equation, and after some algebra, we can get the temperature dependence of the order parameter as follows.

Introducing the variable x and m as $x = \cos\theta_i$, and

$$m = \frac{3AS}{2k_BTV^2} \, . \tag{3.27}$$

the expression for S of (3.26) can be rewritten as:

$$S = \frac{3}{2} \frac{\int\limits_0^1 x^2 e^{mx^2} dx}{\int\limits_0^1 e^{mx^2} dx} - \frac{1}{2} \, . \tag{3.28}$$

Integrating the numerator in parts ($\int_0^1 x^2 e^{mx^2} dx = \frac{e^m}{2m} - \frac{1}{2m}\int_0^1 e^{mx^2} dx$) and making the substitution $y = x\sqrt{m}$, we get:

$$S = \frac{3}{4m}\left[\frac{\sqrt{m}}{D(\sqrt{m})} - 1\right] - \frac{1}{2} \, . \tag{3.29}$$

where D is the so-called Dawson integral, defined as:

$$D(t) = e^{-t^2}\int\limits_0^t e^{y^2} dy \, . \tag{3.30}$$

A convenient way to see how S depends on T is to choose a value for m, and using tabulated values of the Dawson integral or, these days, rather a computer , one can find S from (3.29). Now knowing the pair m and S and using the definition of m from (3.27), we get the corresponding temperature.

To find the temperature where the isotropic phase becomes unstable, we need to analyze the Helmholtz free energy, which is given by:

$$F = U - T\Sigma \, . \tag{3.31}$$

where Σ is the entropy (usually the entropy is denoted by S, but now this is used for the order parameter). The average energy of a molecule is:

$$\langle U_i \rangle = \left\langle -\frac{AS}{V^2}\left(\frac{3}{2}\cos^2\theta_i - \frac{1}{2}\right)\right\rangle = -\frac{AS}{V^2}\left\langle\frac{3}{2}\cos^2\theta_i - \frac{1}{2}\right\rangle = -\frac{AS^2}{V^2} \, . \tag{3.32}$$

where <...> denotes the thermal averaging using the probability function given in (3.24). To calculate the total energy coming from the pair interaction energies, $\langle U_i \rangle$, we have to multiply it with half of the number of molecules, $N/2$, to avoid counting a molecule twice. Therefore:

$$U = -\frac{1}{2}N\frac{AS^2}{V^2} . \tag{3.33}$$

From the definition of the entropy of a molecule ($\Sigma_i = -k_B \ln P_i(\theta_i)$), and from the expression for $P_i(\theta_i)$ of (3.24), we get the total entropy for N molecule as:

$$\Sigma = N\langle \Sigma_i \rangle = N\left(\frac{\langle U_i \rangle}{T} + k_B \ln Z\right) = -N\frac{AS^2}{TV^2} + Nk_B \ln Z . \tag{3.34}$$

Thus, the free energy for N molecule is:

$$F = N\left[\frac{1}{2}\frac{AS^2}{V^2} - k_B T \ln Z\right] . \tag{3.35}$$

Since $Z = 4\pi$ when $S = 0$, the free energy of the isotropic phase is $-Nk_B T \ln(4\pi) = -2.53 Nk_B T$. The nematic phase is stable only when the free energy is less than this value. Using the Equations (3.25), (3.27) and (3.30), the free energy can be expressed as:

$$F = Nk_B T\left[\frac{mS}{3} - \ln\left(\frac{4\pi}{\sqrt{m}}\exp\left(\frac{2m}{3}\right)D\left(\sqrt{m}\right)\right)\right] . \tag{3.36}$$

To evaluate F, one can choose a value for m, find the value of Dawson integral for that value m, then calculate S from this, and finally find F in the units of $Nk_B T$.

Executing this procedure, one finds that for $S < 0.43$, the isotropic phase has a lower free energy and is therefore the stable phase. For values of S larger than 0.43, the nematic phase is stable. This theory also predicts that the order parameter increases continuously from 0.43 with decreasing temperature (see Figure 3.6). Although this prediction is universal, not depending on the materials, and there are no parameters to adjust (unlike the Landau–de Gennes theory that uses four parameters), it gives fairly good agreement with experiments. This shows that the theory captures the most important essence of the material interactions.

Experimentally, one can detect minor deviations from this universal behavior,[29] indicating that the short-range steric interactions are also important. They were also considered, first by Maier and Saupe,[28] and later by others.[30]

A generalization of the Maier–Saupe model to include forces of other symmetry has been given by Freisner,[31] Chandrasekhar and Madhusudana.[32]

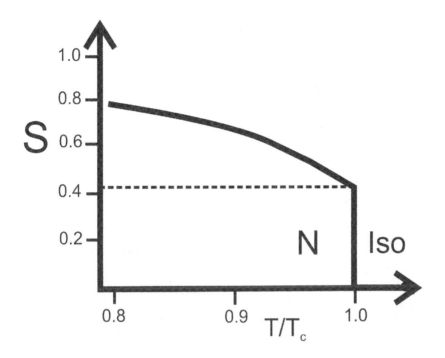

FIGURE 3.6
Schematic representation of the order parameter versus the reduced temperature.

The effect of the density ρ was included in the Maier–Saupe by Tao, Sheng, and Lin,[33] who obtained values of $T_t - T_c = 1K$.

The Maier–Saupe theory was extended by McMillan[34] to describe the smectic A–nematic transition. For this, two order parameters were introduced into the mean-field potential energy function: the usual orientational order parameter S and an order parameter related to the amplitude of the density wave describing the smectic A layers:

$$\sigma = \left\langle \cos\left(\frac{2\pi z_i}{d}\right)\left(\frac{3}{2}\cos^2\theta_i - \frac{1}{2}\right)\right\rangle. \tag{3.37}$$

where z_i is the position of the center of mass of the molecule along the layer normal, θ_i is the angle between the molecular long axis and the director, and d is the layer spacing. The mean-field potential energy also contains both order parameters, and in analogy to the Maier–Saupe potential, is written as:

$$U_i(\theta_i, z_i) = -U_oS[1 + \alpha\cos(2\pi z/d)]\left(\frac{3}{2}\cos^2\theta_i - \frac{1}{2}\right). \tag{3.38}$$

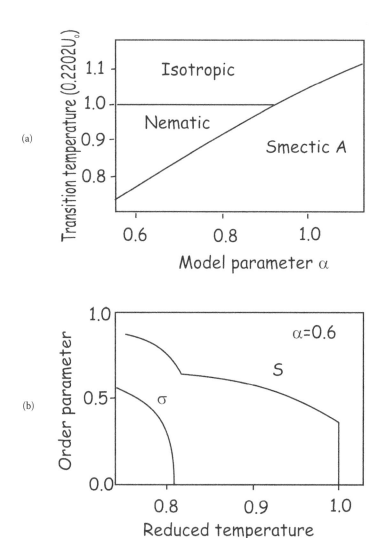

FIGURE 3.7
(a) Phase diagram for the the theoretical model parameter α of the McMillan model. (b) Temperature dependence of the order parameters S and σ.[35]

where α is a parameter describing the strength of the short-range interaction between molecules.

Similar to the Maier–Saupe theory, one can get the expression for the probability function just inserting $U_i(\theta_i, z_i)$ of (3.38) into the expression of P_i given in (3.24). To be self-consistent, the average of the appropriate functions using this probability function must be equal to the order parameters. These two self–consistent equations can be solved numerically. Without going into details of these calculations, which are similar to the steps made in deriving the Maier–Saupe theory, we just state that, depending on the values of α and U_o, three classes of solutions are obtained. First, at high values of α and low values of U_o, both order parameters are nonzero, indicating that the SmA phase is stable. Second, when both α and U_o are intermediate in value, only S is nonzero, indicating that the nematic phase is stable. Third, at low values of α and high values of U_o, both order parameters are zero, indicating that the isotropic phase is the stable one. These three regions are illustrated in Figure 3.7. The statistical theory for the SmA–SmC phase transition was also worked out first by McMillan.[35]

References

1. T.W. Stinson, J.D. Litster, *Phys. Rev. Lett.*, 25, 503 (1970).
2. R. Alben, *Phys. Rev. Lett.*, 30, 778 (1973).
3. G. Melnik, P. Photinos, A. Saupe, *Phys. Rev. A*, 39, 1597 (1989).
4. L.D. Landau, *Collected Papers*, p. 193, Gordon and Breach, New York (1974).
5. T.W. Stinson, J.D. Litster, N.A. Clark, *J. Phys. (Paris) Colloq.*, 33, C1 (1972).
6. L.D. Landau, E.M. Lifshitz, *Statistical Physics*, Pergamon, Oxford (1958).
7. R.G. Priest, *Mol. Cryst. Liq. Cryst.*, 41, 223 (1978).
8. L.J. Yu, A. Saupe, *Phys. Rev. Lett.*, 45, 1000 (1980).
9. P.K. Mukherjee, J. Saha, B. Nandi, M. Saha, *Phys. Rev. B*, 50, 9778 (1994).
10. P.K. Mukherjee, T.R. Bose, D. Ghose, M. Saha, *Phys. Rev. E*, 51, 4570 (1995).
11. P.K. Mukherjee, *J. Phys. Condens. Matter*, 10, 9191 (1998).
12. K. Mukhopadhyay, P.K. Mukherjee, *Int. J. Mod. Phys. B*, 11, 3429 (1997).
13. D. Lacoste, A.W.C. Lau, T.C. Lubensky, *Eur. Phys. J. E*, 8, 403 (2002).
14. Z. Dogic, K.R. Purdy, E. Grelet, M. Adams, S. Fraden, *Phys. Rev. E*, 69, 051702 (2004).
15. P.G. de Gennes, J. Prost, *The Physics of Liquid Crystals*, Clarendon Press, Oxford (1993).
16. Yu. M. Gufan et al., *Phys. Rev. B*, 51,9219 (1995).
17. V. Lorman, B. Mettout, *Phys. Rev. Lett.*, 82, 940 (1999).
18. P.J. Collings, Phase structures and transitions in thermotropic liquid crystals, in *Handbook of Liquid Crystal Research*, Ch. 4, Ed. P.J. Collings, J.S. Patel, Oxford University Press, New York (1997).
19. M. Born, Sitzungberichted. Koenig Preuss. Akad. d. Wissenschaften, *Phys.-Mat. Kl.*, XXX: 613 (1916); also in *Ann. Phys.*, 55, 221 (1918).
20. P. Langevin, *Ann. Chim. Phys. (Paris)*, 5, 70 (1905).
21. P. Weiss, *J. Phys. Rad.*, 6, 661 (1907).
22. G. Szivessy, *Z. Physik.*, 34, 474 (1925).
23. M.F. Grandjein, On the application of magnetism theory to anisotropic liquids, *Comptes Rendes*, 164, 280 (1917).
24. T. Sluckin, D. Dunmur, H. Stegemeyer, *Crystals That Flow — Classic Papers from the History of Liquid Crystals*, Taylor & Francis, Boca Raton, FL (2004).
25. L. Onsager, *Ann. NY Acad. Sci.*, 51, 627 (1949).
26. W. Maier, A. Saupe, *Z. Naturforsch.*, A, 13, 451 (1958); ibid., A13, 564 (1958).
27. W. Maier, A. Saupe, *Z. Naturforsch.*, A, 14, 882 (1959); ibid., A15, 287 (1960).

28. A. Speranza, P. Sollich, arXiv:cond-mat/0203325 v2 9 Aug (2002).
29. I.W. Hamley et al., *J. Chem. Phys.*, 104, 10046 (1996).
30. J.R. McColl, C.S. Shih, *Phys. Rev. Lett.*, 29, 85 (1972).
31. M.J. Freisner, *Mol. Cryst. Liq. Cryst.*, 14, 165 (1971).
32. S. Chandrasekhar, N.V. Madhusudana, *Mol. Cryst. Liq. Cryst.*, 10, 151 (1970).
33. R. Tao, P. Sheng, Z.F. Lin, *Phys. Rev. Lett.*, 70, 1271 (1993).
34. W.L. Macmillan, *Phys. Rev. A*, 4, 1238 (1971).
35. W.L. Macmillan, *Phys. Rev. A*, 8, 1921 (1973).

4

Rheological Properties

The continuum theories of mechanical properties of orientationally ordered soft matter, such as liquid crystals, can be developed in close analogy to solid crystals,[1] but it is necessary to extend the usual formalism. In addition to the usual translational displacements, which describe deformation in solids, one also has to consider the displacements that correspond to rotations of alignment axes.[2] The situation is schematically illustrated in Figure 4.1.

Denoting the displacement vector for the translations by $\mathbf{u}$ and for the rotations by α, and assuming that all displacements are small, the symmetrical part of the strain tensor can be written as:

$$u_{ik} = \frac{1}{2}\left(\frac{\partial u_i}{\partial x_k} + \frac{\partial u_k}{\partial x_i}\right) \tag{4.1}$$

In addition, the asymmetrical "torsion tensor" can be introduced as:

$$\alpha_{ik} = \frac{\partial \alpha_i}{\partial x_k} \tag{4.2}$$

A strain described by u_{ik} leads to a symmetric stress tensor σ_{ik} in materials with elastic response. The force per unit area f on a surface of the surrounding body has the components:

$$f_i = \sigma_{ik} n_k \tag{4.3}$$

Here n is a unit vector pointing outward normal to the considered surface element. As usual in tensor calculus, we understand sum over equal suffixes when they occur in the same term and if not indicated differently. The force per unit volume due to the stress is:

$$\sigma_i = \frac{\partial}{\partial x_k} \sigma_{ik} \tag{4.4}$$

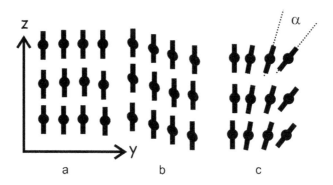

FIGURE 4.1
(a) Not deformed state; (b) deformation by displacements u_z = const. × y; (c) deformation by rotation of alignment α_x = const. × y.

To account for the torques that the particles exert on each other, we define the elements τ_{ik} of the "torque tensor" related to the torsions α_{ik} analogous to σ_{ik}. Accordingly, the torque per unit area is:

$$t_i = \tau_{ik} n_k \tag{4.5}$$

and the torque per unit volume is:

$$\tau_i = \frac{\partial}{\partial x_k} \tau_{ik} \tag{4.6}$$

A distortion of the director is opposed by an elastic force, which is analogous to the elasticity of solids with positional ordering. We consider a weakly distorted situation: $a/l \ll 1$, where a is the molecular dimension (~2 nm), and l is the wavelength of deformation. With these definitions, and assuming that all interactions are of short range, furthermore allowing a rotation of the lattice by the angles $\beta_i = \frac{1}{2}\varepsilon_{ikl}\frac{\partial u_l}{\partial x_k}$, the elastic free energy can be given as:

$$F = F_o + \frac{1}{2}\lambda_{iklm}u_{ik}u_{lm} + \frac{1}{2}\mu_{iklm}\alpha_{ik}\alpha_{lm} + \frac{1}{2}\nu_{ik}(\alpha_i - \beta_i)(\alpha_k - \beta_k)$$
$$+ \kappa_{iklm}^{(1)}u_{ik}\alpha_{lm} + \kappa_{i,kl}^{(2)}(\alpha_i - \beta_i)u_{kl} + \kappa_{i,kl}^{(3)}(\alpha_i - \beta_i)\alpha_{kl} \tag{4.7}$$

In general, there are a total of 21 independent λ, 45μ, 6ν, $54\kappa^{(1)}$, $18\kappa^{(2)}$ and $27\kappa^{(3)}$ coefficients.[2] This means altogether 171 possible parameters, which is impossible to handle both theoretically and experimentally. Fortunately, these numbers reduce for symmetrical systems, since F is invariant under symmetry operations. In addition, some λ and v reduce to zero in liquid crystalline systems, where the lattice is incomplete, or completely absent as for nematic liquid crystals. The terms connected with coefficients μ and κ can become leading term only in liquid crystalline systems.

As structured fluids such as liquid crystals are at least partially fluid, we also need to consider the forces and torques produced by friction. The frictional forces are given by a dissipative stress tensor, which is most conveniently derived from the dissipative function ϕ.[3] It is a homogeneous positive definite quadratic function of the time derivatives of the strains and rotations (the time derivatives of the torsions can be generally ignored) giving:

$$\phi = \frac{1}{2}\eta_{iklm}\dot{u}_{ik}\dot{u}_{lm} + \frac{1}{2}\gamma_{ik}^{(1)}(\dot{\alpha}_i - \dot{\beta}_i)(\dot{\alpha}_k - \dot{\beta}_k) + \gamma_{i,kl}^{(2)}(\dot{\alpha}_i - \dot{\beta}_i)\dot{u}_{kl} \tag{4.8}$$

The coefficients suffix the symmetry relations: $\eta_{iklm} = \eta_{kilm} = \eta_{ikml} = \eta_{lmik}$ and $\gamma_{ik}^{(1)} = \gamma_{ki}^{(1)}; \gamma_{i,kl}^{(2)} = \gamma_{i,lk}^{(2)}$.

The viscosity tensors may have $81 + 27$ elements. However, fortunately only those which are allowed by symmetry will be nonzero. The nonzero elements are determined so that the symmetry of the viscosity tensors must be compatible with the symmetry of the material[4] (for example, a calamitic nematic and smectic A material with uniaxial symmetry will be invariant under $\bar{n} \Leftrightarrow -\bar{n}$).

4.1 Nematic Liquid Crystals

Nematic liquid crystals are 3D anisotropic fluids, and as such they have no translational order, i.e., they do not support extensional or shear strains. For this reason, the rheology of nematic liquid crystals is similar to conventional organic liquids with similar size of molecules. The main difference is due to the anisotropic nature of the materials: the director distortion results in elastic responses, and the magnitude of the viscosity depend on the relative orientation of the director with respect to the velocity gradient.

The macroscopic theory that takes into account the effect of the orientation order was developed by Ericksen,[5] Leslie[6] and Parodi,[7] and usually is referred as ELP theory. A microscopic theory based on correlation functions, which then were "translated" to macroscopic terms and extended to other mesomorphic phases, was developed by the Harvard group.[8] Although usually the ELP theory is accepted,[16] it seems that the two approaches are equivalent.[9] A continuum theory of biaxial nematics was developed by Saupe,[2] who followed the description we give with (4.1)–(4.8). In the uniaxial situation, they reproduce the Leslie–Ericksen and Harvard theories.

The elastic response to the distortion of the director is analogous to the elasticity of solids with positional ordering, and in complete absence of the lattice it can be described only by the term of (4.7) that is quadratic function of the rotation α_i ($\frac{1}{2}\mu_{iklm}\alpha_{ik}\alpha_{lm}$). All other terms are not leading terms anymore. In the case of completely asymmetric hypothetical fluid nematic liquid crystals, this means that "only" 45 elastic constants are needed. However, as has been shown by Frank[11] (see Appendix C.1), in the case of nonchiral and nonpolar nematic liquid crystals of uniaxial symmetry, the distortion free energy, in general, can be written as:[10,11]

$$F_d = \frac{1}{2}K_{11}(\vec{\nabla}\cdot\vec{n})^2 + \frac{1}{2}K_{22}(\vec{n}\cdot\vec{\nabla}\times\vec{n})^2 + \frac{1}{2}K_{33}(\vec{n}\times\vec{\nabla}\times\vec{n})^2$$

$$+\frac{1}{2}\left(K_{22}+K_{24}\right)\cdot\left(\vec{\nabla}\cdot\left\{\vec{n}\vec{\nabla}\cdot\vec{n}+\vec{n}\times\vec{\nabla}\times\vec{n}\right\}\right)$$

$$(4.9)$$

The diagonal elements of the elastic tensor are known as deformation elastic constants for splay (K_{11}), twist (K_{22}) and bend (K_{33}). They reflect configurations in which the director is constant within or parallel to a plane (see Figure 4.2).

The last term in (4.9) is the divergence of a vector and, according to Gauss's theorem,[12] its volume integral can be converted to a surface integral; therefore it describes only contributions to the surface and not to the volume energies.

In the early 1970s Nehring and Saupe proposed[15] that one should add a term linear in second gradients of director to the Frank free energy, namely

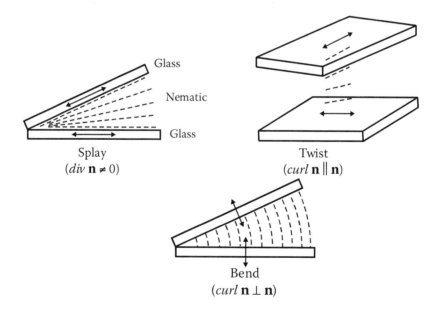

FIGURE 4.2

Illustration of the splay, twist and bend deformations of the thermotropic uniaxial nematic liquid crystals.

$K_{13}\vec{\nabla}(\vec{n}\vec{\nabla}\vec{n})$. This is often called second order or mixed splay-bend term. Microscopic calculations[15] indicate that the mixed splay-bend term is comparable in magnitude to the other terms in the Frank distortion energy; however it is often neglected because in the volume integral of the free energy it transfers to a surface integral, just as the last term in (IV.9). However, near the surfaces this term may influence the equilibrium director through boundary conditions, and thus contribute to the energy.[16]

When discussing bulk energies, this term is usually omitted.[13] For curved surfaces of liquid crystals, e.g., in the description of lyotropic and membrane liquid crystals, however, this term becomes important.[14] The surface term, K_{24}, is usually important in case of saddle-splay (potato chips type) deformations and is called saddle-splay constant.

The free energy must be positive definite, otherwise the undistorted state would not correspond to the minimal energy. Because it is possible to generate pure splay, twist and bend, $K_{ii} > 0$ ($i = 1,2,3$). The saddle-splay deformation involves other deformations, too, so K_{24} is not necessarily positive. It can be shown[15] that $|K_{24}| \leq K_{22}$ and $|K_{11} - K_{22} - K_{24}| \leq K_{11}$.

In parallel plate geometries, the free energy of nematic liquid crystals mostly can be simply expressed as:[16]

$$F_d = \frac{1}{2}\left[K_{11}(\vec{\nabla}\cdot\vec{n})^2 + K_{22}(\vec{n}\cdot\vec{\nabla}\times\vec{n})^2 + K_{33}(\vec{n}\times\vec{\nabla}\times\vec{n})^2\right] \qquad (4.10)$$

The magnitude of the elastic constant can be easily estimated assuming that $K_{ii} \sim U/a$, where U is typical interaction energy between molecules. $U \sim k_B T_{dis}$, where T_{dis} is the temperature where the molecules dissociate (evaporate). $T_{dis} \sim 10^3 K$, so $U \sim 10^{-20}$ J ($\sim$0.1 eV). $a \sim 20$ *Angstrom*; accordingly $K \sim 10^{-20}/2 \times 10^{-9}$ $\sim 5 \times 10^{-12}$ N. Indeed (e.g., for PAA): $K_{11} = 7 \times 10^{-12}$ N, $K_{22} = 4 \times 10^{-12}$ N, $K_{33} = 17 \times 10^{-12}$ N.[17] In general, it is observed that the bend constant is the hardest, especially near to a transition to the smectic phase, where director bend is prohibited due to the constant layer spacing requirements.

Based on the above estimate, we can express the magnitude of distortion energy per molecule:

$$F_d \cdot a^3 \sim \left(\frac{K}{l^2}\right)a^3 \sim U\cdot\left(\frac{a}{l}\right)^2 \ll U \qquad (4.11)$$

This means that distortions do not modify significantly the phase sequences, order parameter, etc.

The most general expression of the viscous stress tensor, $\sigma_{\alpha\beta}^v$, compatible with the symmetry of a uniaxial and incompressible calamitic nematic liquid crystal, was determined first by Leslie as:[6]

$$\sigma_{\alpha\beta}^v = \alpha_1 n_\alpha n_\beta n_\gamma n_\delta A_{\gamma\delta} + \alpha_2 n_\alpha N_\beta + \alpha_3 n_\beta N_\alpha + \alpha_4 A_{\alpha\beta} + \alpha_5 n_\alpha n_\mu A_{\mu\beta} + \alpha_6 n_\beta n_\mu A_{\mu\alpha}$$

$$(4.12)$$

where α_i ($i = 1,2, ...6$) are the Leslie viscosity coefficients.

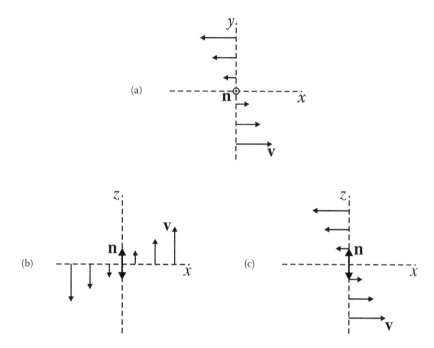

FIGURE 4.3

The three possible orientations of the director in a nematic liquid crystal with respect to the velocity gradient. (a) The director is perpendicular to the shear plane. The corresponding viscosity is η_a. (b) The director is parallel to both the shear plane and the velocity. The viscosity in this situation is η_b. (c) The director is parallel to the shear plane, but is perpendicular to the velocity. In this case the relevant viscosity is η_c. The coordinate systems were chosen so that in each case the director is along the z direction.

In this expression, $A_{\alpha\beta} = \frac{1}{2}(\frac{\partial v_\alpha}{\partial x_\beta} + \frac{\partial v_\beta}{\partial x_\alpha})$ is the symmetric velocity gradient tensor, and N_μ, is the rotational torque with respect to the background fluid, and can be expressed as:

$$\vec{N} = \frac{d\vec{n}}{dt} - \vec{\omega} \times \vec{n} \qquad (4.13)$$

where ω is the local angular velocity of the material.

In addition, it was shown by Parodi[7] that:

$$\alpha_3 + \alpha_2 = \alpha_6 - \alpha_5 \qquad (4.14)$$

This means that five independent nonzero viscosity coefficients exist. Three of them correspond to flow viscosities. They were first measured by Miesovicz[18] and are called Miesovicz viscosities. The three different possibilities and the corresponding viscosity coefficients η_a, η_b and η_c are illustrated in Figure 4.3.

Typically, the viscosities are in the range of 10–100 cP, with η_b being the smallest and η_c being the largest with ratios of about 3.

The Miesovicz viscosities can be related to the Leslie coefficients by comparing (4.12) to the geometries of Figure 4.3.

In case of geometry (a), $A_{yx} = \frac{1}{2}\frac{\partial v}{\partial y}$ and $\sigma_{yx}^v = 2\eta_a A_{yx}$. From (4.12), $\sigma_{yx} = \alpha_4 \cdot A_{yx}$, which gives that:

$$\eta_a = \frac{\sigma_{yx}}{2A_{yx}} = \frac{1}{2}\alpha_4 \tag{4.15}$$

In geometry (b), $A_{zx} = \frac{1}{2}\frac{\partial v}{\partial x}$; $N_x = -\omega_y = A_{xz} = A_{zx}$, and the relevant stress tensor is $\sigma_{xz} = \alpha_3 N_x + (\alpha_4 + \alpha_6)A_{zx}$, giving the relation:

$$\eta_b = \frac{1}{2}(\alpha_3 + \alpha_4 + \alpha_6) \tag{4.16}$$

For (c), $A_{xz} = \frac{1}{2}\frac{\partial v}{\partial z}$; $N_x = -\omega_y = -A_{zx}$, so $\sigma_{zx} = \alpha_2 N_x + (\alpha_4 + \alpha_5)A_{zx}$. This means that:

$$\eta_c = \frac{1}{2}(-\alpha_2 + \alpha_4 + \alpha_5) \tag{4.17}$$

Note here that one of the Leslie coefficients, α_1, does not appear in any of the Miesovicz visocosities. This appears only in such flow measurements, where the director is oblique with respect to the velocity and its gradient. That is why it is difficult to measure. It is also important to note that the measurements of the Miesovicz viscosities have to be carried out in the presence of external fields, which stabilize the director during shear flow. Usually the measurements are done on films between two plates, where displacing one plate with respect to the other provides the shear, and the alignment is given by a magnetic field (in the range of H ~ 1 T) along the director. This is illustrated in Figure 4.4.

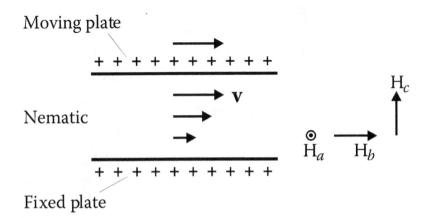

FIGURE 4.4
Usual geometry to measure the Miesowicz viscosities. The film thickness is typically in the range of 20–100 μm, and the strength of the magnetic field is in the range of 0.5–1 T.

The meaning of the remaining two viscosity components can be comprehended by understanding why an external aligning agent is needed to keep the director fixed during the flow. Without the external torque the flow would realign the director. It means that the flow and the director fields are coupled. This is an inherent property of nematic liquid crystals: the coupling reciprocal. The gradient of the velocity field leads to an inhomogeneous rotation of the director (flow alignment), and an inhomogeneous director rotation results in an inhomogeneous flow (backflow effect).

One of the remaining viscosity coefficients is related to the rotation of the director, and the other one is due to the coupling between the flow gradient and director rotation.

The viscosity coefficient related to the director rotation can be obtained from the viscous torque,

$$\vec{\Gamma}^{visc} = \vec{n} \times \hat{\sigma}^v \tag{4.18}$$

In matrix representation it can be written as:

$$\Gamma_i^{visc} = \varepsilon_{ijk}\sigma_{kj} \tag{4.19}$$

where ε_{ijk} is the totally asymmetric Levi–Civita tensor,[19] and σ_{kj}^v has the form given in (4.12). It is easy to see that only the asymmetric part of the viscous stress tensor will give a contribution to the viscous torque, accordingly:

$$\Gamma_i^{visc} = \varepsilon_{ijk}\{(\alpha_2 - \alpha_3)n_j N_k + (\alpha_5 - \alpha_6)n_j A_{kp}n_p\} \tag{4.20}$$

where the first term is the contribution due to director rotation, and $\gamma_1 = \alpha_3 - \alpha_2$ is called rotational viscosity. The physical meaning of the second term is best illustrated considering a laminar flow of the material without an external stabilizing field. Measurements using sandwich cell geometry with simple shear flow indicated that far from the surfaces the director mainly aligns in an angle independent of the shear rate.

Considering the experimental geometry shown in Figure 4.5, we have $A_{xz} = \frac{1}{2}\frac{dv}{dx}$, $N_z = -\omega_y \cdot n_x = -A_{xz}n_x$, $N_x = -\omega_y \cdot n_z = A_{xz}n_z$, and $n_x = \sin\theta$, $n_z = \cos\theta$. Denoting $\alpha_5 - \alpha_6 = \gamma_2$, the viscous torque can be expressed as:

$$\Gamma_{visc} = -\gamma_1(n_z N_x - n_x N_z) - \gamma_2(n_z n_z A_{zx} - n_x n_x A_{xz})$$

$$= -\frac{1}{2}\frac{dv}{dx}[\gamma_1 + \gamma_2(\cos^2\theta - \sin^2\theta)] \tag{4.21}$$

In case of equilibrium and steady director structures, i.e., $\Gamma^{visc} = 0$, the stable orientation is given by $\gamma_1 + \gamma_2(\cos^2\theta - \sin^2\theta) = 0$. This gives the flow alignment angle θ_o as:

$$\cos 2\theta_o = -\frac{\gamma_1}{\gamma_2} \tag{4.22}$$

A stable angle obviously requires that $|-\frac{\gamma_2}{\gamma_1}| \equiv |\lambda| > 1$. Indeed, in most nematics, $|\lambda| > 1$, and there is an equilibrium with flow alignment angles[20] in the range

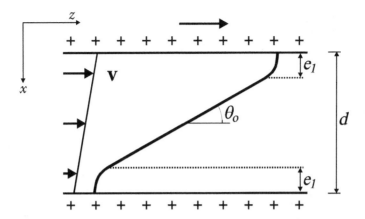

FIGURE 4.5
Illustration of the flow alignment angle in simple shear flow.

of $\theta_o \sim 5\text{--}15°$. It is important to note that θ_o (and consequently the apparent viscosity) does not depend on the shear rate, i.e., the nematic liquid crystal is a Newtonian fluid. However, due to surface effects, as it is illustrated in Figure 4.5, this is true only for thick films, where the surface effects are negligible. Near the substrates, there is a transition layer with thickness: $e_1 \sim \sqrt{\frac{K}{\eta \nabla \nu}}$ [21] (K and η are the relevant distortion elastic constant and shear viscosities, respectively), where the alignment is different form that of the inner area. For this reason, in practice, one always measures a weak shear rate dependence of the apparent viscosity.

Note that for $|\lambda| \le 1$, there is no angle where the hydrodynamic torque would be zero, and a strongly deformed nematic structure forms. Such situation is called "tumbling."[22]

4.2   Cholesteric Liquid Crystals

The flow properties of cholesteric liquid crystals are surprisingly different from those of the nematics. The most important difference is that, in some directions (along the helical axis), the viscosity measured in Poiseuille flow geometries (see Appendix B) is about six orders of magnitude larger than in the isotropic phase, or in the cholesteric phase when the flow direction is perpendicular to the helix axis.[23] In this latter case, the viscosity is similar to that of nematics, although the behavior is somewhat non-Newtonian above a pitch-dependent threshold shear rate. It was found that the shear rate above which the fluid becomes non-Newtonian is inversely proportional to the square of the pitch.[24] The apparent viscosities as the function of shear rate of materials with different pitch values are shown in Figure 4.6.

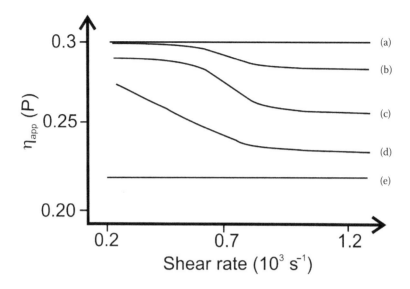

FIGURE 4.6

Apparent viscosity in Poiseuille flow as a function of shear rate. Flow normal to the helical axis at different pitch values (in μm units). (a) 2, (b) 3, (c) 6, (d) 3. (After Chandau et al.[24])

The general theory of shear flow normal to the helical axis has been discussed by Leslie.[25] This basically uses concepts similar to those described by the Ericksen–Leslie–Parodi theory of nematics. The main difference is that the twist term of the deformation free energy will be:

$$\frac{1}{2} K_{22} (\bar{n} \cdot cur\, \bar{n} + q_o)^2 \tag{4.23}$$

where $q_o = 2\pi/P_o$ is the wave-number of the helix in the ground state. It gives that, at low rates of shear, the preferred direction appears to adopt twisted configurations. This modification follows directly from Frank's description (see Appendix C.1) by dropping the nonchirality requirement, which means that K_{12} not zero bringing a term to the distortion free energy that is linear function of the twist (a feature that has no analogy in solid elasticity). The terms containing K_2 and K_{12} can be lumped to one twist term given in (4.23). At high shear rates the flow dictates a uniform orientation thorough the liquid crystal, except in thin layers at the plates, in which the preferred direction changes to that required by the boundary condition. The shear stress necessary to achieve the uniform orientation, i.e., the one needed to overcome the elastic stress of twist deformation, is in the order of $K_{22}(\frac{2\pi}{P})^2$, where K_{22} is the twist elastic constant, and P is the pitch of the helix. In general, the apparent viscosity decreases with increasing shear rates and approaching the nematic value (P = ∞) (see Figure 4.6). This behavior is consistent with the picture that the shear continuously deforms the helix structure, which imposes an extra dissipation until it unwinds the helix. Concerning the pitch dependence, a periodic behavior of the apparent viscosity was found experimentally.[26] This is due to an interplay between the pitch and the film thickness, L; when L/P is an integer the ground state of the helix is pure sinusoidal. The shear-induced deformation of such structure leads to smaller dissipation, i.e., smaller viscosity, than of the ground state when the helix is already distorted. The pitch dependence was given by Kini[27] as:

$$\eta_{app} \approx \frac{\alpha_4(\alpha_3 + \alpha_6)}{2\alpha_4 + \alpha_3 + \alpha_6 - (\alpha_3 + \alpha_6)(P \cdot \sin(2\pi L/P)/2\pi L)} \tag{4.24}$$

Although experimentally the pitch dependence is difficult to measure, the temperature dependence of the apparent viscosity indeed was found[28] to show oscillating behavior, which is due to the temperature dependence of the pitch. Details of the Leslie and other theories and of the relevant experiments are discussed in the book of Chandrasekhar.[29]

The explanation of the flow along the helix axis is much more challenging. This was done by Helfrich, based on a new concept called "permeation."[34] The idea is the following. Imagine that we have a normal Poiseuille flow in a tube, i.e., the speed is larger inside than at the walls. This flow would evidently lead to a distortion of the helix. This distortion, however, requires a threshold torque, so Helfrich assumed that at low shear rates the flow occurs so that the helical structure remains intact. The velocity profile is flat and not parabolic (plug flow) (see Figure 4.7).

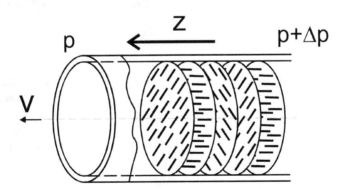

FIGURE 4.7
Plug flow of cholesteric liquid crystal along the helix axis.

Under these circumstances, the translational motion along the capillary will be related to the rotational motion of the director.

$$\vec{\Gamma}^v = \gamma_1 \frac{d\varphi}{dt} \tag{4.25}$$

Where $\vec{\Gamma}^v$ is the viscous torque and γ_1 is the rotational viscosity. With v being the velocity, one has:

$$\frac{d\varphi}{dt} = 2\pi v/P \tag{4.26}$$

The energy density dissipated by the rotation in unit time is:

$$-\gamma_1\left(\frac{d\varphi}{dt}\right)^2 = -\gamma_1\left(\frac{2\pi v}{P}\right)^2 \tag{4.27}$$

If the Poiseuille shear can be neglected, the dissipated energy must be balanced by the energy gained by the motion in the pressure gradient, dp/dz, i.e.,

$$\gamma_1\left(\frac{2\pi}{P}v\right)^2 = \frac{dp}{dz}v \tag{4.28}$$

The quantity of the fluid flowing through in the cross section per second is:

$$Q = -\frac{\pi R^2\left(dp/dz\right)}{\gamma_1\left(2\pi/P\right)^2} \tag{4.29}$$

where R is the radius of the capillary. Applying Poiseuille's law (see B.20),

$$\eta_{app} = -\frac{\gamma_1\pi^2 R^2}{2P^2} \tag{4.30}$$

In the experimentally typical cases $R \sim 500$ μm and $P \sim 1$ μm, which gives the experimentally observed high viscosity at low pressures. Later it was shown that the essential features of the Helfrich model can be explained also on the basis of the Ericksen–Leslie theory.[30]

The above situation is true for weak stress gradients. When the stress gradient becomes larger than that needed to realign the helical axis, the flow will be normal to the helix, and the apparent viscosity becomes similar to that of nematics (see Figure 4.8).

This shows that cholesteric liquid crystals can be characterized by a very strong non-Newtonian behavior, which resembles to the behavior of Bingham fluids (Appendix B).

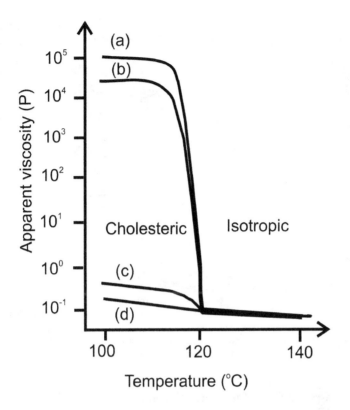

FIGURE 4.8

Temperature dependence of apparent viscosity of cholesteryl acetate. Capillary shear rate. Curves labeled with different symbols correspond to shear rates 5000 (a), 1000 (b), 100 (c) and 10 (d) in s[-1]. (After Porter et al.[23])

4.3 Rheology of Smectic Liquid Crystals

The strange behavior of smectic liquid crystals can be beautifully illustrated by a demonstration that William Doane, Director Emeritus of the Liquid Crystal Institute of Kent State University, used to show in the 1980s at high schools or business meetings. This is illustrated in Figure 4.9. One fills a small amount of liquid crystal material that has isotropic–nematic–smectic A phase sequence, with the smectic A phase conveniently falling in the room temperature range. When the sample is heated to the isotropic phase and then cooled in a uniform magnetic field of a few tenths of Tesla, we end up with an aligned sample with uniform layer structure. The quality of the alignment can be easily verified by the transparency of the sample: higher transparency indicates better alignment. When the sample is taken out of the magnetic field, and poured by flipping the tube around the axis normal to the direction of the previously

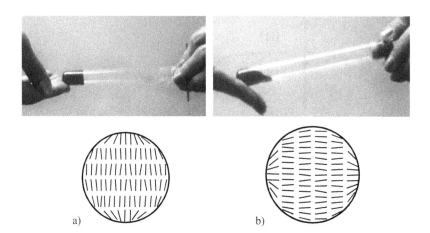

FIGURE 4.9

Illustration of anisotropic flow properties of smectic A liquid crystals. Sticks are parallel to the director. (a) The director is in the plane of the flipping. In this case the material acts as a solid. Lower row: The director and layer structure of the cross section of the tube in the view when facing the opening of the tube. The smectic layers are in horizontal position. (b) The director is normal to the plane of the rotation of the sample tube. The material behaves as a fluid. Lower row: The director and layer structure of the cross section of the tube in the view when facing the opening of the tube. The smectic layers are in vertical position.

applied magnetic field, the material behaves as a solid, and no flow is observed. When rotated by 90 degrees around its symmetry axis, so flipped around the axis of previously applied magnetic field, the fluid flows like a normal liquid. This looks like a magic, but can be easily understood keeping in mind the concept of permeation introduced by Helfrich.

In order for the flow to occur, some part of the molecules need to move downward. In the case corresponding to Figure 4.9a, such a motion would require permeation of the molecules through the layers, whereas in the situation corresponding to Figure 4.9b, the molecules can move along the layers, which does not require permeation. The situation therefore is very similar to plug flow in cholesterics, except that now the gravitational force replaces the pressure gradient and that the permeation is much harder. To understand quantitatively this and other magic of the smectic liquid crystals, we need to understand their rheology. Introduction of one-dimensional positional order makes the description of elasticity and flow properties very complicated. Here we would like to understand the most important concepts, outline the basic steps of the theories and give the relevant references where the details can be found.

4.3.1 Elasticity of Smectic A Phase

To proceed from simple to more complicated, let us first assume that the smectic layers (Figure 4.10) are incompressible, i.e., only those deformations are allowed that leave the smectic layer spacing constant. It is obvious that

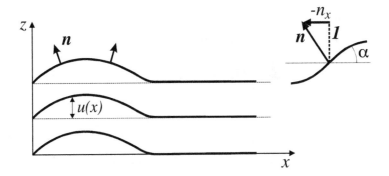

FIGURE 4.10
Illustration of the stack of smectic layers.

neither the director bend nor the twist is compatible with the constant layer spacing requirement.

This means that the only term remaining in the free energy density is:

$$F_d = \frac{1}{2} K_{11} (\vec{\nabla} \cdot \vec{n})^2 \qquad (4.31)$$

In case of sufficiently strong director bend, we will create defects (see Chapter 6): edge dislocations will invade the smectic in an attempt to keep the layer thickness constant.

If we consider a stack of layers, which are originally flat and at equilibrium, we can describe a deformed state by introducing $u(x,y)$, which gives the displacement relative to the original position along the z axis, which is along the undistorted layer normal (see Figure 4.10).

For small distortions $\partial u / \partial x = \tan \alpha \approx -n_x$ (similarly: $\partial u / \partial y = -n_y$). Accordingly,

$$F_d = \frac{1}{2} K_{11} (\vec{\nabla} \cdot \vec{n})^2 = \frac{1}{2} K_{11} \left(\frac{\partial n_x}{\partial x} + \frac{\partial n_y}{\partial y} \right)^2 = \frac{1}{2} K_{11} \left(\frac{\partial^2 u}{\partial x^2} + \frac{\partial^2 u}{\partial y^2} \right)^2 \qquad (4.32)$$

The layers will be compressed or dilated if $\partial u / \partial z \neq 0$. We assume that there is a linear relation between the stress and strain across the layers. The corresponding term in the elastic energy then will be $\frac{1}{2} B (\frac{\partial u}{\partial z})^2$, where B is the layer compression modulus (experiments[32,33] show that $B \sim 10^6$ N/m^2). With this,

$$F_d = \frac{1}{2} B \left(\frac{\partial u}{\partial z} \right)^2 + \frac{1}{2} K_{11} \left(\frac{\partial^2 u}{\partial x^2} + \frac{\partial^2 u}{\partial y^2} \right)^2 \qquad (4.33)$$

Note, if the displacement varies along z, two additional terms proportional to $\partial^2 u / \partial z^2$ will appear, too.[16] They, however, practically always can be neglected.

The equilibrium structures are determined by the condition that $\int F_d dV$ should be minimal. This is the problem of variation calculus, and the solution is given by the Euler–Lagrange equation. Assuming uniformity in y direction:

$$\frac{\partial}{\partial z}\frac{\partial F_d}{\partial(\partial u/\partial z)} - \frac{\partial^2}{\partial x^2}\frac{\partial F_d}{\partial(\partial^2 u/\partial x^2)} = 0 \tag{4.34}$$

Comparing (4.35) with (4.34) we get:

$$B\frac{\partial^2 u}{\partial z^2} - K_{11}\frac{\partial^4 u}{\partial x^4} = 0 \tag{4.35}$$

In case of periodic deformation in x direction we can search the solution in the form:

$$u = u_o(z)\cos kx \tag{4.36}$$

Inserting this into (4.36) we get:

$$B\cos(kx)\frac{\partial^2 u_o(z)}{\partial z^2} - K_{11}k^4 u_o(z)\cdot\cos(kx) = 0 \tag{4.37}$$

Introducing the permeation length,

$$\lambda = \left(K_{11}/B\right)^{1/2} \tag{4.38}$$

we get

$$(\partial^2 u_o(z)/\partial z^2)/u_o(z) = \lambda^2 k^4 \tag{4.39}$$

It is easy to see that this equation is satisfied by the solution:

$$u = e^{-\lambda k^2 z}\cos kx \tag{4.40}$$

From the definition of the characteristic length λ in Eq.(4.39), and from the typical distortion modulus $K \sim 10^{-11}$ N, we get that the permeation length is $\lambda \sim 3$ nm, i.e., comparable to the layer spacing, as it was indeed predicted theoretically.[34] From Eq.(4.41) we can introduce a second characteristic length, the penetration length:

$$l = 1/(\lambda k^2) \tag{4.41}$$

which gives the length range in which an undulation imposed by a surface penetrate into the volume (Figure 4.11). Assuming a surface waviness with periodicity of $10\,\mu m$, l becomes as large as 1 mm. This is indeed a macroscopic parameter, and this distinguishes smectics from both ordinary crystals and nematic liquid crystals, where distortions relax in the medium in length scales comparable to those of distortions. This is perhaps the most striking general result of smectic elasticity.

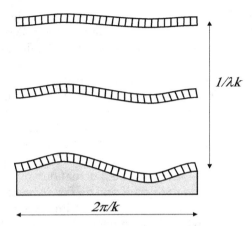

FIGURE 4.11
Illustration of the penetration of a surface-induced deformation.

This analysis can be also extended to study layer fluctuations $u(q)$, where q is the wave-number of the periodic deformations. In this case, it is useful to express the free energy (4.34) in the Fourier space of a unit volume:

$$F_{sm} = \frac{B}{2} \int \frac{d^3q}{(2\pi)^3} (q_z^2 + \lambda^2 q_\perp^4) | u(q)|^2 \qquad (4.42)$$

where q_z is the wavenumber of the deformation along the smectic layer normal (solid direction), and $q_\perp$ is the wavenumber of deformation along the smectic layers, i.e., in the fluid direction. Utilizing the theorem of equipartition of energy, we can write:

$$\left\langle | u(q)|^2 \right\rangle = k_B T/B(q_z^2 + \lambda^2 q_\perp^4) \qquad (4.43)$$

where $k_B = 1.38 \times 10^{-23}$ J/K is the Boltzmann constant. Introducing the cut-off wave vector q_c, beyond which the elastic theory breaks down ($q_c \sim 1/l$, where l is the size of the building units), we get for smectics:

$$\langle u^2(r) \rangle = \frac{k_B T}{(2\pi)^3 B} \int_{1/L}^{1/l} \frac{d^3q}{q_z^2 + \lambda^2 q_\perp^4}$$

$$= \frac{k_B T}{(2\pi)^3 B} \int_{L^{-1}}^{l^{-1}} \frac{dq_z 2\pi q_\perp dq_\perp}{q_z^2 + \lambda^2 q_\perp^4} = \frac{k_B T}{(2\pi)^2 B} \int_{L^{-1}}^{l^{-1}} dq_z \left(\int_{L^{-1}}^{l^{-1}} \frac{q_\perp dq_\perp}{q_z^2 + \lambda^2 q_\perp^4} \right)$$

$$= \frac{k_B T}{(2\pi)^2 B} \int_{L^{-1}}^{l^{-1}} dq_z \frac{1}{2q_z} \tan^{-1}\left(\frac{q_\perp^2}{q_z} \right) \cong k_B T \log(L/l)/4\pi\lambda B \qquad (4.44)$$

It is seen that for smectics the fluctuation logarithmically diverges with L.[†]
This was first shown by Peierls and Landau and is called Landau–Peierls
instability.[35] This is behavior is very different from the fluctuations in crystal
and columnar liquid crystals (3D and 2D elasticity, respectively) where the
amplitude of the fluctuations remain finite even for infinite samples,
as $\langle u^2 \rangle_{crystal} = \frac{k_B T}{2\pi B}(1/l - 1/L)$, and $\langle u^2 \rangle_{columnar} = \frac{k_B T}{(2\pi)^2 B}(l^{-1}\lambda^{-1})^{1/2}$. Inserting typical
values, $k_B T \sim 4 \cdot 10^{-21}$J, $\lambda \sim 3$nm; B$\sim 10^9$J/m^3 (crystals) $\sim 10^6$J/m^3 (LC-s); q$_c \sim 10^{10}$
m^{-1} (for crystals) $\sim 10^9$m^{-1} (for liquid crystals) to these expressions, we get
for crystal, columnar and smectic samples is in the cm ranges $\sqrt{\langle u^2 \rangle} \simeq 0.04$ *nm*,
$\sqrt{\langle u^2 \rangle} \simeq 0.15nm$ and $\sqrt{\langle u^2 \rangle} \simeq 0.4nm$, respectively. This means that even in
smectic phases in everyday situations the fluctuations are still much smaller
than the layer spacing. Although the Landau–Peierls instability practically
does not destabilize the smectic phase, it has important influence on the
rheology of the smectics, which is often referred to as breakdown of conventional elasticity.[36,37]

In writing the expression for the energy density of Eq. (4.34), we neglected
that, in the second order, the bending of the layers alters the effective layer
spacing along z and makes a second-order contribution. With this correction
and up to the second order, the energy density can be written as:

$$F_d = \frac{1}{2} B \left[\frac{\partial u}{\partial z} - \frac{1}{2}(\nabla_\perp u)^2 \right]^2 + \frac{1}{2} K_{11}(\Delta_\perp u)^2 \qquad (4.45)$$

Here $\nabla_\perp$ and $\Delta_\perp$ are the gradient operator and *Laplacian*, respectively, in the
xy plane. Using a renormalization procedure, Grinstein and Pelcovits
showed[36] that the effect of the higher-order term is that the elastic constants
become dependent on the wave-numbers of the fluctuations. This effect
becomes increasingly important for asymptotically small q, where the fluctuations diverge as shown in (4.45). According to Grinstein and Pelcovist,
in this limit the layer compression modulus vanishes as $B(q)$-|ln q|$^{-4/5} \to 0$,
which means that at very large scale it behaves as a fluid nematic. Interestingly, however it was also shown that in this limit the splay elastic constant
diverges as $K_{11}(q) \sim |\ln q|^{2/5} \to \infty$, which means that this fluid is rigid against
splay. Although these corrections are practically small, they are measurable.

It is interesting to note that the requirement of undistorted helix of cholesterics is analogous to the constant layer spacing requirements of smectic
liquid crystals. For this reason the concept of permeation is also important
in smectics, and to some extent their mechanical properties are similar. The
"only" difference is the length scale of the periodicity: a few nanometers in
smectics and micrometers in cholesterics. Based on this analogy, Lubensky[31]
and de Gennes[16] constructed a coarse-grained version of the continuum

---

[†]Note that for crystals and columnar liquid crystals the fluctuations do not depend on the size
of the sample.

theory of cholesterics. They found that the deformation of the helix gives a contribution F_{cg} to the free energy as:

$$F_{cg} = \frac{1}{2} B \left(\frac{P}{P_0} - 1 \right)^2 + \frac{1}{2} \frac{\tilde{K}}{r^2} \tag{4.46}$$

where P is the local value of the pitch, and r is the curvature radius of plane of the helix axis. It was shown that $B = K_{22} q_0^2$, and $\tilde{K} \approx \frac{3}{8} K_{33}$.[16]

4.3.2 Flow Properties of Smectic A Liquid Crystals

The existence of crystalline order in at least one direction reflected most dramatically in flow properties, just as illustrated by the demonstration shown in Figure 4.9a. Instead of a complete overview of the theories,[38] we give other demonstrative examples.

Just like in nematics, there are three basic geometries corresponding to different orientation of the layer normal and the pressure gradient. These three geometries are shown in Figure 4.12.

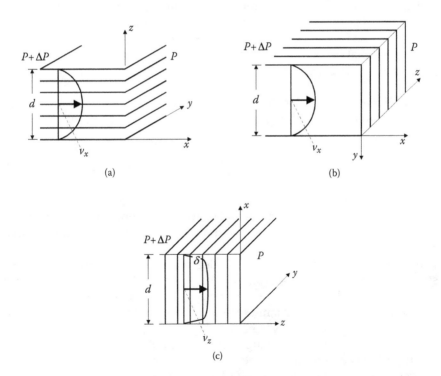

(a)

(b)

(c)

FIGURE 4.12
Characteristic flow geometries for smectic A phases. A pressure head Δp promotes the flow. (a) Velocity parallel and gradient is perpendicular to the layers. (b) Velocity and the gradient are parallel to the layers. (c) Both the velocity and the gradient are perpendicular to the layers.

When the velocity field probes liquid directions (Figure 4.12a and b) the velocity profile is parabolic, just like in Poiseuille flow geometries of isotropic fluids. When the velocity probes a solid direction, a plug flow appears, as in cholesterics, with flow along the helical axis. The velocity profile in plug flow is constant, except in a narrow region δ, which is of the order of molecular length. In geometry (a), experimental observations are basically consistent with the expectations that the flow is Newtonian, with relevant viscosity corresponding to Miesowicz viscosity η_c.[39] The problem doing experiments in this geometry is that the film thickness has to be kept constant in the whole sample with an accuracy comparable to the layer spacing, which is not possible with standard viscometric devices. If a thickness variation,

$$\Delta h > 2\pi \lambda = 2\pi \sqrt{K_{11}/B} \qquad (4.47)$$

would exist, a system would be dilated or compressed, and a buckling instability would arise, leading to arrays of focal conic defects (see Chapter 6).[40] Mechanical instabilities under dilative or compressive stresses are summarized by Ribotta and Durand.[41] The problems partially can be overcome by making alternating vibrations of the top plate of geometry (a), instead of a steady flow. If the amplitude of the vibration is only 10 nm, the plate parallelism is greater than 10^{-4} rad (which is easily done), and thickness of 100 μm, the residual strains would be[42] $\delta h/h \sim 10^{-8}$. Even in this case, the residual normal stress would be $\sigma_{zz} = B(\delta h/h) \sim 10^{-2} N/m^2$, which is still comparable to the viscous stress $\sigma_{xz} = \eta_c(\partial v_x/\partial z)$ (η_c ~0.1 *Pas and v* ~ 10 nm x 1 *kHz* ~ $10^{-6}s^{-1}$).

In geometry (b) one does not find a purely viscous behavior, but rather that of a Bingham fluid,[39] and the shear stress then can be described as:

$$\sigma_{xy} = \frac{\eta}{2}\partial_x v_y + \sigma_o \qquad (4.48)$$

Experimentally, η was found to be in the order of nematic viscosity, with value depending on the shear rate and history, indicating the effect of defects on the flow properties. It was also found that $\sigma_o \sim 1N/m^2$, which is comparable to $\sigma_{zz} = B(na/L) \sim n \cdot 3N/m^2$ with n ~ 3, where a is the layer thickness, and L is the width of the film (~1 cm). Because $\sigma_{zz} > \sigma_o$, the effect of residual stress may explain the observations. We will see a bit later, however, that more subtle explanation of the existence of σ_o may be connected to the breakdown of the hydrodynamics.[43] Results obtained under periodic shears, however, show pure viscous behavior with no frequency dependence. This shows the importance of working with small amplitudes, which reduces the residual effects due to the thickness variations and of the nonlinear effects.

In geometry (c), close to the boundary in a range controlled by the characteristic length δ, which is in the order of molecular size,[44] the speed varies from zero given by the boundary condition to the uniform velocity of the bulk dictated by the plug flow. In this range, the viscous friction is important, and is called proximal region.[45] In the bulk a uniform velocity profile establishes where the velocity will be proportional to the pressure gradient, i.e., the flow can be described by the equation:

$$v_z = \lambda_p \partial P / \partial z \qquad (4.49)$$

This range has been called the "distal" region. Although the layer compression modulus is about two orders of magnitude larger than in the short-pitch cholesterics (see (4.31)), permeation-type flow similar to that observed in cholesterics was observed near the nematic phase,[40] showing that the apparent viscosity maybe smaller than in the cholesteric, probably due to defects that cause plastic behavior. This and measurements under periodic deformations indicate the importance of layer defects, which are hard to regulate.

In addition to the simple shear, the rheological properties can also be characterized by studying the flow of the fluid around fixed obstacles. When the flow plane is parallel to the smectic layers the flow around the sphere involves mainly only fluid directions (see Figure 4.13), and the force acting on a sphere is purely viscous.

The force embedded in a flowing smectic A liquid crystal was calculated by de Gennes.[45] In the absence of emission of dislocations, the force $\vec{F}$ acting on a sphere of radius R has the same form as of Stokes law of an ordinary fluid (see Appendix B), except that the multiplying factor 6 becomes 8.

$$\vec{F} = 8\pi \eta R \vec{v} \qquad (4.50)$$

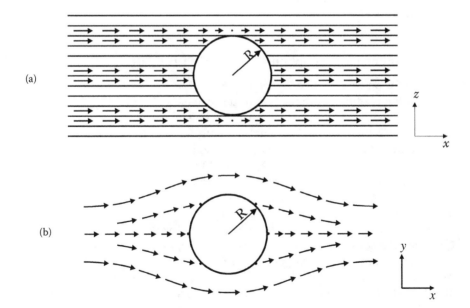

FIGURE 4.13
Flow past a sphere. (a) Flow in the xz plane. (Note that $v_z = 0$ everywhere). (b) The flow in the xy plane.

Here $\eta \approx \eta_c$ is the relevant Miesovicz viscosity, and $\vec{v}$ is the velocity of the sphere. The situation is quite different for semi-infinite[47] and finite barriers,[48] where permeation is necessarily involved in the flow, leading to undulation of layers and arrays of defects. Since it is very difficult to trap and stabilize a sphere with a size smaller than the film thickness, as far as we know, the flow of smectics around a sphere has never been studied experimentally. However, experiments when finite particles are moving in the liquid crystal medium, which is at rest far from the particles has been carried out very recently, and it was indeed found that the flow of beads in smectic A and smectic C liquid crystals is purely viscous at sufficiently high speeds.[49] Such technique is analogous to the one-bead micro-rheology[50] developed recently to monitor the mechanical properties of viscoelastic soft materials, especially biological systems.[51]

The Peierls–Landau instability that results in logarithmical divergences of the fluctuations may also have effect on the frequency dependence of the viscosity, as was shown by Mazenko et al.,[37] who examined the anharmonic fluctuation effects to the hydrodynamics of the smectics. To understand the effect of the long-wavelength fluctuations on the viscosity, we have to recall that a force normal to the layers due to the distortion u is given as $g = -\partial F_d/\partial u$. As we have seen in (4.46), g contains the nonlinear term $\partial_z(\nabla_\perp u)^2$, yielding a nonlinear contribution to the stress. Taking into account that the viscosity at frequency ω is the Fourier transformation of a stress autocorrelation function,[52] it was shown by Mazenko et al.[37] that $\delta\eta/B \sim 10^{-2}/\omega$.[16] Taking $B \sim 10^6$ *N/m²*, $\omega \sim 10^7$ rad s⁻¹, we get $\delta\eta \sim 10^{-3}$ *Pas*, which is hard to measure, but seems to be confirmed by experiments using ultrasonic attenuation methods.[53]

4.3.3   Continuum Description of the SmC Phase

The director deformations described by K_{ii} that do not lead to layer compressions, in the continuum range where the wavelengths λ of the deformation are much larger than the molecular dimensions ($\lambda \gg 10$ nm) can be induced by stress $K(2\pi/p)^2 < 10^5$ N/m². This is usually smaller than of the layer compression modulus $B \sim 10^6$ N/m². For this reason, deformations that do not lead to layer compression (such as splay in SmA) are usually called soft deformations, whereas those that require layer compression (such as bend and twist in SmA) are the so-called hard deformations. In SmC there will be six soft[54] and three hard[55] deformations, so it is basically impossible to take into account all elastic terms while keeping the transparent physics. (In the chiral smectic C* materials, additional three terms are needed, as shown by de Gennes.[34]) Fortunately, however, the larger number of soft deformations enable for the material to avoid the hard deformations, which makes it possible to understand most of the elastic effects, even in SmC materials.

In the following, we will mainly concentrate on the soft deformations, assuming undistorted layers, and only outline the hard deformations.

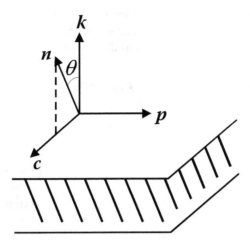

FIGURE 4.14

Local structure of smectic C liquid crystals. k is the local smectic layer normal, c is the c-director, $\vec{k}$ is the layer normal, $\vec{n}$ is the director, and θ is the tilt angle.

To start out, we need to introduce the concept of two directors, c and k as shown in Figure 4.14. Note that the c-director is a real vector, i.e., ($\vec{c} \neq -\vec{c}$). For mathematical convenience, we also introduce a third unit vector, $\vec{p} = \vec{k} \times \vec{c}$. Since a change in the tilt angle θ leads to a variation of the layer spacing, we require that θ be constant.

From Figure 4.14 we have that:

$$\vec{n} = \vec{k} \cos \theta + \vec{c} \sin \theta \equiv a\vec{k} + b\vec{c} \tag{4.51}$$

To express director splay, twist and bend with the new variables is a simple formality. Accordingly:

$$\nabla \cdot \vec{n} = a\nabla \cdot \vec{k} + b\nabla \cdot \vec{c} \tag{4.52}$$

$$\nabla \times \vec{n} = a\nabla \times \vec{k} + b\nabla \times \vec{c} \tag{4.53}$$

$$\vec{n} \cdot \nabla \times \vec{n} = (a\vec{k} + b\vec{c})(a\nabla \times \vec{k} + b\nabla \times \vec{c})$$
$$= a^2\vec{k} \cdot \nabla \times \vec{k} + b^2\vec{c} \cdot \nabla \times \vec{c} + ab(\vec{k} \cdot \nabla \times \vec{c} + \vec{c} \cdot \nabla \times \vec{k}) \tag{4.54}$$

$$\vec{n} \times \nabla \times \vec{n} = (a\vec{k} + b\vec{c}) \times (a\nabla \times \vec{k} + b\nabla \times \vec{c})$$
$$= a^2\vec{k} \times \nabla \times \vec{k} + b^2\vec{c} \times \nabla \times \vec{c} + ab(\vec{k} \times \nabla \times \vec{c} + \vec{c} \times \nabla \times \vec{k}) \tag{4.55}$$

In the case of incompressible and undistorted layers $\nabla \cdot \vec{k} = 0$ and $\nabla \times \vec{k} = 0$. With these, furthermore, utilizing that c and k are perpendicular unit vectors (for example, $(\vec{c} \times \nabla \times \vec{c})^2 = (\vec{k} \cdot \nabla \times \vec{c})^2$), in the soft deformation limit we get that:

$$
\begin{aligned}
F_d &= \frac{1}{2} K_{11}(\nabla \cdot \vec{n})^2 + \frac{1}{2} K_{22}(\vec{n} \cdot \nabla \times \vec{n})^2 + \frac{1}{2} K_{33}(\vec{n} \times \nabla \times \vec{n})^2 \\
&= \frac{1}{2} B_1(\vec{k} \cdot \nabla \times \vec{c})^2 + \frac{1}{2} B_2(\nabla \cdot \vec{c})^2 + \frac{1}{2} B_3(\vec{c} \cdot \nabla \times \vec{c})^2 - B_{13}(\vec{k} \cdot \nabla \times \vec{c}) \cdot (\vec{c} \cdot \nabla \times \vec{c})
\end{aligned}
\tag{4.56}
$$

where the B_i constants are related to the Oseen–Frank constants[56,57] as follows: $B_1 = K_{22} \sin^2 \theta \cos^2 \theta + K_{33} \sin^4 \theta$ describes the bend of the c-director; $B_2 = K_{11} \sin^2 \theta$ describes the splay of c-director; $B_3 = K_{22} \sin^4 \theta + K_{33} \sin^2 \theta \cos^2 \theta$ corresponds to the c-twist, and $B_{13} = (K_{33} - K_{22}) \sin^3 \theta \cos \theta$ is the coupling between the c-director bend and twist.

It is interesting to analyze the SmA → SmC phase transition. For small tilt angles, we have

$$
B_1 = K_{22}\theta^2 + K_{33}\theta^4 ; \; B_2 = K_{11}\theta^2 ; \; B_3 = K_{22}\theta^4 + K_{33}\theta^2 \text{, and } B_{13} = (K_{33} - K_{22})\theta^3.
$$

This shows that at the transition only the twist K_{22} contributes to the c-bend B_1, and only the bend constant K_{33} contributes to the c-twist constant B_3. This illustrates how twist and bend are mixed in smectic structures.

So far, only a few experiments have been performed to test the B_i constants. Similar for the measurement of the K_{ii} constants, a kind of Freedericksz transition may be used. Experimentally,[58] B_2 was always found to be smaller than K_{11}, which is in agreement with the above relations between the B_i and K_{ii} terms.

The soft layer distortions s1, s2, and s3 and together with the hard layer compression, give F_{ld} can be described by the rotation vector components,

$$
F_{ld} = \frac{1}{2} B\gamma^2 + \frac{1}{2} A_{11} \left(\frac{\partial \Omega_x}{\partial x} \right)^2 + \frac{1}{2} A_{12} \left(\frac{\partial \Omega_y}{\partial x} \right)^2 + \frac{1}{2} A_{21} \left(\frac{\partial \Omega_x}{\partial y} \right)^2
\tag{4.57}
$$

where $\gamma \equiv \partial u/\partial z$, $\Omega_x = \partial u/\partial y$ and $\Omega_y = -\partial u/\partial x$.

Finally, there are the terms describing the coupling between the distortions of the smectic layers and of the c-director field as:[57]

$$
F_{l-c} = C_1 \frac{\partial \Omega_x}{\partial x} \frac{\partial \Omega_z}{\partial x} + C_2 \frac{\partial \Omega_x}{\partial y} \frac{\partial \Omega_z}{\partial y}
\tag{4.58}
$$

This expression is not very transparent and has been rarely used in explaining any experimental situation. Since, in SmC, the director can move around a cone, in most cases it has enough freedom to avoid the hard deformations, and one usually ends up dealing only with soft deformations. In fact, in most cases, even these are complicated to work with, and this theory was refined and reformulated by several authors.[59,60]

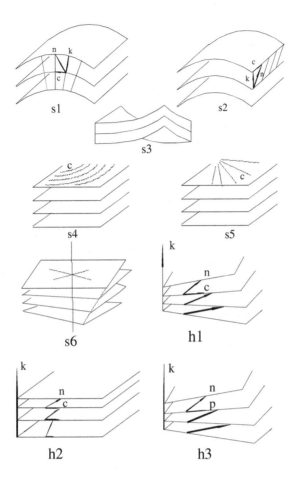

FIGURE 4.15

Illustration of the soft and hard deformations of smectic layer structures (after. S.T. Lagerwall[57]). (s1) Layer bend (splay in k) with the bending axis along p described by A_{12}; (s2) the corresponding layer bend with bending axis along c described by A_{21}; (s3) a saddle-splay deformation of the layers with the two bending axes making 45° to c and p. Corresponding elastic constant A11; (s4) a bend in the c-director (elastic constant B1); (s5) a splay in the c-director (elastic constant B2); (S6) a twist in the c-director corresponding to B3; (h1) a layer splay inducing a splay-bend deformation in the n field; (h2) layer compression or dilation, which varies along k and thereby it induces a bend in the n director; (h3) layer splay perpendicular to the tilt plane, which induces a twist-bend in the n field along p.

A demonstration of the six soft and three hard deformations of a smectic C structure without the coupling terms B_{13}, C_1, C_2 (which are difficult to visualize) is shown in Figure 4.15.

4.3.4 Continuum Description of the Chiral Smectic C Phase

We have seen that the soft elastic energy of a smectic C material can be derived from the Oseen–Frank energy of nematic liquid crystals (see (4.56)).

Similarly, one can construct the soft elastic contribution of the free energy density, starting out from the energy density of the cholesteric liquid crystals. In 1958, Frank extended the nematic free energy expression to the chiral nematic case,[61] introducing the term $(\vec{n} \cdot \nabla \times \vec{n} + q)^2$. Based on this, and guided by Meyer's recognition[62] that the SmC* is a space filling twist-bend structure, Lagerwall and Dahl[56] introduced the corresponding term $(\vec{n} \times \nabla \times \vec{n} - \vec{B})^2$ to account for spontaneous bend. This allows us to treat the smectic C* phase in the nematic description. Accordingly we can write the soft free energy density of the SmC* in the form:

$$F_{dC^*} = \frac{1}{2} K_{11}(\nabla \cdot \vec{n})^2 + \frac{1}{2} K_{22}(\vec{n} \cdot \nabla \times \vec{n} - q)^2 + \frac{1}{2} K_{33}(\vec{n} \times \nabla \times \vec{n} - \vec{B})^2 \quad (4.59)$$

With $\vec{B} = -\beta \vec{k} \times \vec{n} = -\beta b \vec{k} \times \vec{c}$, including only the first order terms, the free energy density can be written as:

$$F_{dC^*} = \frac{1}{2} B_1(\vec{k} \cdot \nabla \times \vec{c})^2 + \frac{1}{2} B_2(\nabla \cdot \vec{c})^2 + \frac{1}{2} B_3(\vec{c} \cdot \nabla \times \vec{c})^2 - B_{13}(\vec{k} \cdot \nabla \times \vec{c}) \cdot (\vec{c} \cdot \nabla \times \vec{c})$$

$$+ D_1 \vec{k} \cdot \nabla \times \vec{c} - D_3 \vec{c} \cdot \nabla \times \vec{c}$$

$$(4.60)$$

where

$$D_1 = K_{22} q \sin \theta \cos \theta + K_{33} \beta \sin^2 \theta$$
$$D_3 = K_{22} q \sin^2 \theta - K_{33} \beta \sin \theta \cos \theta$$
$$(4.61)$$

The new chiral terms are of first order and therefore describe spontaneous deformations. The D_1 term means that there is a tendency for the c-vector to bend in the plane of the layer, which means that a coupled ferroelectric polarization (see next chapter) will have a tendency for a spontaneous splay. The D_3 term is a spontaneous twist, describing the helical order of the SmC* ground state. In the ground state, there is a pure twist in c. This means that $\vec{c} \cdot \nabla \times \vec{c} = -q$, and consequently, $\vec{k} \cdot \nabla \times \vec{c} = 0$. This reduces (4.60) near the equilibrium to:

$$F_{dC^*}(q) = F_{do} + \frac{1}{2} B_3 q^2 - D_3 q \quad (4.62)$$

Hence,

$$\partial F_{dC^*}/\partial q = B_3 q - D_3 = 0 \quad (4.63)$$

which gives the wave number $q = D_3/B_3$. If we notice that $\vec{k} \cdot \nabla \times \vec{c} = -(\vec{c} \times \nabla \times \vec{c}) \cdot \vec{p}$, where $\vec{c} \times \nabla \times \vec{c}$ is a vector in the $\vec{p}$ direction, we can rewrite (4.60) as:

$$F_{dC^*} = (D_1/B_1)^2 - (D_3/B_3)^2 + \frac{1}{2}B_1(\vec{c} \times \nabla \times \vec{c} \cdot \vec{p} - D_1/B_1)^2 + \frac{1}{2}B_2(\nabla \cdot \vec{c})^2$$

$$+ \frac{1}{2}B_3(\vec{c} \cdot \nabla \times \vec{c} + D_3/B_3)^2 + B_{13}(\vec{c} \times \nabla \times \vec{c}) \cdot \vec{p} \cdot (\vec{c} \cdot \nabla \times \vec{c})$$

(4.64)

All these terms are easy to interpret. The B_1 term expresses the fact that there is a spontaneous bend D_1/B_1 in this plane. The second term describes the splay. It does not couple to other deformations, and there is no spontaneous splay. The third term describes the twist between c in successive layers and recalls the fact that in a chiral material there is a spontaneous twist equal to $-D_3/B_3$. Finally, the cross term, B_{13}, describes the bend-twist coupling.

4.3.5 Flow Behavior of Smectic C and Smectic C* Liquid Crystals

To describe theoretically the flow of SmC and SmC* materials, we neglect the possibility of transportation of material through the layers, and rely on the linear and angular momentum balances, following essentially the same steps described for nematics. Subtle differences are in assumption of the behavior of the layer normal **k** and the c-director **c**, furthermore whether to allow small or large director deformations. Due to the much lower symmetries, however, the number of viscosity and elastic constants are much larger in the SmC and SmC* materials than in nematics. These differences and similarities are most clearly discussed by Carlsson, Leslie and Clark.[63] Their theory contains 20 viscosity coefficients, which is an order of magnitude more than we can normally handle if we wish to compare experiments with theory. If we assume that **k** and **c** vary independently, we reduce this number to "only" 12 constants (just as the number in case of biaxial nematics). However, under this assumption there would be no flow alignment, which does not correspond to the observations. So we have to live with 20 viscosity coefficients. They contain one coefficient, which is independent of the director (isotropic-like), four that are independent of the c-director (smectic A-like), four that are independent of the layer normal (nematic-like), and 11 coupling constants involving both **k** and **c**. Such coupling constants have no counterpart in any other theory.

Similar to nematic liquid crystals, there is a rotational viscosity related to the rotation of the c-director, which in this case means a rotation of the director around a cone as shown in Figure 4.16.

Also, just as we learned for nematics, shear may induce an alignment, except that now there are two shear alignment geometries, as shown in Figure 4.17.

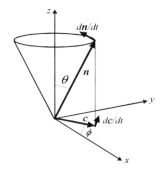

FIGURE 4.16
Illustration of the motion around the cone with rotational viscosity. The variable ϕ describes the motion of the c director around the layer normal.

When the shear is within the smectic layers, we have a nematic-like behavior, and the torque balance simplifies to the following equation:

$$-2\lambda_5 \cdot \dot{\phi} + \left[\lambda_2(\sin^2 \phi - \cos^2 \phi) - \lambda_5 \right]\frac{dv}{dy} = 0 \qquad (4.65)$$

where λ_2 and λ_5 are the relevant viscosity components.[64] In the steady state, when $\dot{\phi} = 0$, the flow alignment angle becomes:

$$\cos 2\phi_m = -\frac{\lambda_5}{\lambda_2} \qquad (4.66)$$

This is analogous to the flow alignment angle in nematics (see (4.22)), which again shows that the behavior of the c-director within the layers is similar to the meaning of the director in the nematic phase.

Another situation occurs when the shear gradient is between the smectic layers (Figure 4.17c). In this case the relevant torque balance reads as:

$$-2\lambda_5 \cdot \dot{\phi} + (\tau_1 - \tau_5)\sin\phi \cdot \frac{v_0}{h}\cos\omega t = 0 \qquad (4.67)$$

In the stationary case ($d\phi/dt = 0$), the solution is $\phi_m = 0$ or $\phi_m = \pi$. Stability analysis shows that the former solution is stable if $\tau_5 > \tau_1$.

Only a few shear experiments have been performed in SmC materials.[65] In one case, it was shown that the c-director becomes parallel to the flow direction.[66] Experimental data about the SmC viscosity coefficients is very rare (mainly because there are so many that people find it impossible to measure

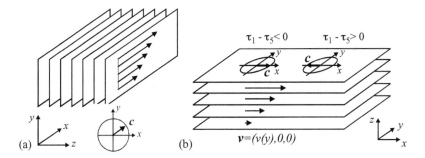

FIGURE 4.17

Flow alignment geometries. (a) Shear within the plane; (b) shear between the planes.

all of them). For example, several measurements of the elastic constant B_3 are available,[67] although measurements on the constants B_1 and B_2 are scarce.[68]

In addition, it is frequently observed that shear forces may distort the layer structure, and layer realignments can occur. In fact, shear is the most effective method for alignment of the layers in sandwich cells. Shear-induced alignment is especially important when there is no nematic phase above the smectic A phase, so the surface alignment techniques would not work. Studies in the 1980s[69,70] and 1990s[71] showed that proper shear techniques can provide uniform bookshelf alignment up to 100 μm film thickness, especially if the shear is applied during the I–SmA transition, when the shear acts on the smectic nuclei, the so-called "batonnets" that are elongated SmA objects floating in the isotropic liquid. The shear stress normal to the layers imposes a permeative motion for SmA domains. This hard deformation, however, can easily avoided by rotating the object so that the shear plane becomes parallel to smectic layers. In this case, only soft deformations will appear. However, this requirement can be fulfilled both by bookshelf (corresponding to the geometry of Figure 4.17a) and by homeotropic alignments (corresponding to the geometry of Figure 4.17b) with respect to the boundary plates (or by any tilted bookshelf alignment). Experimentally, it was observed that, at low frequencies and moderate shear rates, planar alignment is stabilized, whereas high shear rates often prefer a homeotropic geometry.[72] The phase diagram of a typical shear-induced alignment is shown in Figure 4.18.

Stability of the planar geometry was attempted to be explained theoretically by arguing that thermal fluctuations of the layers are diffused differently in the homeotropic and planar configurations.[73] Convection leads to a greater suppression in the homeotropic orientation, which leads to a smaller entropy, what is energetically less favorable. This theory, however, cannot explain the high shear rate–high frequency situation when homeotropic alignment becomes stable again.

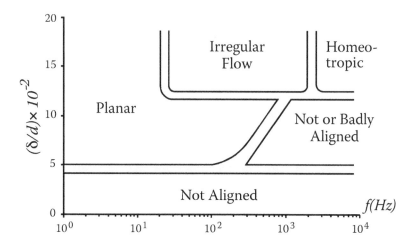

FIGURE 4.18
Schematic representation of shear-induced texture transformations.

4.4   Rheology of Columnar Liquid Crystals

In first approximation, the rheology of columnar liquid crystals can be treated very similarly to the smectic phases, except that the dimensionality of the fluid directions is different. The difference between systems of one or two fluid directions is much smaller that between those of zero or three. In fact the rheology of columnar and smectic phases is treated by a unified theory by de Gennes and Prost.[16] For example, similar to the three SmA flow geometries, we can distinguish three analogous geometries in the columnar phases (see Figure 4.19), and we can use the same arguments concerning the Poiseuille or plug flow situations.

Experimentally, there are numerous studies on the viscoelastic properties of discotic columnar phases[74,75] and for pyramidal or bow-shaped columnar phases.[76] In discotic columnar liquid crystals, assuming no intercolumnar molecular positional correlation, we would measure pure viscous behavior in geometry Figure 4.19c. However experimentally, the material even in this geometry was found to be too stiff, indicating[74] an apparent curvature constant $K_{app} \sim 10^{-6}$ N, which is a million times larger than that deduced by X-ray observations[77] or by surface tension and compressibility measurements.[78] Rayleigh scattering measurements[79] also indicated that the columnar mesophase behaves as a 3D crystal, thus leading to the large K_{app} when the columns are bent over distances larger than 0.3 μm. A possible mechanism to explain these observations was proposed by Prost,[75] who showed that a large density of column ends (so-called lock-in fault lines or in-grain boundaries) can make the material stiff. The shear stress can be approximated

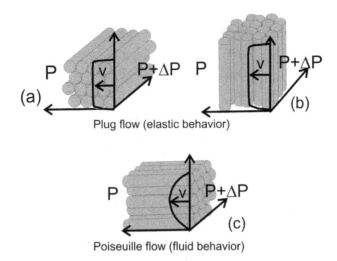

FIGURE 4.19
Characteristic flow geometries for hexagonal columnar phases.

as K_{app}/h^2 ($h \sim 5$ μm is the film thickness) corresponding to $G \sim 10^5$ N/m², and on average, a micron size of grains. So far, however, there is no evidence of these defects. In pyramidal columnar phases, the apparent shear elastic constant was even an order of magnitude larger ($G \sim 10^6$ N/m²), which would correspond to grain sizes of about 0.3 μm. In addition, it was observed that the stresses relax in a few tens of milliseconds, indicating no permanent bonds or grain boundaries between the columns. For this reason, it was assumed that the shear strains are due to entanglements between the end chains of the neighbor molecules, so that the relaxation times correspond to the lifetime of the entanglements.

It is important to note that, just as there is a major difference between the fluctuations of the columnar and smectic phases, the distortions relax at different lengths scales. In smectics the distortion relaxes in much larger scales than that of the distortions, whereas in columnars they relax in length scales comparable to that of the deformations.[16]

4.5 Lyotropic Lamellar Systems

Lyotropic lamellar systems are very similar to thermotropic smectics, and their elastic free energy is identical to that of SmA given in Eq. (4.34). In lyotropic lamellar systems, the origin of the layer compression modulus B is the steric repulsive interaction energy.[80] This can be visualized as illustrated in Figure 4.20. If a stack of membranes, formed, e.g., by lipid bilayers, is placed between two parallel walls, violent thermal out-of-plane fluctuations of the membranes exert a pressure p on the walls.

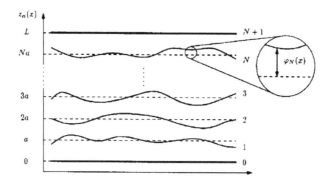

FIGURE 4.20
The stack of lyotropic lamellae stabilized by the mechanism suggested by Helfrich.[80]

For individual lamellae, which interact via collisions, the entropic confinement pressure is:

$$p \sim \frac{k_B T}{L_p^2 d} \qquad (4.68)$$

where L_p is the characteristic distance between collisions (the "patch" length), and d is the layer spacing, from which the layer compression modulus can be estimated to be $B \sim -d\partial p/\partial d$. It was shown that:

$$L_P = cd \sqrt{\frac{K_{11}d}{k_B T}} \qquad (4.69)$$

where $c = \sqrt{4\pi/3}$ according to Helfrich[80] (or $\sqrt{32/3\pi}$ according to Golubovic and Lubensky[81]). Combining (4.68) and (4.69) yields the compression modulus as:[80]

$$B = \frac{6\delta_n n}{n+1} \frac{(k_B T)^2}{K_{11} d^4} \qquad (4.70)$$

where, for $n \to \infty$, $\delta_\infty = 3\pi^2/128 \sim 0.23$. However, strong coupling perturbation calculations[82] gave that $\delta_\infty \sim 0.1$ and, for a single membrane, $\delta_1 = \pi^2/128$; both results are consistent with those given by Monte–Carlo simulations.[83]

Again, similar to thermotropic smectic materials, lyotropic lamellar systems can also undergo a variety of reorientational transitions. They can form multilamellar vesicles, or "onions," either above a critical strain rates[84] or for very small strain rates.[85] At even higher shear rates, these onions can form onion crystals[86] or oscillatory states.[87] Theoretically, in flow the transverse membrane fluctuations are suppressed, leading to fewer collisions and, hence, more space between layers. If there is no permeation or other mechanism to change the layer number on the experimental time scale, initially the layer spacing remains fixed. At sufficiently large strains, an instability will relieve the strain in favor of undulations (similar to Helfrich–Hurault effect[88] of thermotropic smectic A) that keep the number of layers constant in spite of the change in the reduced pressure.[89] This then could be a precursor to either a stable modulated phase, or instability to onion formation observed experimentally.

4.6 Soap Membranes

The study of hydrodynamics of soap films basically stems back to 1959 when G.I Taylor[90] investigated the propagation of waves in plane jets of pure water. He observed that, in the most easily excited mode, the sinusoidal transverse motions of the two surfaces are in phase with each other (Figure 4.21a). For this mode, the restoring force is due to surface tension γ and their velocity is:

$$v_{AS} = \sqrt{\frac{2\gamma}{\rho h}} \tag{4.71}$$

where ρ is the density of the water, and h is the thickness of the film. In the second mode, called peristaltic by Taylor, both surfaces move with opposite phases (Figure 4.21b). This mode involves viscous motion of the fluid from the nodes to the antinodes. It is dispersive and propagates at:

$$v_S = k_o \sqrt{\frac{\gamma h}{2\rho}} \tag{4.72}$$

where k_o is the wave vector.

Expeimental propagation of waves in soap films has been investigated by Vrij and coworkers in a series of papers.[91] The undulation mode was found to be underdamped, whereas the overdamped peristaltic mode (Figure 4.21c)

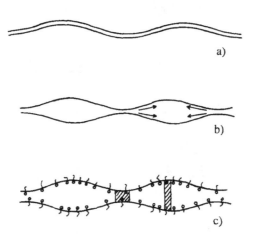

a)

b)

c)

FIGURE 4.21
Waves in a liquid film. (a) Antisymmetrical mode; (b) symmetrical mode; (c) elastic mode.

involves Poiseuille flow of the fluid core with the flow velocity zero within 2Å of the surfactant solution interface, indicating rigid interfacial water layer.[92] It can be shown[93] that the viscous terms corresponding to the symmetrical mode of pure water described by Taylor are not relevant for soaps, due to Marangoni elasticity. The waves due to the Marangoni elasticity propagate at a speed:

$$v_M = \sqrt{\frac{2E_M}{\rho h}} \tag{4.73}$$

Using $E_M \sim 8 \times 10^{-3}$ N/m, we find $v_M \sim 13 m/s$ for a 1 μm film.

A soap film is easily dragged into motion by an airflow parallel to its surface. If the airflow is steady, the resulting thickness profiles will be related to the pressure distribution. The film then provides an excellent visualization of the flow. The observed interference fringes in a circular shear flow are shown in Figure 4.22a; in the central region and around each vortex core, concentric fringes are observed.[94] It can be understood considering that the radial pressure gradient is balanced by the centrifugal force $\omega^2 r^2 /2$, where r is the distance from the axis of the rotation. With centrifugal forces, the thickness of the soap film is analogous to the equilibrium thickness in the gravitational field. The vortices are due to the von Kármán vortices forming in the rotating air stream.

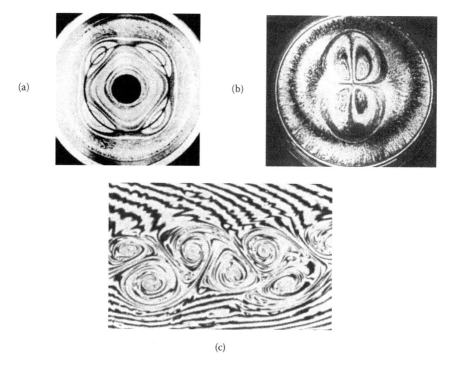

(a)

(b)

(c)

FIGURE 4.22

(a) Interference fringes of a film in a circular shear flow showing four corotating vortices. (b) The circulations observed in films when the air jet impinges perpendicularly to its surface. (It shows two inward and two outward flows.) (c) Unstable von Kármán wake pairs.

Observations of soap bubbles show that they are very sensitive to drafts of the surrounding air. The interaction of the three-dimensional motion of air with the film has been investigated in an arrangement where a thin air jet impinges on a plane film.[94,95] The soap film shows a slight bulge at the point where the air jet impinges on its surface. The observation of the film in monochromatic light shows that the axisymmetry of the system is broken, and the axis of the jet becomes a saddle point separating four regular steady circulations (see Figure 4.22b). They result from two effects: the deflected air flow drives the film in a radial outward motion, but the local increase of the surface tension due to stretching and thinning of the film tends to create a radial inward motion of the film. The two opposite effects break the axisymmetry of the system. The air motion creates in the film plane two strong thin opposite radial jets, while the surface tension gradient generates large recirculations of fluid towards the center. This experiment clearly illustrates the stabilizing influence of the Gibbs elasticity.

Experiments when a flow is generated directly in the film have been undertaken by Couder et al.[94,96] These experiments show that grid turbulence can be generated by towing a comb in the film. The regularly spaced cylinders around the tips of the comb create parallel interacting wakes in which the vortices grow in time by a pairing process. The growth of the vortice size is linear in time, which is predicted for two-dimensional decaying turbulence.[97] Another version of this experiment was undertaken by Goldburg et al.,[98] who studied combs in flowing soap. In some other experiments, tiny objects, like discs[99] or circular cylinders,[100] were moving with respect to the soap film (in the first case the soap being at rest, in the second one the soap film was moving in a tunnel with the help of film-pulling device). It was observed that, depending on the velocity, various types of von Kármán wakes appear at Reynolds numbers larger than 40. At low velocities, pairing and merging of likewise vortices was observed (see Figure 4.22c). This phenomenon is known as the elementary process of inverse energy cascade and characterizes two-dimensional turbulence. At larger velocities, singular structures of opposite sign of vortices formed. The formation of these couples is very fast, and the wakes move into quiescent zones of the fluid, thus diffusing turbulence. Both processes also appeared in numerical simulations, and are intrinsic features of two-dimensional flows.[101,99] Recently flexible threads[102] were placed in flowing soap film. It was observed that, when the filament length is larger than a critical value, it could stay either in a stretched-straight or in an oscillatory state. In the first case, the von Kármán vortices shed from its free end. In the oscillatory flapping state, flow structures modulated by the filament passed along the flow. This experiment can be considered as a model system to study the behavior of 1D flag in 2D fluid.

A similarly interesting study of behavior with reduced dimensionality is the popular demonstrative experiment showing the effect of surface tension (Plateau[103] or Boys[104]). We knot both ends of a fine thread (e.g., hair) to form a closed loop, which we place in the soap film. Breaking the enclosed part of the film gives a circular shape to the loop obtaining a circular hole.

Placing the film in vertical frame creates two competing forces on the hole: its weight $m \cdot g$ (where m is the mass of the thread and of the attached meniscus), and buoyancy force equal to the weight of the excluded volume of fluid. The resulting force is:

$$F = mg - \rho g \iint_{loop} h(z)\,dy\,dz \tag{4.74}$$

When the loop is placed at the bottom of the frame, where the film is thickest, there is a buoyant force that causes it to move upward. As it rises the surrounding film becomes thinner, and the loop loses its buoyancy until it stops where F = 0. This is exactly similar to the case when a balloon is rising in the stratified atmosphere where the density decreases with height. This, therefore, is the experimental demonstration of the similarity of the equilibrium of the atmosphere. Figure 4.23a illustrates the rising of a hole with Reynolds number Re = 200.

For this value an alternate shedding of vortices is observed ("von Kármán street"). During the ascent the hole has an oscillatory motion, where it simultaneously rotates and slips sideways in reaction to the alternate emission of vortices.

A situation when a bubble is rising in a normal fluid is illustrated in Figure 4.23b with the motion of thinner zones of soap fluid in a vertical soap fluid frame. It is possible to create zones of different two-dimensional density touching the film with pure soap surfactants. The locally increased soap concentration causes the surrounding film to stretch rapidly characterized by a smaller surface tension. This process creates a thinner zone with a constant thickness h_2, which could be as small as that of the black films (4.5 nm or 30 nm) or to the silvery white (~100 nm). The limit between the zone of area S and the film is sharply defined with the width of the rim comparable to the film thickness h_1. If this thin zone is created near the bottom of the frame and surrounded by thicker fluid, it will experience a buoyant force:

$$F = \rho g \left[h_2 S - \iint_S h_1(z)\,dy\,dz \right] \tag{4.75}$$

and rise up in the frame until it reaches the level where $h_1(z) = h_2$. The velocity of its motion and the shape during the ascent is the two-dimensional equivalent of the motion and the shape of rising bubbles in a classical fluid. The shape of these zones was studied extensively.[105]

In normal fluids the shape of rising bubbles and the nature of their wakes are usually described as the function of a set of different parameters representing the relative importance of buoyancy, inertia, viscosity and surface tension. The most useful parameters are the Reynolds number, Eötvös number and Morton number, which in 2D can be defined as follows.

$$Re = \frac{\rho h_2 D_e v}{\eta_S}; \quad Eo = \frac{g \rho D_e^2}{\gamma}; \quad Mo = \frac{g \rho \eta_S (h_2 - h_1)}{\rho^2 h_2^2 \lambda^2} \tag{4.76}$$

(a)

(b)

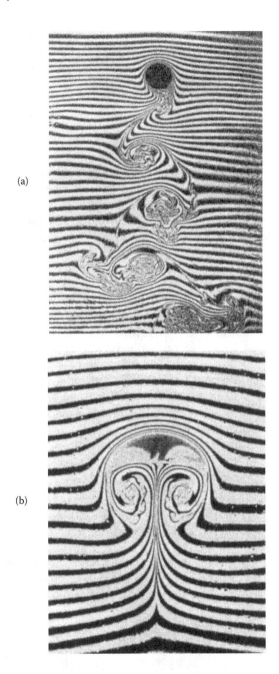

FIGURE 4.23
(a) The motion of a hole of diameter d = 1.9 cm rising in film with velocity 9 cm/s, where the Reynolds number is Re ~ 200. (b) The motion of a rising black zone.

Here D_e is the diameter of the surface equivalent circle, $\rho(h_2 - h_1)$ is the difference in density between two fluids, γ is the interfacial tension, $\lambda = \gamma(h_2 - h_1)$ is the line tension, and η_s is the surface viscosity of the film.

4.7 Rheology of Foams

Nowadays an increasing attention is given to the dynamics of the foams. This is not an easy subject, but the well-defined structure of the foam gives us hope to successfully understand the flow properties. The present stage of the physics, including the rheology of the fluid foams, is substantially described by the book of Weaire and Hutzler.[106] Our summary is based on this book and on more recent reviews.[107,108]

A fluid foam has a low yield stress, above which it flows as viscous fluid. In the elastic regime the individual cells are deformed without significant rearrangements, and the material can be well characterized by a well-defined shear modulus G. The plastic effects which give rise to a yield stress are due to the topological rearrangements. At low shear rates, these rearrangements can be pictured as instantaneous adjustments between two equilibrium structures. For large strain rates, the microscopic rearrangements are irreversible, so that pattern does not return to the initial stage. In this range the foam behaves plastically, characterized by the plastic viscosity η_p. A sequence of topological changes in the plastic regime of a perfect hexagonal honeycomb structure is illustrated Figure 4.24.

In general, the fluid soaps behave as Bingham fluids, and their shear strain follows the Bingham model:

$$S = S_y + \eta_p \dot{\varepsilon} \tag{4.77}$$

where S_y is the yield stress, and $\dot{\varepsilon}$ is the strain rate. Such behavior is represented in Figure 4.25.

The shear modulus and yield stress are strongly dependent on the liquid fraction of the foam. Both become smaller at larger liquid concentrations.

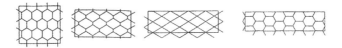

FIGURE 4.24
Illustration of the sequence of topological changes leading to the yielding behavior. Note, these topological changes correspond to two-dimensional first-order phase transitions illustrated in Figure 3.1.

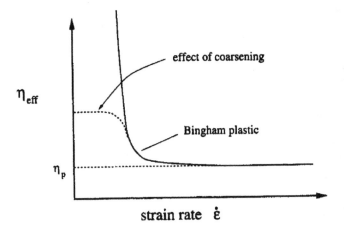

FIGURE 4.25
Typical variation of the effective viscosity with the strain rate.

Reliable measurement of rheological properties of foams is not an easy task. For example, in a Couette viscosimeter the foam is not homogeneously deformed: part of it remains in elastic solid, while another part flows. Flow in pipe, for example, may consist of a plug flow in some places and fluid type in other parts.

Recently a relatively successful approach is to study the two-dimensional foam steadily flowing through a construction[108] (see Figure 4.26).

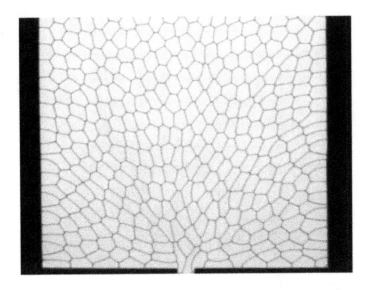

FIGURE 4.26
Two-dimensional foam flowing downward through a construction. The wide field of view shows only the end of the 1-m-long horizontal channel.

4.8   Rheology at Surfaces

Anisotropic fluids are often studied in small quantities, mostly sandwiched between two plates, where the plates impose restrictions on the alignment and motion of the molecules (surface anchoring). Due to the long-range orientational order, these boundary effects have long-range effect on the alignment of anisotropic fluids.

First we describe how the interaction of the orientationally ordered fluid materials with the surface can be treated mathematically, and then we review the most common boundary conditions and their possible physical mechanisms.

4.8.1   Surface Anchoring

The total free energy of an anisotropic fluid sample of volume V limited by a surface Ω, assuming constant orientational order ($S =$ constant), can be written as:

$$G[\vec{n}(\vec{r})] = \iiint_V F(n_i, n_{i,j})dV + \oiint_A \Phi_s(n_i)d\Omega \qquad (4.78)$$

The first term is related to the bulk distortion energy, determined by the elastic deformation and external bulk field interactions. The second term is the surface contribution of the limiting surface Ω. The actual director field can be obtained by minimizing the energy. In a simple sandwich cell geometry (see Figure 4.27), the director ($\vec{n}(\vec{r})$) can be expressed with the tilt angle as $\theta = \cos^{-1}(\vec{n} \cdot \vec{k})$, where $\vec{k}$ is the unit vector of the surface normal.

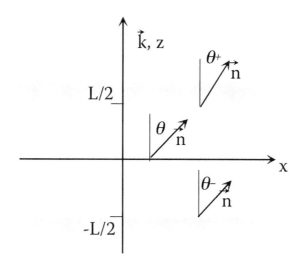

FIGURE 4.27
Director field of an anisotropic material sandwiched between two parallel surfaces assuming uniform director field in the xy plane. The tilt angle θ is the tilt angle between n and the z axis. θ^+ and θ^- are the surface tilt angles.

In this case, the total free energy per unit surface area is:

$$\Phi[\theta(z)] \equiv \frac{G}{\Omega} = \int\limits_{-L/2}^{L/2} F(\theta,\theta')dz + \Phi_s^-(\theta^-) + \Phi_s^+(\theta^+) \tag{4.79}$$

where $\theta'(z)=d\theta(z)/dz$ and $\Phi_s^\pm(\theta^\pm)$ represent the surface energies per unit areas; furthermore $\theta^\pm = \theta(\pm L/2)$ are the surface tilt angles. Taking into account that $F(\theta,\theta')=(\frac{\partial F}{\partial\theta}\frac{d\theta}{dz}+\frac{\partial F}{\partial\theta'}\frac{d\theta'}{dz})dz$, and utilizing the identity: $\int \frac{\partial F}{\partial\theta'}\frac{\partial\theta'}{dz}dz = \frac{\partial F}{\partial\theta'}\cdot\frac{\partial\theta}{dz}\Big]_{-L/2}^{L/2}$ $-(\frac{d}{dz}\frac{\partial F}{\partial\theta'})\frac{d\theta}{dz}dz$, the first variation of $\Phi[\theta(z)]$ becomes:

$$\delta\Phi[\theta(z)] = \int\limits_{-L/2}^{L/2}\left(\frac{\partial F}{\partial\theta}-\frac{d}{dz}\frac{\partial F}{\partial\theta'}\right)\theta'(z)dz + \left(\frac{-\partial F}{\partial\theta'}+\frac{\partial\Phi_s^-}{\partial\theta^-}\right)\theta'(-L/2)+\left(\frac{\partial F}{\partial\theta'}+\frac{\partial\Phi_s^+}{\partial\theta^+}\right)\theta'(L/2)$$

$$\tag{4.80}$$

where $\theta'(z)$ is continuous in first-order derivative.[109] The requirement that $\delta\Phi[\theta(z)] = 0$ for all $\theta'(z)$ implies that in the bulk $\theta(z)$ satisfies the Euler–Lagrange equation:

$$\frac{\partial F}{\partial\theta} - \frac{d}{dz}\frac{\partial F}{\partial\theta'} = 0 \tag{4.81}$$,

and at the boundaries ($z=\pm L/2$), the conditions are:

$$\frac{-\partial F}{\partial\theta'} + \frac{\partial\Phi_s^\pm}{\partial\theta^\pm} = 0 \tag{4.82}$$

In case of nematic liquid crystals, in one elastic constant approximation,

$$F(\theta,\theta') = \frac{1}{2}K\theta'^2 + F_e(\theta) \tag{4.83}$$

where $F_e(\theta)$ takes into account the interaction with external bulk forces. The surface interactions with $W^\pm >0$ can be approximated by the Rapini–Papoular expression:[110]

$$\Phi_s^\pm = \frac{1}{2}W^\pm\sin^2(\theta^\pm - \theta_e^\pm) \tag{4.84}$$

where $\theta_e^\pm$ are the surface easy tilt angles, i.e., the directions imposed by external surface conditions. The RP expression is valid only for small deviations from the easy direction. For large enough angular deviations, higher harmonic terms appear, and one can observe a surface disordering (change in S at the surface).[111]

With these simplifying assumptions the equations (4.81)–(4.82) to be solved become:

$$K\theta'' - \frac{dF_e}{d\theta} = 0 \quad -L/2 \le z \le L/2 \tag{4.85}$$

which represents the bulk equilibrium of the mechanical torque (note: ordinary liquids cannot transmit torque, but anisotropic fluids can[112]), and

$$-K\theta' + \frac{1}{2}W^{\pm}\sin 2(\theta^{\pm} - \theta_e^{\pm}) = 0 \quad \text{at} \quad z = \pm L/2 \tag{4.86}$$

which represents the equilibrium between the torques at the surfaces. For small difference between the easy axis and the actual surface orientation:

$$\frac{K}{W^{\pm}} \equiv b = \frac{\theta^{\pm} - \theta_e}{\theta'} \tag{4.87}$$

which has a dimension of length and is called extrapolation length. As illustrated in Figure 4.28, it basically gives the distance from the surface where θ would reach the easy axis value without disturbing the $\theta' = const$ dependence valid in the bulk.

Qualitative estimates by dimensional analysis tell us that $W \approx \frac{U_{ai}}{a^2}$, where U_{ai} is the anisotropic part of the interaction between the wall and the anisotropic molecules of average size of a. Previously, we have seen that $K \approx \frac{U}{a}$, where U is the interaction energy between two alike molecules, approximated

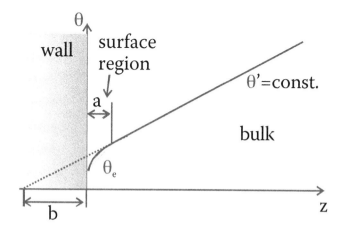

FIGURE 4.28
Explanation of the extrapolation length and the surface region of an anisotropic fluid near an interface.

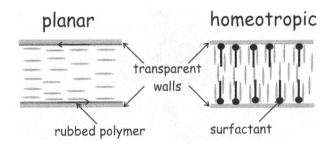

FIGURE 4.29
Two types of uniform anisotropic textures obtained between parallel glass walls.

by the energy needed to cause transition to the isotropic phase. Combining these, we get that:

$$\frac{K}{W} \equiv b \approx a \cdot \frac{U}{U_{ai}} \tag{4.88}$$

Experimentally, we distinguish between two practically important situations: strong and weak anchoring.

The anchoring is said to be strong when $U_{ai} \geq U$ *i.e.* $b \leq a$. This means that:

$$\frac{\Phi_{surf}}{\Phi_{bulk}} = \frac{\frac{1}{2}W(\Delta\theta_s)^2}{\frac{1}{2}K\int \theta'^2 dL} \approx \frac{W}{KL} \cdot \left(\frac{\Delta\theta_s}{\theta'}\right)^2 \leq \frac{1}{bL}b^2 = \frac{b}{L} \leq \frac{a}{L} \tag{4.89}$$

In the continuum limit $(a/L) \ll 1$, so Φ_{surf} can be neglected, and $\Delta\theta_s = \theta_e - \theta_o = 0$.

In the limit of weak anchoring, $U_{ai} \ll U$, which means that $b \gg a$, i.e., $\Delta\theta_s \sim b \cdot \frac{\theta(L)}{L}$, i.e., external constraints disrupt the surface alignment.

We note that real situations can be extremely complex, but it remains true that the surface energies can be omitted, and the surface acts as boundary condition of the problem if $U_{ai} \sim U$.

4.8.2 Surface Alignments

The principal alignments are the so-called homeotropic[†] and homogeneous (planar) alignments as shown in Figure 4.29. In the former case, the director is constrained to be perpendicular to the substrates, whereas in the latter situation, the director is parallel to the substrates. In the parallel alignment any direction along the substrates can be equal (degenerate planar anchoring),

---

[†] In display technology the homeotropic anchoring is often called vertical alignment.

or can have an easy direction that would dictate the alignment along the substrate.

The homeotropic alignment, when the director is normal to the boundary plates, is usually achieved by surfactants and detergents, such as lecithin or some silanes (for example, octadecyl trietoxy silane), or by special polyimide coating. Generally, liquid crystals align homeotropically at air interface, although in some materials the molecules are tilted with respect to the surface normal. If there is no bias in any direction along the surfaces, we end up with a degenerate alignment, termed conical anchoring.

Planar anchorings can be achieved by several ways:

1. The oldest technique utilizes the alignment on the surface of a single crystal. If the interface corresponds to a well-defined crystallographic plane, the easy axis becomes parallel to some of the crystallographic axes, but there is no rule for the selection.[113]

2. Surface rubbed by paper or cloth makes easy axis parallel to the rubbing. This was already noticed by O. Lehmann,[114] but the precise recipe giving reproducible planar textures was given only by Chatelain.[115] Although the exact mechanism of the alignment is not known, most widely accepted is the Berreman's model[116] that is based on calculating the elastic energy difference between the alignments when the director is along and perpendicular to the wave vector of the surface deformation (see Figure 4.30). In both cases, the molecules are parallel to the surface, but in the first case (a) there is a deformation, which decays with increasing z, whereas in the second case (b) the director is parallel to the grooves, which (in absence of the positional order) does not cause any director deformation. This latter configuration, therefore, is obviously preferred over the first one. This difference in the free energies can be interpreted as an effective anchoring energy.

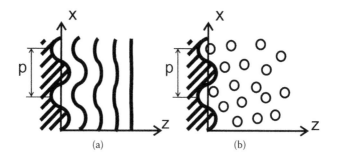

(a)   (b)

FIGURE 4.30
Illustration of Berreman's model for planar anchoring by rubbed polymer. (a) Deformation of the molecular distribution when the director is oriented perpendicular to the ridges of the surface. (b) No deformation is present for directors parallel to the ridges and troughs on the surface.

The anchoring energy estimated from this model,[117] has a good order of magnitude (typically in the order of 10^{-2} to 10^{-4} J/m^2); however, it cannot explain all details of the experimental results. Other theories have also been published to explain the surface alignment by rubbing. For example, the surface tension model,[118] the van der Waals's force model,[119] excluded volume effect,[120] the "frictional rubbing and hot spot model,"[121] but none of them can explain all results.

3. Obliquely evaporated SiO, which forms a saw-tooth profile at the surface, was first reported by Janning.[122] Similar to the Berremann's model in these cases, the easy axis is determined by the evaporation direction. We note that, in special evaporation directions and SiO thickness ranges, this method could provide bistable anchoring conditions[123] leading to bistable nematic devices.[124]

4. Recently, much interest is devoted to control surface alignment of liquid crystals with the help of polarized light, because it enables high-density optical storage and the construction of sophisticated electro-optic devices. For this, photosensitive polymer layers are used. One scheme is to illuminate the layer by polarized UV radiation, which either creates direction-sensitive cross-linking or selectively breaks up polymer chains.[125] A different approach is to incorporate azo dyes into a polymer matrix,[126] in which the polarized light induces an anisotropic orientational distribution of the dye molecules, the preferred direction being perpendicular to the electric vector of the optical field.

5. Surface memory effects refer to phenomena where an initially isotropic surface can be rendered anisotropic in contact with an anisotropic medium. This anisotropic medium can be a crystal[127] or smectic liquid crystal.[128] It is believed that the surface memory effect arises from the interfacial (smectic liquid) crystal–polymer coupling that leads to a deformed surface.

For relatively new and extensive reviews of surface alignment effect of liquid crystals, we can recommend Uchida and Seki's[129] and Barbero and Durand's[130] reviews.

References

1. See, for instance, L.D. Landau, E.M. Lifschitz, *Theory of Elasticity*, 2nd ed., Pergamon, New York (1970).
2. A. Saupe, *J. Chem. Phys.*, 75(10), 5118 (1981).
3. See for example, L.D. Landau, E.M. Lifschitz, *Statistical Physics*, 2nd ed., Pergamon, New York (1970).

4. This is the Neumann–Curie principle, developed by Franz Neumann in 1833 and Pierre Curie in 1894; see, e.g., in J.F. Nye, *Physical Properties of Crystals,* Oxford University Press, Oxford (1957).

5. J.L. Ericksen, *Trans. Soc. Rheol.,* 5, 23 (1961); J.L. Ericksen, *Adv. Liq. Cryst.,* 2, 233 (1976).

6. F.M. Leslie, *Arch. Ration. Mech. Anal.,* 28, 265 (1968); F.M. Leslie, *Mol. Cryst. Liq. Cryst.,* 63, 111 (1981)

7. O. Parodi, *J. Phys. (Paris),* 31, 581 (1970).

8. D. Forster, T. Lubensky, P. Martin, J. Swift, P. Pershan, *Phys. Rev. Lett.,* 26, 1016 (1971); P.C. Martin, O. Parodi, P.J. Pershan, *Phys. Rev. A,* 6, 2401 (1972).

9. T.C. Lubensky, unpublished.

10. C.W. Oseen, *Trans. Faraday Soc.,* 29, 883 (1933); H. Zocher, ibid., 29, 945 (1933).

11. F.C. Frank, *Discuss. Faraday Soc.,* 25, 19 (1958).

12. See for example, G.B. Arfken, H.J. Weber, *Mathematical Methods for Physicists,* 4th ed., Academic Press, New York (1995).

13. P.G. de Gennes, *The Physics of Liquid Crystals,* Clarendon Press, Oxford (1974).

14. J. Nehring, A. Saupe, *J. Chem. Phys.,* 54, 337 (1971), ibid., 55, 5527 (1972).

15. J.L. Ericksen, *Phys. Fluids,* 9, 1205 (1966).

16. P.G. de Gennes, J. Prost, *The Physics of Liquid Crystals,* 2nd ed., Oxford Science Publications (1993).

17. V. Zvetkov, *Acta. Phys. Chem., USSR,* 6, 866 (1937); A. Saupe, *Z. Naturforsch.,* 15A, 815 (1960); G. Durand, L. Leger, F. Rondelez, M. Veissie, *Phys. Rev. Lett.,* 22, 227 (1969).

18. M. Miesowicz, *Nature,* 17, 261 (1935); M. Miesowicz, *Nature,* 158, 27 (1946).

19. See for example, G.B. Arfken, H. Weber, *Mathematical Methods for Physicists,* 4th ed., Academic Press, San Diego, CA (1995).

20. J. Fisher, J. Wahl, *Mol. Cryst.,* 22, 359 (1973).

21. J.L. Ericksen, *Trans. Soc. Rheol.,* 13, 9 (1969).

22. P.E. Cladis, S. Torza, *Phys. Rev. Lett.,* 35, 1283 (1975); P. Manneville, *Mol. Cryst. Liq. Cryst.,* 70, 223 (1981); T. Cralsson, *Mol. Cryst. Liq. Cryst.,* 104, 307 (1984).

23. R.S. Porter, E.M. Barrall, J.F. Johnson, *J. Chem. Phys.,* 45, 1452 (1966).

24. S. Candau, P. Martinoty, F. Debeauvais, *C.R. Acad. Sci.,* B277, 769 (1973).

25. F.M. Leslie, *Mol. Cryst. Liq. Cryst.,* 7, 407 (1969).

26. S. Battacharya, C.E. Hong, S.V. Letcher, *Phys. Rev. Lett.,* 41, 1736 (1978).

27. U.D. Kini, *J. de Physique,* 40, C3 (1979).

28. S. Bhattacharaya, C.E. Hong, S.V. Letcher, *Phys. Rev. Lett.,* 41, 1736 (1978).

29. S. Chandrasekhar, *Liquid Crystals,* 2nd ed., Cambridge University Press, Cambridge (1992).

30. U.D. Kini, G.S. Raganath, S. Chandrasekhar, *Pramana,* 5, 101 (1975).

31. T.C. Lubenski, *Phys. Rev. A.,* 6, 452 (1972).

32. D. Guillon, J. Stamatoff, P.E. Cladis, *J. Chem. Phys.,* 76, 2056 (1982).

33. P. Martinoty, J.L. Gallani, D. Collin, *Phys. Rev. Lett.,* 81, 144 (1998).

34. W. Helfrich, *Phys. Rev. Lett.,* 23, 372 (1969).

35. L.D. Landau, E.M. Lifshitz, *Statistical Physics,* Part I, 3rd ed., Pergamon, Oxford (1980).

36. G. Grinstein, R. Pelcovits, *Phys. Rev. Lett.,* 47, 856 (1981); *Phys. Rev. A,* 26, 915 (1982).

37. G.F. Mazenko, S. Ramaswamy, J. Toner, *Phys. Rev. Lett.,* 49, 51 (1982); *Phys. Rev. A,* 28, 1618 (1983).

38. P.C. Martin, O. Parodi, P.S. Pershan, *Phys. Rev. A,* 6, 2401 (1972); G.F. Mazenko, S. Ramaswamy, J. Toner, *Phys. Rev. A,* 28, 1618 (1983).

39. S. Battacharya, S.V. Letcher, *Phys. Rev. Lett.*, 44, 414 (1980).
40. L. Leger, A. Martinet, *J. Phys. (Paris)*, 37 (Suppl. C3), 89 (1976).
41. R. Ribotta, G. Durand, *J. Phys. (Paris)*, 38, 179 (1977).
42. M. Cagnon, G. Durand, *Phys. Rev. Lett.*, 45, 1418 (1980).
43. S. Ramaswamy, *Phys. Rev. A*, 29, 1506 (1984).
44. N.A. Clark, *Phys. Rev. Lett.*, 40, 1663 (1978); R. Bartolino, G. Durand, *J. Phys. (Paris)*, 42, 1445 (1981); P. Oswald, M. Kleman, *J. Phys. (Paris)*, 43, L411 (1982).
45. P.G. de Gennes, *Physics of Fluids*, 17, 1645 (1974).
46. P.G. de Gennes, *Phys. Fluids*, 17, 1645 (1974).
47. N.A. Clark, *Phys. Rev. Lett.*, 40, 1663 (1978).
48. H.G. Walton, I.W. Stewart, M.J. Towler, *Liq. Cryst.*, 20, 665 (1996).
49. G. Liao, I.I. Smalyukh, J.R. Kelly, O.D. Lavrentovich, A. Jákli, Electronic liquid crystal communications, http://www.e-lc.org/dics/2005_02_25_10_57_49.
50. For a review see T. Gisler, D.A. Weitz, *Curr. Opin. Colloid Interface Sci.*, 3, 586 (1998); F.C. MacKintosh, C.F. Schmidt, ibid., 4, 300 (1999).
51. A. Palmer, T.G. Mason, J. Xu, S.C. Kuo, D. Wirtz, *Biophys. J.*, 76, 1063 (1999).
52. R. Kubo, *Rep. Prog. Phys.*, 29, 255 (1966).
53. S. Battacharya, J.B. Ketterson, *Phys. Rev. Lett.*, 49, 997 (1982); J.L. Gallani, P. Martinoty, *Phys. Rev. Lett.*, 54, 333 (1985).
54. A. Saupe, *Mol. Cryst. Liq. Cryst.*, 7, 59 (1969).
55. Orsay Group, *Solid. State Commun.*, 9, 653 (1971).
56. S.T. Lagerwall, I. Dahl, *Mol. Cryst. Liq. Cryst.*, 114, 151 (1984).
57. S.T. Lagerwall, *Ferroelectric and Antiferroelectric Liquid Crystals*, Wiley-VCH, Weinheim (1999).
58. G. Pelzl, P. Schiller, D. Demus, *Liq. Cryst.*, 2, 131 (1987).
59. I. Dahl, S.T. Lagerwall, *Ferroelectrics*, 58, 215 (1984).
60. T. Carlsson, I.W. Stewart, F.M. Leslie, *Liq. Cryst.*, 9, 661 (1991).
61. F.C. Frank, *Discuss. Faraday Soc.*, 25, 19 (1958).
62. R.B. Meyer, *Mol. Cryst. Liq. Cryst.*, 40, 33 (1977).
63. T. Carlsson, F.M. Leslie, N.A. Clark, *Phys. Rev. E*, 51, 4509 (1995).
64. F.M. Leslie, I.W. Stuart, M. Nakagawa, *Mol. Cryst. Liq. Cryst.*, 198, 443 (1991).
65. T. Scharf, Chirale smecktische C Phasen in periodischen Scherfeldern, Ph.D. dissertation, Martin Luther Universitat, Halle (1997).
66. P. Pieranski, E. Guyon, P. Keller, *J. Phys. (Paris)*, 36, 1005 (1975).
67. F. Gouda, K. Skarp, G. Andersson, H. Kresse, S.T. Lagerwall, *Jpn. J. Appl. Phys.*, 28, 1887 (1989); A. Levstik, Z. Kutnjak, C. Filipic, I. Levstik, Z. Bregar, B. Zeks, T. Carlsson, *Phys. Rev. A*, 42, 2204 (1990).
68. G. Pelzl, P. Schiller, D. Demus, *Liq. Cryst.*, 2, 131 (1987).
69. N.A. Clark, S.T. Lagerwall, *Appl. Phys.*, 36, 899 (1980).
70. A. Jákli, I. Jánossy, L. Bata, Á. Buka, *Cryst. Res. Technol.*, 23, 949 (1988).
71. C.R. Safinya, E.B. Sirota, R.J. Plano, *Phys. Rev. Lett.*, 66, 1986 (1991); C.R. Safinya, E.B. Sirota, R.F. Bruinsma, C. Jeppesen, R.J. Plano, L.J. Wenzel, *Science*, 261, 588 (1993).
72. A. Jakli, R. Bartolino, N. Scaramuzza, R. Barberi, *Mol. Cryst. Liq. Cryst.*, 178, 21–32 (1990).
73. M. Goulian, S.T. Milner, *Phys. Rev. Lett.*, 74, 1775 (1995).
74. M. Cagnon, M. Gharbia, G. Durand, *Phys. Rev. Lett.*, 53, 938 (1984); M. Gharbia, M. Cagnon, G. Durand, *J. Phys. Lett.*, 46, L-683 (1985).
75. J. Prost, *Liq. Cryst.*, 8, 123 (1990).
76. A. Jákli, M. Müller, G. Heppke, *Liq. Cryst.*, 26, 945 (1999).

77. A.M Levelut, *J. Chem. Phys.*, 80, 149 (1983).
78. L. Sallen, P. Oswald, J.C. Geminard, J. Malthete, *J. Phys. II. Fr.*, 5, 937 (1995).
79. M. Gharbia, T. Otthmann, A. Gharbi, C. Destrade, G. Durand, *Phys. Rev. Lett.*, 68, 2031 (1992).
80. W. Helfrich, *Z. Naturforsch.*, 33A, 305 (1978).
81. L. Golubovic, T.C. Lubensky, *Phys. Rev. B*, 39, 12110 (1989).
82. M. Bachmann, H. Kleinert, A. Pelster, *Phys. Rev. E*, 63, 051709 (2001).
83. W. Janke, H. Kleinert, M. Meinhart, *Phys. Rev. B*, 217, 525 (1989).
84. D. Roux, F. Nallet, O. Diat, *Europhys. Lett.*, 24, 53 (1993); O. Diat, D. Roux, F. Nallet, *J. Phys. II*, 3, 1427 (1993).
85. M. Bergmeier, H. Hoffmann, C. Thunig, *J. Phys. Chem. B*, 101, 5767 (1997); A. Leon, D. Bonn, J. Meunier, A. Al-Kahwaji, O. Greffier, H. Kellay, *Phys. Rev. Lett.*, 84, 1335 (2000).
86. O. Diat, D. Roux, F. Nallet, *Phys. Rev. E*, 51, 3296 (1995).
87. A.S. Wunneburger, A. Colin, J. Leng, A. Arneodo, D. Roux, *Phys. Rev. Lett.*, 86, 1374 (2001).
88. W. Helfrich, *Appl. Phys. Lett.*, 17, 531 (1970); J.P. Hurault, *J. Chem. Phys.*, 59, 2086 (1973).
89. S.W. Marlow, P.D. Olmsted, *Eur. Phys. J. E*, 8, 485 (2002).
90. G.I. Taylor, *Proc. Royal Soc. London A*, 253, 296 (1959).
91. A. Vrij, *Adv. Colloid Interface Sci.*, 2, 39 (1968); J. Lucassen, M. Van de Tempel, A. Vrij, F. Th. Hesselink, *Proc. K. Ned. Akad. Wet.*, B 73, 109 (1970).
92. C.Y. Young, N.A. Clark, *J. Chem. Phys.*, 74 (7), 4171 (1981).
93. A.I. Rusanov, V.V. Krotov, *Prog. Surf. Membrane Sci.*, 13, 415 (1979).
94. Y. Couder, J.M. Chomaz, M. Rabaud, *Physica D*, 37, 384 (1989).
95. J.A. Kitchener, G.F. Cooper, *Quart. Rev.*, 1, 71 (1959).
96. Y. Couder, *J. Phys. Lett.*, 45, 353 (1984).
97. G.K. Batchelor, *Phys. Fluids (suppl. II)*, 12, 233 (1969).
98. W.I. Goldburg, A. Belmonte, X.L. Wu, I. Zusman, *Physica A*, 231 (1998).
99. Y. Couder, C. Basdevant, *J. Fluid Mech.*, 173, 225 (1986).
100. M. Gharib, P. Derango, *Physica D*, 406 (1989).
101. H. Aref, E.D. Siggia, *J. Fluid Mech.*, 109, 435 (1981).
102. J. Zhang, S. Childress, A. Libchaber, M. Shelley, *Nature*, 408, 836 (2000).
103. J. Plateau, Statique experimentale et theoretique des liquids soumis aux seules forces moleculaires, Gauthier–Villars, Paris (1873).
104. C.V. Boys, *Soap, Bubbles and the Forces Which Mould Them*, Society for Promoting Christian Knowledge, London, (1890), and Anchor Books, New York.
105. J.F. Harper, *Adv. Appl. Mech.*, 12, 59 (1972); P.P. Wegener, J.Y. Parlange, *Ann. Rev. Fluid Mech.*, 5, 79 (1973); R. Clift, J.R. Grace, M.E. Weber, *Bubbles, Drops and Particles*, Academic Press, New York (1978).
106. D. Weaire, S. Hutzler, *The Physics of Foams*, Clarendon Press, Oxford (1999).
107. D. Weaire et al., *J. Phys. Condens. Matter*, 15, S65 (2003).
108. M. Aubouy et al., *Granular Matter*, 5, 67 (2003).
109. L. Elsgolts, *Differential Equations and the Calculus of Variations*, Moscow, Mir. (1977).
110. A. Rapini, M. Papoular, *J. Phys. Colloq.*, 30, C4 (1969).
111. M. Nobili, G. Durand, *Phys. Rev. A*, 46, R6147 (1992).
112. J. Prost, P.G. de Gennes, *The Physics of Liquid Crystals*, 2nd ed. (1994).
113. F. Grandjean, *Bull. Soc. Fr. Miner.*, 29, 164 (1916).
114. O. Lehmann, *Verh. Naturwiss. Vereins. Karlsruhe*, 19, Sonderdruck 275 (1906).
115. P. Chatelain, *P. Bull. Soc. Fr. Miner.*, 66, 105 (1943).

116. D. Berreman, *Phys. Rev. Lett.*, 28, 1683 (1972).
117. D. Berreman, *Mol. Cryst.*, 23, 215 (1973); U. Wolff, W. Grubel, A. Kruger, ibid., 23, 187 (1973).
118. H. Mada, *J. Chem. Phys.*, 75, 372 (1981); J.C. Dubois, M. Gazard, A. Zann, *J. Appl. Phys.*, 47, 1270 (1976).
119. P.G. de Gennes, *C.R. Acad. Sci. Paris*, t-271, B-4469 (1970); H. Mada, *Mol. Cryst. Liq. Cryst.*, 53, 127 (1979).
120. K. Okano, *Jpn. J. Appl. Phys.*, 22, L343 (1983).
121. H. Mada, T. Sonoda, *Jpn. J. Appl. Phys.*, 32, L1245 (1993).
122. J.L. Janning, *Appl. Phys. Lett.*, 21, 173 (1972).
123. R. Barberi, G. Durand, *Appl. Phys. Lett.*, 58, 2097 (1991); R. Barberi, M. Gicondo, G. Durand, *Appl. Phys. Lett.*, 60, 1085 (1992).
124. R. Barberi, M. Gicondo, Ph. Martinot-Lagarde, G. Durand, *Appl. Phys. Lett.*, 62, 3270 (1993).
125. A.G. Dyadyusha, T. Marussii, Y. Reznikov, A. Khiznyak, V. Resetnyak, *JETP Lett.*, 56, 17 (1992); M. Schadt, K. Schmidt, V. Koozinkov, V. Chigrinov, *Jpn. J. Appl. Phys.*, 31, 2155 (1992); J.L. West, X. Wang, Y. Ji, J.R. Kelly, *SID Digest*, XXVI, 703 (1995).
126. W.M. Gibbons, P.J. Shannon, S.T. Sun, B.J. Swetlin, *Nature*, 351, 49 (1991); W.M. Gibbons, T. Kosa, M. Palffy-Muhoray, P.J. Shannon, S.T. Sun, *Nature*, 377, 43 (1995).
127. G. Friedel, *Ann. Phys. (Paris)*, 18, 273 (1922).
128. N.A. Clark, *Phys. Rev. Lett.*, 55, 292 (1985).
129. T. Uchida, H. Seki, *Liquid Crystals: Applications and Uses*, Vol. 3, Ch. 5, Ed. B. Bahadur, World Scientific, Singapore (1992).
130. G. Barbero, G. Durand, Surface anchoring of nematic liquid crystals, in *Liquid Crystals in Complex Geometries*, Ch. 2, Ed. G. Crawford, S. Zumer, Taylor & Francis, Bece Reton, FL (1996).

5

Optics of Anisotropic Materials

Light is an electromagnetic wave, and its propagation can be derived from the well-known Maxwell equations, which in their differential form read as:

$$\vec{\nabla} \times \vec{H} - \frac{\partial \vec{D}}{\partial t} = \vec{j} \tag{5.1}$$

$$\vec{\nabla} \times \vec{E} + \frac{\partial \vec{B}}{\partial t} = 0 \tag{5.2}$$

$$\vec{\nabla} \cdot \vec{D} = \rho \tag{5.3}$$

$$\vec{\nabla} \cdot \vec{B} = 0 \tag{5.4}$$

Here $\vec{j} = \hat{\sigma}\vec{E}$ is the current density, and ρ is the free charge density.
The electric displacement $\vec{D}$ and electric field $\vec{E}$ are connected through the dielectric tensor as:

$$\vec{D} = \varepsilon_o \hat{\varepsilon} \vec{E} \tag{5.5}$$

The magnetic induction $\vec{B}$ and magnetic field $\vec{H}$ are related through the magnetic permeability tensor as:

$$\vec{B} = \mu_o \hat{\mu} \vec{H} \tag{5.6}$$

Here $\mu_o = 4\pi \times 10^{-7}$ N/A^2 is the permeability of the vacuum.
First we assume that there are no free charges ($\rho = 0$), and the material is insulating (DC conductivity $\sigma_{DC} = 0$). Assuming also that the material is nonmagnetic ($\mu = 1$), and taking the time derivative of (5.1) and using the relation (5.3), we get:

$$\vec{\nabla} \times \frac{\partial \vec{H}}{\partial t} = \varepsilon_o \hat{\varepsilon} \frac{\partial^2 E}{\partial t^2} \tag{5.7}$$

Applying the *curl* operation on Eq.(5.2) and using Eq. (5.6), we arrive at:

$$\vec{\nabla} \times \vec{\nabla} \times \vec{E} = -\mu_o \varepsilon_o \hat{\varepsilon} \frac{\partial^2 \vec{E}}{\partial t^2} \tag{5.8}$$

This is the basic wave equation. Note that $\mu_o \varepsilon_o$ has a dimension of $s^2 m^{-2}$, i.e., the inverse of square of speed. It is customary, therefore, to replace it with $1/c^2$, where $c = \frac{1}{\sqrt{\varepsilon_o \mu_o}} = 3 \cdot 10^8 \, m/s$ is the velocity of light in vacuum.

In the following, we will deal only with monochromatic plane wave, where the electric field component of the wave can be written as:

$$\vec{E} = \vec{A} \cdot \cos(\omega t - \vec{k} \cdot \vec{r} + \delta) \tag{5.9}$$

where $\vec{A}$ is the amplitude, ω is the angular frequency, $\vec{k}$ is the wave vector of the light (for isotropic media, $\vec{k}$ is perpendicular to the electric vector, i.e., $\vec{k} \cdot \vec{E} = 0$), and δ is the phase shift).

Different plane-polarized lights in isotropic media are illustrated in Figure 5.1. For light propagating along z:

$$E_x = A_x \cos(\omega t - k \cdot z + \delta_x)$$
$$E_y = A_y \cos(\omega t - k \cdot z + \delta_y) \tag{5.10}$$

Here we have used two independent and positive amplitudes, A_x and A_y, and added two independent phases, δ_x and δ_y to reflect the mutual independence of the two components. At a definite frequency the x and y components can oscillate independently.

Without loss of generality we can take z = 0, so

$$E_x = A_x \cos(\omega t + \delta_x)$$
$$E_y = A_y \cos(\omega t + \delta_y) \tag{5.11}$$

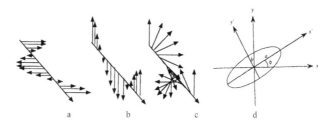

FIGURE 5.1

Illustration of differently polarized lights in isotropic medium. (a) and (b) Linearly polarized light in horizontal and vertical planes, respectively. (c) A left-handed circular polarized light. (d) An elliptically polarized light at fixed z coordinate.

We define the relative phase as: $\delta = \delta_y - \delta_x;\quad -\pi < \delta < \pi$.
In the general case, eliminating ωt in (5.11) after some algebra we get:

$$\left(\frac{E_x}{A_x}\right)^2 + \left(\frac{E_y}{A_y}\right)^2 - 2\frac{\cos\delta}{A_x A_y}E_x E_y = \sin^2\delta \tag{5.12}$$

By suitable rotation of the coordinate system, this can be transferred to the equation of an ellipse:

$$\left(\frac{E_{x'}}{a}\right)^2 + \left(\frac{E_{y'}}{b}\right)^2 = 1 \tag{5.13}$$

where the axes a and b of the ellipse can be expressed by A_x, A_y and δ as

$$a^2 = A_x^2\cos^2\phi + A_y^2\sin^2\phi + 2A_x A_y\cos\delta\cos\phi\sin\phi$$
$$b^2 = A_x^2\sin^2\phi + A_y^2\cos^2\phi + 2A_x A_y\cos\delta\cos\phi\sin\phi \tag{5.14}$$

where

$$\tan 2\phi = \frac{2A_x A_y}{A_x^2 - A_y^2}\cos\delta \tag{5.15}$$

Here ϕ is the angle between the long axis of the ellipse and the x axis (see Figure 5.1). It is useful to define the ellipticity as $e = \pm\, b/a$, where the signs determine the handedness of the light as + (−) correspond to left (right) handed rays.

In the special cases when $\delta = 0$ or π, we speak about linear polarization states, where the electric field vector vibrates sinusoidally along a constant direction in the xy plane defined by the ratio of two components (often called plane-polarized light).

When $\delta = \pm\pi/2$, the ellipse becomes a circle, and we speak about circularly polarized light.

In the following we split our description of the optics of anisotropic materials to two parts. First we investigate the optical properties of achiral and nonhelical systems; then we will see how these results will be modified in chiral and helical systems, which are especially important in biological materials.

5.1 Achiral Materials

To describe the propagation of light in an anisotropic medium, we write the wave equation of (5.8) in the form:

$$\vec{\nabla} \times \vec{\nabla} \times (\hat{\varepsilon}^{-1}\vec{D}) = -\frac{1}{c^2}\frac{\partial^2\vec{D}}{\partial t^2} \tag{5.16}$$

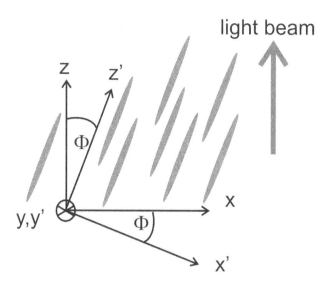

FIGURE 5.2
Coordinate system to describe light propagation in a uniaxial system.

We consider a plate of anisotropic and nonhelical material. Two orthogonal coordinate systems (x,y,z) and (x',y',z') are used to characterize the system (Figure 5.2). In the laboratory frame the light propagates along the $+z$ axis. The x',y',z' system is aligned along the principal directions of the medium. Provided that the medium is uniaxial, the optical axis coincides with the z' axis. Without loss of generality we choose y to be parallel to y', and the primed system is rotated around y by an angle ϕ.

The transformation matrix from the primed to the unprimed system, and the inverse matrices read as:

$$\hat{U} = \begin{pmatrix} \cos\Phi & 0 & \sin\Phi \\ 0 & 1 & 0 \\ -\sin\Phi & 0 & \cos\Phi \end{pmatrix} \quad \hat{U}^{-1} = \begin{pmatrix} \cos\Phi & 0 & -\sin\Phi \\ 0 & 1 & 0 \\ \sin\Phi & 0 & \cos\Phi \end{pmatrix} \quad (5.17)$$

The dielectric tensor and its inverse are diagonal in the primed system and are expressed as:

$$\hat{\varepsilon}' = \begin{pmatrix} \varepsilon_1 & 0 & 0 \\ 0 & \varepsilon_2 & 0 \\ 0 & 0 & \varepsilon_3 \end{pmatrix} \quad \hat{\varepsilon}'^{-1} = \begin{pmatrix} \dfrac{1}{\varepsilon_1} & 0 & 0 \\ 0 & \dfrac{1}{\varepsilon_2} & 0 \\ 0 & 0 & \dfrac{1}{\varepsilon_3} \end{pmatrix} \quad (5.18)$$

Transforming the dielectric tensor and its inverse to the laboratory system we get:

$$
\hat{\varepsilon} = \hat{U}\hat{\varepsilon}'\hat{U}^{-1} =
\begin{pmatrix}
\varepsilon_1 \cos^2 \Phi + \varepsilon_3 \sin^2 \Phi & 0 & (\varepsilon_3 - \varepsilon_1)\sin \Phi \cos \Phi \\
0 & \varepsilon_2 & 0 \\
(\varepsilon_3 - \varepsilon_1)\sin \Phi \cos \Phi & 0 & \varepsilon_1 \sin^2 \Phi + \varepsilon_3 \cos^2 \Phi
\end{pmatrix}
$$

$$(5.19)$$

$$
\hat{\varepsilon}^{-1} = \hat{U}\hat{\varepsilon}'^{-1}\hat{U}^{-1} =
\begin{pmatrix}
\dfrac{1}{\varepsilon_1}\cos^2 \Phi + \dfrac{1}{\varepsilon_3}\sin^2 \Phi & 0 & \left(\dfrac{1}{\varepsilon_3} - \dfrac{1}{\varepsilon_1}\right)\sin \Phi \cos \Phi \\
0 & \dfrac{1}{\varepsilon_2} & 0 \\
\left(\dfrac{1}{\varepsilon_3} - \dfrac{1}{\varepsilon_1}\right)\sin \Phi \cos \Phi & 0 & \dfrac{1}{\varepsilon_1}\sin^2 \Phi + \dfrac{1}{\varepsilon_3}\cos^2 \Phi
\end{pmatrix}
$$

The *curl* and *curl curl* operators in matrix forms in the specific Cartezian coordinate systems are:

$$
curl =
\begin{pmatrix}
0 & -\dfrac{\partial}{\partial z} & \dfrac{\partial}{\partial y} \\
\dfrac{\partial}{\partial z} & 0 & -\dfrac{\partial}{\partial x} \\
-\dfrac{\partial}{\partial y} & \dfrac{\partial}{\partial x} & 0
\end{pmatrix}
\qquad
curlcurl =
\begin{pmatrix}
-\dfrac{\partial^2}{\partial y^2} - \dfrac{\partial^2}{\partial z^2} & \dfrac{\partial^2}{\partial x \partial y} & \dfrac{\partial^2}{\partial x \partial z} \\
\dfrac{\partial^2}{\partial x \partial y} & -\dfrac{\partial^2}{\partial x^2} - \dfrac{\partial^2}{\partial z^2} & \dfrac{\partial^2}{\partial y \partial z} \\
\dfrac{\partial^2}{\partial x \partial z} & \dfrac{\partial^2}{\partial y \partial z} & -\dfrac{\partial^2}{\partial x^2} - \dfrac{\partial^2}{\partial y^2}
\end{pmatrix}
$$

$$(5.20)$$

The light propagates in the z direction, and we assume that it consists of plane waves extending far in the x and y directions. Except near the rim of the beam, the wave must be independent of the x and y coordinates, so we can assume that $\partial/\partial x = 0$ $\partial/\partial y = 0$. Under these circumstances, the *curl curl* operator is simplified to:

$$
curlcurl =
\begin{pmatrix}
-\dfrac{\partial^2}{\partial z^2} & 0 & 0 \\
0 & -\dfrac{\partial^2}{\partial z^2} & 0 \\
0 & 0 & 0
\end{pmatrix}
\qquad (5.21)
$$

This means that we can write the wave equation (5.16) for $\vec{D}$ in matrix form as:

$$
\begin{pmatrix}
-\dfrac{\partial^2}{\partial z^2} & 0 & 0 \\
0 & -\dfrac{\partial^2}{\partial z^2} & 0 \\
0 & 0 & 0
\end{pmatrix}
\begin{pmatrix}
\dfrac{1}{\varepsilon_1}\cos^2\Phi + \dfrac{1}{\varepsilon_3}\sin^2\Phi & 0 & \left(\dfrac{1}{\varepsilon_3}-\dfrac{1}{\varepsilon_1}\right)\sin\Phi\cos\Phi \\
0 & \dfrac{1}{\varepsilon_2} & 0 \\
\left(\dfrac{1}{\varepsilon_3}-\dfrac{1}{\varepsilon_1}\right)\sin\Phi\cos\Phi & 0 & \dfrac{1}{\varepsilon_1}\sin^2\Phi + \dfrac{1}{\varepsilon_3}\cos^2\Phi
\end{pmatrix} \quad (5.22)
$$

$$
\begin{pmatrix} D_x \\ D_y \\ D_z \end{pmatrix} + \dfrac{1}{c^2}\dfrac{\partial^2}{\partial t^2}\begin{pmatrix} D_x \\ D_y \\ D_z \end{pmatrix} = 0
$$

Multiplying the matrices, we find that (5.22) splits up to three equations:

$$
\dfrac{\partial^2}{\partial z^2}c^2\left[\dfrac{\cos^2\Phi}{\varepsilon_1}+\dfrac{\sin^2\Phi}{\varepsilon_3}\right]D_x + \dfrac{\partial^2}{\partial z^2}c^2\left(\dfrac{1}{\varepsilon_3}-\dfrac{1}{\varepsilon_1}\right)\sin\Phi\cos\Phi\,D_z - \dfrac{\partial^2}{\partial t^2}D_x = 0 \quad (5.23)
$$

$$
\dfrac{\partial^2}{\partial z^2}\dfrac{c^2}{\varepsilon_2}D_y - \dfrac{\partial^2}{\partial t^2}D_y = 0 \quad\quad\quad (5.24)
$$

$$
\dfrac{\partial^2}{\partial t^2}D_z = 0 \quad\quad\quad (5.25)
$$

The last equation (5.25) shows that the oscillating component of D_z is zero. Because we do not need to care about the static fields, the second term of (5.23) disappears, and the corresponding equation reduces to:

$$
\dfrac{\partial^2}{\partial z^2}c^2\left[\dfrac{\cos^2\phi}{\varepsilon_1}+\dfrac{\sin^2\phi}{\varepsilon_3}\right]D_x - \dfrac{\partial^2}{\partial t^2}D_x = 0 \quad\quad\quad (5.26)
$$

This equation involves only D_x, and its solution is a plane wave polarized in x direction traveling with the velocity:

$$
v_{xz} = c\sqrt{\dfrac{\cos^2\Phi}{\varepsilon_1}+\dfrac{\sin^2\Phi}{\varepsilon_3}} \quad\quad\quad (5.27)
$$

Defining the refractive index as $n = c/v$, we get that:

$$
n_{xz} = \dfrac{\sqrt{\varepsilon_1}\sqrt{\varepsilon_3}}{\sqrt{\varepsilon_3\cos^2\Phi + \varepsilon_1\sin^2\Phi}} \quad\quad\quad (5.28)
$$

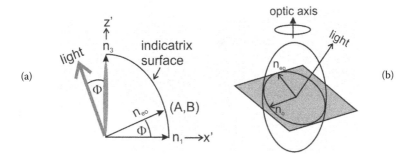

FIGURE 5.3
(a) The geometry of the indicatrix in uniaxial media. (b) Illustration how the construction indicatrix makes it easy to determine the refractive indices.

This is the refractive index experienced by the electric vector of the light that has a component along the director, and is called extraordinary index.

Formula (5.24) relates to D_y and gives a wave with a velocity polarized in the y direction $v_{yz} = c/\sqrt{\varepsilon_2}$ corresponding to the ordinary mode. In the local system the principal refractive indices in general are given by $n_i = \sqrt{\varepsilon_i}$ with $i = 1, 2, 3$.

It is interesting to point out that the Φ-dependent refractive index, given in (5.28), geometrically describes an ellipse as illustrated in Figure 5.3.

From Figure 5.3a we can see that:

$$\sin\phi = \frac{B}{n_{xz}} \quad and \quad \cos\phi = \frac{A}{n_{xz}} \tag{5.29}$$

Substituting this into (5.28), we find that:

$$\frac{A^2}{n_1^2} + \frac{B^2}{n_3^2} = 1 \tag{5.30}$$

This means that the end point of the refractive index follows an ellipse, which is the arc in Figure 5.3a. The indicatrix in a real 3D system is an ellipsoid. The use of the indicatrix model makes it easy to visualize the propagation of the light in an anisotropic material. To find the refractive indices of a light beam, insert a plane through the center of the indicatrix and perpendicular to the propagation of the light (see Figure 5.3b). The plane will intersect the indicatrix surface along an ellipse. One axis of this ellipse (the long axis in case of positive birefringence, and the short axis for the negative one) will have a length that shows the extraordinary refractive index. In addition, its direction will coincide with the polarization of the extraordinary light wave. In the same manner, the other axis of the ellipse corresponds to the ordinary wave.

5.1.1   Birefringence

In case of uniaxial materials, such as conventional nematic liquid crystals, $\varepsilon_1 = \varepsilon_2 = \varepsilon_\perp$, and $\varepsilon_3 = \varepsilon_{\parallel}$. Accordingly, we can write the dielectric tensor in the optical frequency range as:

$$\hat{\varepsilon} = \begin{pmatrix} n_\perp^2 & 0 & 0 \\ 0 & n_\perp^2 & 0 \\ 0 & 0 & n_{\parallel}^2 \end{pmatrix} \tag{5.31}$$

The above mathematical derivations can be summarized as follows. If the anisotropy of a material persists in a range longer than the wavelength of the light, the material will be optically anisotropic. In this case, the propagation through the material depends on the direction of propagation and of the vibration of the electric vector. The speed of the light is determined by the refractive index $\hat{n}$, which for uniaxial materials will be different along $n_{\parallel}$ and perpendicular to the director $n_\perp$. The numerical difference $\Delta n = n_{\parallel} - n_\perp$ between these refractive indices (or in short: refringences) is called "birefringence." The birefringence depends on the wavelength of the light (dispersion), which in the visible and near infrared region typically can be approximated by the Cauchy formula: $\Delta n = A + B/\lambda^2$. Calamitic thermotropic liquid crystals in the visible range typically have positive Δn, being in the range of 0 and 0.45. Very low birefringence is characterized for materials made from bicyclohexanes, whereas the most birefringent nematic is made of biphenyl-diacetylenics. Generally the anisotropy is high if two or more benzene rings are connected by conjugated bonds. For example, in MBBA (CH_3-Φ-C=N-Φ-C_4H_9), where Φ denotes benzene ring, the birefringence is Δn = 0.19. The contribution of alkyl groups, which often terminate the molecules, depends on the number of carbon atoms. Very often a so-called "odd–even" effect is observed: the difference between the longitudinal and transversal polarizability usually increases by about 0.83×10^{-40} Fm² as the number of carbon atoms increases from even to odd, and decreases by 0.28×10^{-40} Fm² when the number increases from odd to even.

The birefringence of lyotropic liquid crystals is almost an order of magnitude smaller due to the water content and because of the hydrocarbon chain, which takes relatively large part of the amphiphilic molecules, contributes to birefringence only weakly.

A birefringence with the induced optical axis parallel to $\vec{E}$ can be also induced in any isotropic medium if a uniform electric field is applied at right angles to the light beam. This is the so-called Kerr effect. The induced birefringence is proportional to E^2, and it is customary to express it as:

$$\Delta n(E) = \lambda B E^2 \tag{5.32}$$

Here B is the Kerr constant, and λ is the wavelength of the light. The Kerr effect is due to partial alignment of the molecular dipoles $\vec{\mu}$ parallel to the field.

Liquid crystals well above the transition to the nematic phase behave like ordinary liquids, with Kerr constant comparable to of nitrobenzene ($+4 \times 10^{-12}$ m V^{-2}). As the transition approaches, the absolute value of the Kerr constant increases strongly (B is either positive or negative), and the induced birefringence approaches values of about 10^{-5}–10^{-6}.

The anomalous Kerr effect near the isotropic–nematic transition can be qualitatively understood as follows. Approaching the *I–N* transition nematic-like aggregates form, and the sizes of the aggregates increase close to the transition. The growth of the Kerr constant can be calculated by taking into account the contribution of nematic aggregates to the orientation order parameter to the dielectric properties of the isotropic phase. This contribution was calculated in the frame of the Landau theory. Accordingly, the field induced orientation order is:[1]

$$S(E) = \rho \Delta \varepsilon E^2 / 12 \pi a_0 (T - T^*) \tag{5.33}$$

where ρ is the number density of the molecules, and a_o is the parameter of the Landau expansion. The field-induced order can be related to the Kerr constant as:

$$B = \Delta n^o S(EI / \lambda E^2) \tag{5.34}$$

Experimentally, the Kerr effect was intensively studied in the 1970s.[2] The relaxation times associated with the order parameter fluctuations increase when approaching T_{N-I} (critical slowing down) but still in the submicrosecond range.

For nonpolar molecules, B is always positive. In case of polar molecules, the sign of B depends on the sign of Δn, which is related to the direction of the molecular dipole with respect to the molecular long axis. $B > 0$ if $\beta < 54°$ (for example, for PAA $\beta > 54°$ and $\Delta \varepsilon < 0$, whereas for APAAB $\beta < 54°$ and $\Delta \varepsilon > 0$).

5.1.2 Optical Retardation

Because the electric vector of the light has components normal to the beam, in anisotropic materials they feel different electronic polarizabilities, i.e., different refractive indices. For this reason, the speed of the light of different polarization directions will be different: $v_o = c/n_\perp$ describes the speed of the "ordinary" wave, and $v_e = c/n_\parallel$ relates to the "extraordinary" beam, which exists only in anisotropic materials. Due to the differences of the speeds, there will be a phase difference between the ordinary and extraordinary waves. Since the wavevector $\vec{k}$ of the light relates to the wavelength, λ as $|\vec{k}| \equiv k = n\frac{\omega}{c} = n \cdot \frac{2\pi}{\lambda}$, the difference between the phases (the so called retardation) of the ordinary and extraordinary waved can be expressed as:

$$\Gamma = \frac{2\pi}{\lambda}(n_s - n_f)d = \frac{2\pi}{\lambda}\Delta n \cdot d \tag{5.35}$$

Here n_s (n_f) denotes the fast and slow indices (for anisotropic materials with rod-shaped molecules $n_s = n_e$ and $n_f = n_o$), d is the thickness of the birefringent film, and λ is the wavelength of the light. The component that feels larger refractive index will be slower, i.e., retarded with respect to the other polarization directions.

Coming back to Eq. (5.25), in uniaxial liquid crystals D_z is zero, i.e., the waves are transverse if one looks at the $\vec{D}$ field. Knowing $\hat{\varepsilon}^{-1}$ from (5.19), we can calculate the direction of the corresponding electric field by using (5.18). If $\vec{D}$ is along y, $\vec{E}$ will be parallel to it, just as in ordinary isotropic materials. However, if $\vec{D}$ is along x, $\vec{E}$ will have components both in x and z. This is one of the extraordinary features of this wave. The angle between $\vec{D}$ and $\vec{E}$ is given as:

$$\eta = \arctan\left(-\frac{\varepsilon_{31}^{-1}}{\varepsilon_{11}^{-1}}\right) = \arctan\left\{-\frac{\left(\dfrac{1}{\varepsilon_3}-\dfrac{1}{\varepsilon_1}\right)\sin\phi\cos\phi}{\dfrac{1}{\varepsilon_1}\cos^2\phi+\dfrac{1}{\varepsilon_3}\sin^2\phi}\right\} \tag{5.36}$$

The same angle will appear between the so-called Poynting vector, which is defined as $\vec{E}\times\vec{H}$, and the wave vector (the normal of the wave planes). The Pointing vector is associated with the energy transport and coincides with the ray direction. This means that we see the doubling of the picture when viewed through the birefringent material with optical axis oblique with the viewing direction. Such a situation was first observed on Iceland spar (calcite crystals) in the seventeenth century by a Danish professor, Rasmus Bartholin as illustrated in Figure 5.4.[3] The double image of an object behind a thick enough birefringent slab, was explained first by Christian Huygens in 1690.[4]

To estimate the maximum of η (at $\phi = 45°$), we take $n_o = 1.5$ and $n_e = 1.7$, which gives $\eta \sim 3°$. In most cases, η is smaller and does not result in observable doubling in the image if one looks through a uniformly aligned liquid crystal slab of typically 10 µm thickness. Besides, mostly $\phi = 0$ (homeotropic alignment) or $\phi = 90°$ (planar alignment), which gives $\eta = 0$.

FIGURE 5.4
Illustration of the double refraction by viewing a text behind a calcite crystal.

When it matters, the double refraction will also modify the Snell's law that describes the refraction of a light obliquely incident on the surface of bire-fringent plate even when the optical axis is parallel or perpendicular to the interface. Let $\bar{k}_i$ be the wave vector of the incident wave; furthermore, let $\bar{k}_o$ and $\bar{k}_e$ be the wave vectors of the refracted ordinary and extraordinary waves. The boundary conditions require that all wave vectors should be in the plane of incidence with the same components. This means that:

$$k_i \sin \theta_i = k_o \sin \theta_o = k_e \sin \theta_e \tag{5.37}$$

This resembles Snell's law, but here $\bar{k}_e$ varies with the direction between the director $\bar{n}$, leading to a quadratic equation.

5.1.3 Light Propagation through Uniaxial Materials (Jones Method)

In studying anisotropic materials, such as liquid crystals, soaps and free-standing films, most often we use microscopic observations, when the size of the light beam is much larger than the studied area, so the incoming beam is a plane wave. In this case the Jones formalism[5] provides a rela-tively simple and straightforward method to calculate the propagation of light in an anisotropic medium, especially for the description of nonchiral materials.

Neglecting reflection and absorption, the incident beam of light with a polarization state can be conveniently described by the Jones vector,[6] which is expressed in terms of complex amplitudes as a column vector:

$$\bar{J} = \begin{pmatrix} A_x e^{i\delta_x} \\ A_y e^{i\delta_y} \end{pmatrix} = \begin{pmatrix} J_x \\ J_y \end{pmatrix} \tag{5.38}$$

Here $\bar{J}$ is a complex vector in an abstract mathematical space related to the electric field as $E_x(t) = \text{Re}[J_x e^{i\omega t}] = \text{Re}[A_x e^{i(\omega t + \delta_x)}]$. This representation con-tains complete information about the plane-polarized light with equation (5.11). If we are interested only in the polarization state, we can normalize the Jones vector as $\bar{J}^* \cdot \bar{J} = 1$ (* denotes the complex conjugation).

Such normalized Jones matrices can be expressed for the different polar-ization states as:

- Linearly polarized light: $\binom{\cos\phi}{\sin\phi}$, where ϕ is the angle between x and the oscillation directions;
- Circularly polarized light: $\frac{1}{\sqrt{2}}\binom{1}{\pm i}$, where $-(+)$ denotes right (left) circularly polarized lights;
- Elliptic polarization: $J(\phi, \delta) = \binom{\cos\phi}{e^{i\delta}\sin\phi}$.

Decomposing the light into a linear combination of the "fast" (typically ordinary for elongated molecules) and "slow" (typically extraordinary for disc-shape molecules) modes is done by a coordinate transformation:

$$\begin{pmatrix} J_s \\ J_f \end{pmatrix} = \begin{pmatrix} \cos\phi & \sin\phi \\ -\sin\phi & \cos\phi \end{pmatrix} \begin{pmatrix} J_x \\ J_y \end{pmatrix} \equiv U(\phi) \begin{pmatrix} J_x \\ J_y \end{pmatrix} \tag{5.39}$$

Comparing it with (5.38), we can write the Jones vector of a plane wave with wavelength λ emerging an anisotropic medium of thickness d as:

$$\begin{pmatrix} J_s' \\ J_f' \end{pmatrix} = \begin{pmatrix} \exp\left(-in_s\dfrac{2\pi}{\lambda}d\right) & 0 \\ 0 & \exp\left(-in_f\dfrac{2\pi}{\lambda}d\right) \end{pmatrix} \begin{pmatrix} J_s \\ J_f \end{pmatrix} \tag{5.40}$$

The phase retardation is the difference of the exponents in (5.40) in agreement with the definition of (5.35).

The birefringence of a typical liquid crystal plate is small compared to the refractive index, i.e., the absolute change in the phase caused by the retarder (where one of the beam is retarded) is much larger than the retardation. The absolute phase change is:

$$\psi = \frac{1}{2}(n_s + n_f)\frac{2\pi}{\lambda}d \tag{5.41}$$

With these,

$$\begin{pmatrix} J_s' \\ J_f' \end{pmatrix} = e^{-i\psi} \begin{pmatrix} e^{-i\Gamma/2} & 0 \\ 0 & e^{i\Gamma/2} \end{pmatrix} \begin{pmatrix} J_s \\ J_f \end{pmatrix} \tag{5.42}$$

The Jones vector of the polarization state of the emerging beam in the x-y coordinate system is given by transforming the vector back from the s-f coordinate system as:

$$\begin{pmatrix} J_x' \\ J_y' \end{pmatrix} = \begin{pmatrix} \cos\phi & -\sin\phi \\ \sin\phi & \cos\phi \end{pmatrix} \begin{pmatrix} J_s' \\ J_f' \end{pmatrix} \tag{5.43}$$

Combining (5.39), (5.42) and (5.43) we get:

$$\begin{pmatrix} J_x' \\ J_y' \end{pmatrix} = U(-\phi)W_oU(\phi) \begin{pmatrix} J_x \\ J_y \end{pmatrix} \tag{5.44}$$

where

$$W_o = e^{-i\psi} \begin{pmatrix} e^{-i\Gamma/2} & 0 \\ 0 & e^{i\Gamma/2} \end{pmatrix} \tag{5.45}$$

is the Jones matrix of a birefringent slab of the retardation Γ.

The phase factor $e^{-i\psi}$ can be neglected if interference effects due to multiple reflections are not important.

The intensity of the light transmitted by the birefringent liquid crystal film can also be calculated with the help of the Jones vector by defining it with the electric vector of the light as $\vec{E} = \begin{pmatrix} E_x \\ E_y \end{pmatrix}$, where $I = \vec{E}^* \cdot \vec{E} = |E_x|^2 + |E_y|^2$ is the intensity of the light. The Jones vector of the emerging beam is: $\vec{E}' = \begin{pmatrix} E_x' \\ E_y' \end{pmatrix}$, so the transmittance of the birefringent slab is:

$$T \equiv \frac{|E_x'|^2 + |E_y'|^2}{|E_x|^2 + |E_y|^2} .$$

Transmittance of the anisotropic films is usually studied by placing the film between crossed polarizers. Polarizers are uniaxially anisotropic in their absorption, i.e., the complex refractive indices are different, so that:

$$n_o^* = n_o - i\kappa_o, \quad n_e^* = n_e - i\kappa_e \tag{5.46}$$

where κ_o and κ_e are extinction coefficients. Examples of polarizers include tourmaline, tin oxide crystals and Polaroid sheets. Polaroid film, which is the most widely used, consists of long-chain polymers, treated with light-absorbing dyes, and stretched so that the chains are aligned. Light vibrating parallel with the chains is absorbed, while light perpendicular to the chains is transmitted. We note that liquid crystals doped with dichroic dyes[†] can also be used as polarizers, which can even be switched to a nonpolarizer by switching to a homeotropic alignment by electric field.

The quality of the polarizers is characterized by their extinction ratios (T_2/T_1), which measures the ratio of the transmissions with polarizations perpendicular/parallel to the transmission axis. Examples for good polarizers are $T_1 = 0.8$ and $T_2 = 0.0008$, which means an extinction ratio of about 10^{-3}.

In the Jones matrix representation an ideal linear polarizer, neglecting the absolute phase accumulated as a result of the finite optical thickness, can be given as:

$$P_x = \begin{pmatrix} 1 & 0 \\ 0 & 0 \end{pmatrix} \quad and \quad P_y = \begin{pmatrix} 0 & 0 \\ 0 & 1 \end{pmatrix} \tag{5.47}$$

where P_x (P_y) represent polarizers oriented with the transmission axis parallel to the laboratory axis x (y). The intensity of a light going through a birefringent plate sandwiched between a pair of crossed polarizers can be calculated by the Jones matrix method. Without loosing generality, we can set the polarizer, for example, in vertical and the analyzer in horizontal direction.

[†] A dichroic dye absorbs light when aligned normal to the beam and transmits when aligned parallel.

If the optical axis of the birefringent film makes an angle ϕ with x, we can express the Jones matrix as:

$$\hat{W} = \begin{pmatrix} \cos\phi & -\sin\phi \\ \sin\phi & \cos\phi \end{pmatrix} \begin{pmatrix} e^{-i\Gamma/2} & 0 \\ 0 & e^{i\Gamma/2} \end{pmatrix} \begin{pmatrix} \cos\phi & \sin\phi \\ -\sin\phi & \cos\phi \end{pmatrix}$$

$$= \begin{pmatrix} \cos^2\phi e^{-i\Gamma/2} + \sin^2\phi e^{i\Gamma/2} & \sin\phi\cos\phi(e^{-i\Gamma/2} - e^{i\Gamma/2}) \\ \sin\phi\cos\phi(e^{-i\Gamma/2} - e^{i\Gamma/2}) & \sin^2\phi e^{-i\Gamma/2} + \cos^2\phi e^{i\Gamma/2} \end{pmatrix}$$

(5.48)

Applying the Jones matrix of a polarizer in x direction, we get the Jones vector of the emerging beam as:

$$\vec{E} = \begin{pmatrix} 1 & 0 \\ 0 & 0 \end{pmatrix} \hat{W} \frac{1}{\sqrt{2}} \begin{pmatrix} 0 \\ 1 \end{pmatrix} = \frac{1}{\sqrt{2}} \begin{pmatrix} 1 & 0 \\ 0 & 0 \end{pmatrix} \begin{pmatrix} i\sin 2\phi \sin\dfrac{\Gamma}{2} \\ \sin^2\phi e^{-i\Gamma/2} + \cos^2\phi e^{i\Gamma/2} \end{pmatrix}$$

$$= \begin{pmatrix} \dfrac{i}{\sqrt{2}}\sin 2\phi \sin\dfrac{\Gamma}{2} \\ 0 \end{pmatrix}$$

(5.49)

This gives the transmitted intensity.

$$I = \frac{1}{2}\sin^2 2\phi \sin^2\frac{\Gamma}{2}$$ (5.50)

This formula shows that the brightness of a uniformly aligned birefringent slab is determined by two factors: the angle between the crossed polarizers and the optical axis, and the retardation. The former one is largest when the optical axis is rotated by 45° with respect the crossed polarizers. The latter part has a maximum, when $\Gamma/2 = \pi/2$. This is the special case of the so-called half-wave retarder $(d = \frac{1}{2}\cdot\frac{\lambda}{\Delta n})$, which for a material with $\Delta n = 0.1$ is formed when $d = \lambda/(2\Delta n) = 2.5\mu m$ for $\lambda = 0.5$ μm light wavelength. This condition is desirable to achieve in liquid crystal displays. Another special example of wave plates is the so-called quarter-wave plate $(d = \frac{1}{4}\cdot\frac{\lambda}{\Delta n})$, which in general converts a linearly polarized light into an elliptically polarized light and vice versa.

5.2 Helical and Chiral Structures

5.2.1 Optical Activity

Optical activity is a phenomenon when the transparent material rotates the polarization direction of light transversing through the material. As proposed first by Fresnel in 1825, optical activity arises from circular double refraction, in which the independent plane wave solutions of the Maxwell's equations (i.e., the eigenwaves of propagation) are right- and left-handed circularly polarized waves. Denoting the refractive indices associated with these two waves by n_r and n_l, respectively, if a light linearly polarized in x direction with amplitude D_o enters the medium at z = 0, it will be represented by the sum of left- and right-handed circularly polarized beams (LCP and RCP, respectively) with amplitude $D_o / \sqrt{2}$. At distance z in the medium, the resultant wave will be:

$$\vec{E} = \frac{D_o}{\sqrt{2}} e^{i\omega t} (\vec{R} e^{-i\omega z n_r / c} + \vec{L} e^{-i\omega z n_l / c}) = D_o \vec{P} \exp\left\{ i\omega \left(t - \frac{z(n_r + n_l)}{2c} \right) \right\} \quad (5.51)$$

where $\vec{R} = \frac{1}{\sqrt{2}} \binom{x}{-iy}$ and $\vec{L} = \frac{1}{\sqrt{2}} \binom{x}{iy}$ are the Jones vectors, and

$$\vec{P} = \vec{x} \cdot \cos\left[\frac{\omega(n_l - n_r)}{2c} z \right] + \vec{y} \cdot \sin\left[\frac{\omega(n_l - n_r)}{2c} z \right] \quad (5.52)$$

is the polarization vector of the emerging beam. This shows that the resultant wave is linearly polarized, with its plane of polarization turned in counterclockwise sense from x through $\alpha \sim \omega z(n_l - n_r)/2c$. The rotation angle per unit length α/z characterizes the ability of the material to rotate the light, and is called rotatory power (ORP), and it can be given as:

$$ORP = \frac{\pi}{\lambda}(n_r - n_l) \quad (5.53)$$

For example, in the liquid crystal, cholesteryl 2–2-ethoxy-ethoxy ethyl carbonate (CEEC), at λ = 650 nm the ORP = 285°/mm, which from (5.60) gives that $|n_r - n_l|$ = 1.03 × 10⁻³. This shows that optical rotation is an extremely sensitive way to measure circular double refraction.

Typically, ORP decreases with increasing wavelength. For example, in quartz ORP varies between 17 and 49, in AgGaS$_2$ between 430 and 950 (in units of degree/mm), and in the liquid crystal cholesteryl oleyl carbonate it varies between 167 and 4167.

When the molecules are dissolved in isotropic three-dimensional fluids, or when they are heated to their isotropic phase, they show optical activity only if they are chiral, since their structure is random. By definition, an optically active material is right-handed or dextro (D) if the sense of rotation of the plane of polarization is counterclockwise when viewed by an observer facing the approaching light beam. For D materials $n_r < n_l$, that is, the plane of the polarization turns in the same sense as the circularly polarized wave, which travels with larger velocity. For a left-handed or levo (L) material, the sense of rotation of plane of polarization is clockwise when viewed by an observer facing the approaching light beam. (see Figure 5.5).

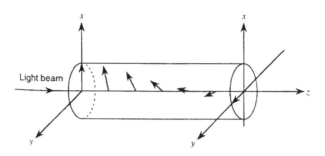

FIGURE 5.5
Illustration of the optical activity of a "levo" material.

There are materials that are both optically active and birefringent, for example quartz crystals, and chiral liquid crystals such as N*, SmC*, or twist grain boundary (TGB) phases. In the liquid crystal phases the optical activity in measurements can be neglected with respect the effect of the birefringence. However, there are a number of optically isotropic liquid crystal (OILC) materials[7] where the optical activity shows up without birefringence. Examples for these materials include the blue phases, which are formed by chiral molecules and first observed (but not identified) by Reinitzer in 1888.[8] These phases can be regarded as three-dimensional counterparts of the cholesteric phase, exhibiting cubic lattices without any long-range positional order.[9] Recently a smectic blue phase[10] was discovered as well, which exhibits quasi-long-range smectic order correlated with the three-dimensional orientation order.[11] Optically isotropic phases with tetrahedratic symmetry were also considered[12] and were predicted[13] to appear in liquid crystals of bent-core ("banana-shaped") molecules.[14,15] Experimentally, optical activity of a thin film can be observed by rotating one polarizer of the microscope from the crossed position. For a number of bent-core liquid crystals, it was observed that some domains of the transformed state became brighter on clockwise rotation, whereas others brighten under counterclockwise rotation due to the optical activity of the domains. The presence of optically active domains in bent-core liquid crystals were observed in a number of cases[16,17,18] and may indicate spontaneously twisted (propeller-type configurations) of the molecules.

5.2.2 Light Propagating through a Helical Medium

Description of the light propagating in a helical medium is much more complicated than in the uniform birefringent medium. A general treatment that can result in numerical solution was worked out by Berreman.[19] Here we describe only the theory restricted to light propagation along the helix axis, since in this case analytical solutions can be obtained. The subject has been treated by Maugin,[20] Oseen[21] and de Vries.[22] Originally the theory was developed for the cholesteric phase, but later it was shown by Parodi that it can be adopted to tilted chiral smectic materials, too.[23] In spite of the constraint that the light should propagate along the helix, it is useful, because this case is common in experiments, for example in free-standing smectic and soap films. Sometimes samples between glass plates also assume this configuration, if the plates are treated for homeotropic alignment.

The optical properties of cholesterics are discussed in several mono-graphs.[24] For simplicity, we describe the propagation of light in SmC* system, because this will also give the results valid for the cholesteric by setting the tilt angle to 90°. Just as we did considering the propagation of light for the uniform structure, we start out with the wave equation (5.16). We will use three coordinate systems to describe the helicoidal geometry (Figure 5.6). The first system is the x,y,z Cartezian laboratory frame, where the x and y axes are along the smectic layers, and z is along the helical axis. The second system, the χ,ψ,ζ-system, is twisted to match the helicoidal medium. The ζ axis coincides with the z axis. The angle g between the two coordinate systems depends on z as $\gamma = 2\pi z/L = qz$, where L is the helical pitch. In the following, we assume that the helix is right-handed, so that q and L are positive. The third, the χ',ψ',ζ' system makes the same spiral as the second system, but it is also locally tilted by an angle θ, which fixes this system to the director. In this local system, the dielectric tensor and its inverse are diagonal and have the same forms as (5.18).

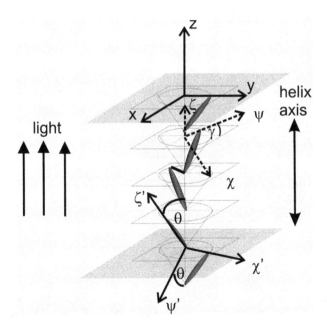

FIGURE 5.6
The coordinate systems used to describe propagation of light in helical medium.

Following the same process as for the uniform director structure described through (5.18)–(5.22), complicated by the fact that now we have to make transformations between two systems, we arrive[25] at an equation for the wave number k

$$k^4 - \left\{ (\varepsilon_{eo} + \varepsilon_2) \frac{\omega^2}{c^2} + 2q^2 \right\} k^2 + q^4 - (\varepsilon_{eo} + \varepsilon_2) q^2 \frac{\omega^2}{c^2} + \varepsilon_{eo} \varepsilon_2 \frac{\omega^4}{c^4} = 0 \quad (5.54)$$

where $\frac{1}{\varepsilon_{eo}} = \frac{1}{\varepsilon_1} \cos^2 \theta + \frac{1}{\varepsilon_3} \sin^2 \theta$. The four roots of the solution can be given as:

$$k = \pm \sqrt{ \frac{\varepsilon_{eo} + \varepsilon_2}{2} \frac{\omega^2}{c^2} + q^2 \pm \sqrt{ \left(\frac{\varepsilon_{eo} - \varepsilon_2}{2} \frac{\omega^2}{c^2} \right)^2 + 2(\varepsilon_{eo} + \varepsilon_2) q^2 \frac{\omega^2}{c^2} } } \quad (5.55)$$

Expressing k in terms of helical pitch L and the wavelength λ of light in vacuum, the solution with + signs will be called "regular" (k_r), because it travels in positive z direction for all values of L and λ. The solution with − sign in front of the first square root in (5.55) would give an equal wave propagating in the opposite direction. The other wave with a minus sign in front of the second square root describes a helical wave with the same handedness as of the helix of the material. It is "irregular" (k_{ir}), because it is imaginary for wavelengths:

$$\sqrt{\varepsilon_2} < \frac{\lambda}{L} < \sqrt{\varepsilon_{eo}} \quad (5.56)$$

This means that a band gap exists in the range of wavelengths $n_o p < \lambda_o < n_e p$ where n^2 is negative and the index is imaginary, so the propagation is forbidden. In a finite sample, this results in the reflectance of the circularly polarized light, if the handedness is the same as of the helix, and transmission for a light circularly polarized with opposite handedness. A typical reflectance spectrum of a circularly polarized light at normal incidence on a cholesteric liquid crystal with a helical axis parallel to the incoming light is shown in Figure 5.7.

We note that the iridescent colors of butterfly wings, fish scales and bird feathers originate in the interference of light after multiple reflections in the more or less regularly layered thin film structure of the material, which is a very similar phenomenon to that causing the color of the cholesteric liquid crystals.

In the reflection band, the nonpropagating modes are the so-called evanescent waves, whose electric vector is exponentially damped ($E \propto e^{-|k|z}$). We note that, in the infinite pitch limit, the regular wave transforms to the extraordinary wave, and the irregular wave turns into an ordinary one.

Outside the bandgap the eigenmodes become elliptically polarized, and the ellipticity increases as the pitch becomes tighter. For a right-handed

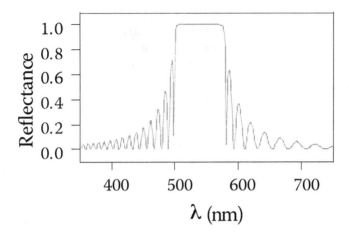

FIGURE 5.7
Typical reflection band of a mono-domain cholesteric liquid crystal at normal incidence.

medium the regular wave forms a left-handed screw, whereas the irregular wave is right-handed in the laboratory frame. One can show that, in the short pitch limit, i.e., far from the reflection band, the effective refractive indices of the regular as well as the irregular waves will be equal: $n_{eff,L\to0} = \sqrt{(n_{eo}^e + n_o^2)/2}$. In this limit the waves are circularly polarized.

5.3 Experimental Methods to Measure Refractive Index and Birefringence

The classical method to measure the index of refraction of transparent materials is based on total internal refraction. The instrument based on this method is the Abbe refractometer designed by Ernst Abbe in the early 1900s. This typically yields index accuracy of two units in the fourth decimal place, but it works only for materials where the refractive index is smaller than the glass prism (about n = 1.89). To measure anisotropic fluids with the Abbe refractometer, the material has to be uniformly (usually homeotropically) aligned between the prisms. Ordinary and extraordinary rays can be selected by using polarization filters polarized perpendicular and along the optic axis, respectively.

To measure the refractive indices of anisotropic fluids, one also can take advantage of crystal structures that uniformly aligned can be achieved in wedge-shape cells. Illuminating the cell with a laser light, the direction of the emerging light will deflect from the normally incident incoming light by an angle θ. Applying Snell's law, one can see from Figure 5.8a that $n \cdot \sin\alpha = 1 \cdot \sin\theta$. For the real wedges $\sin\alpha \sim \alpha$ and $\sin\theta \sim \theta$, which gives that the deviation from the original direction $\beta = \theta - \alpha = (n-1)\alpha$, where α is the wedge angle. This is valid for homeotropic alignment, where $n = n_o$. For planar alignment, the light splits into two beams, with directions $\beta_o = (n_o - 1)\alpha$ and $\beta_e = (n_e - 1)\alpha$. Note: the glass plate separating the liquid crystal film and the air will shift the beam, but would not change the direction. Measuring the laser spots on a screen far enough from the sample, one may resolve the difference between the extraordinary and ordinary beams, and with a polarizer one can identify which one is the extraordinary and ordinary spot. In this way, one can measure the extraordinary and ordinary indices by about 0.1–1% precision.

On wedge cells one also can measure the birefringence by watching the film in transmission under monochromatic plane wave illumination. In this case, fringes appear in the transmitted image, as illustrated in Figure 5.8b. It is due to the phase difference between the extraordinary and ordinary waves described by the retardation: $\Gamma = 2\pi\Delta nd/\lambda$.

Dark fringes appear when $\Gamma = \pi \cdot m$ (m is integer), i.e., where $\frac{\Delta n d_1}{\lambda} = m$, and $\frac{\Delta n d_2}{\lambda} = m+1$ with $d_2 - d_1 = \Delta d$, giving that $\frac{\Delta n \cdot \Delta d}{\lambda} = 1$, i.e. $\Delta n = \frac{\lambda}{\Delta d}$. The lateral distance between two neighbor fringes is Δx. From Figure 5.8 we can see that $\Delta x = \frac{\Delta d}{d} \cdot x$, which can be related to the birefringence as: $\Delta x = \frac{\lambda x}{d \Delta n}$.

Basically, the same principle can be used to measure Δn in samples of uniform thickness by observing the transmitted spectra under white light illumination. The transmitted intensity between crossed polarizers for a given wavelength λ_o becomes zero for $\Gamma/2 = \pi \bmod \pi$, which gives $\Delta n = l\lambda_o/d$ where $l = 1, 2, \ldots$. The measurements can be conveniently done by placing UV–VIS spectrometer (e.g., from OceanOptics) in the eye-piece of a polarizing

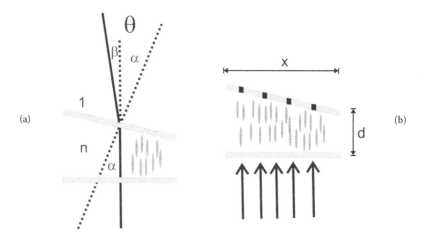

FIGURE 5.8
Experimental setups to measure birefringence using wedge cell. (a) Refraction of laser light beam. (b) Formation of fringes under monochromatic plane wave illumination.

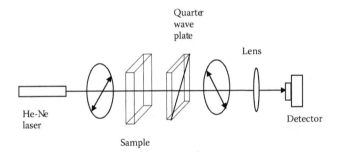

FIGURE 5.9
Senarmont technique to measure birefringence.

microscope. For $\Delta n < 0.2$, which is typical for liquid crystals, in films with thickness of $d \sim 5$ µm, λ_o is in the visible range for $l = 1$.

A more complicated but much more sensitive method of measuring birefringence is based on the Senarmont technique, which is illustrated in Figure 5.9. The light exiting the anisotropic material is generally elliptically polarized; however, a $\lambda/4$ plate converts the light back to a linearly polarized ray, characterized by a polarization direction making an angle θ with respect to $\bar{n}$. This angle θ is related to the retardation as $\cos\theta = \cos\frac{\Gamma}{2}$ and can be found by rotating the analyzer to extinction. Knowing the sample thickness and wavelength used for illuminations, we therefore can calculate Δn from θ.

All the above measurement techniques require uniform alignment. However, the birefringence and much more information can also be obtained by polarizing microscopy (PM). Furnished with the theories explained above, one can easily distinguish phases and judge alignments and molecular structures by simply looking at the polarizing microscopic textures. PM is traditionally the best established method of studying liquid crystals, and in fact, it started with Otto Lehman in 1890.[26] Although the smallest objects that can be resolved in PM are only in the range of 1 µm, which is about 500 times larger than the length of typical liquid crystalline molecules, we can see the defects (details in Chapter 6), which are characteristic of the phases. The schematics of a polarizing microscope are shown in Figure 5.10.

In PM the textures are mostly observed in transmission under collimated (parallel) illuminated light beams, where the light enters the film normal to its surfaces (orthoconic observations). The microscope in orthoconic observations gives a magnified, inverted 2D image of the texture.

PM observations can also be carried out in reflection, which is useful when the material is reflecting (e.g., some cholesterics), or opaque (e.g., polymer dispersed liquid crystals), or if the material is placed on a mirror.

Polarizing microscopes can be also used to determine the magnitude and the sign of birefringence.

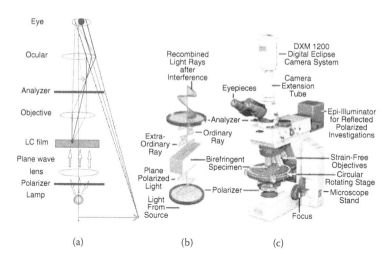

FIGURE 5.10
Principle of polarizing microscope. (a) Main elements with the sketch of the light path. (b) Illustration of the path of the polarized light through the liquid crystal wave plate. (c) A picture of a real polarizing microscope with the main elements (pictures b and c are slightly modified version of pictures in http://www.microscopyu.com).

The sign of the birefringence can be tested by inserting a retardation plate with known slow and fast directions (e.g., Gypsum plate) between the objective and analyzer. If the slow axis of the birefringent liquid crystal plate is parallel to the slow axis of the Gypsum plate, the birefringence increases. If the slow axis of the LC is perpendicular to the slow axis of the Gypsum plate, the resulting birefringence decreases. Most polarizing microscopes come with a wave plate (labeled by λ) oriented 45° away from the crossed polarizers, i.e., more or less along the optic axis if the birefringent film appears to be bright between crossed polarizers. In this case, the conoscopic quadrants become, for example, bluish and yellowish. For uniaxially positive materials the direction of the slow axis bisects the blue quadrants, and for negative uniaxial sample the slow axis bisects the yellow quadrant.

Alternatively, the addition of the wave plate allows mapping the director alignment across the film, if we know the sign of the birefringence of the material. This is very important in the case of liquid crystals, where the director does not necessarily align uniformly (e.g., parallel to the rubbing direction of the aligning polymer surface). Such situation is quite common in smectics, columnars and liquid crystals of biological importance.

The magnitude of the birefringence of an isotropic material can be estimated when observed and/or photographed in a polarized light microscope. A relationship between interference color and retardation can be graphically illustrated in the classical Michel–Levy interference color chart, presented in Figure 5.11. This graph plots retardation on the abscissa and specimen thickness on the ordinate. Birefringence is determined by a family of lines that emanate radially from the origin, each with a different measured value of

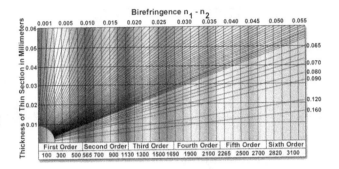

FIGURE 5.11
Michael–Levi chart for estimating the birefringence of a birefringent film of known thickness.

birefringence corresponding to thickness and interference color. The Michel–Levy chart is utilized by comparing the highest-order interference colors displayed by the specimen in the microscope to those contained on the chart. Once the appropriate color has been located, the nearest vertical line along the interference color is followed to the nearest horizontal line representing the known thickness. Birefringence is determined by selecting the diagonal line crossing the ordinate at the intersection of the specimen interference color and thickness value.

In contrast to the orthoconic observations, where we see the 2D image of the sample, in conoscopy[27] the transmission properties of birefringent films are tested by looking through the samples simultaneously at a wide range of incidences. This is achieved by converging the parallel incident light to a focus (see Figure 5.12).

To visualize that picture, we simply need to remove the eyepiece, or to introduce a so-called Bertrand Lens above the analyzer. In this case an interference pattern appears which divides the field of view to four quadrants by the so-called Maltese crosses (Figure 5.13). For uniaxial materials the arms of the cross (so-called *isogyres*) are parallel to the polarizers, and the center of the pattern is dark.

In addition to the cross, one may see one or more colored (dark for monochromatic light) rings. This can be explained as the more light is inclined, the larger is the distance d it travels in the sample. For monochromatic light, there will be a dark ring when the condition $\Delta nd = l\lambda$ is fulfilled. These rings are often not seen in thin samples, since the first ring will be located outside the aperture of the optical system. When the birefringent film is illuminated by white light, the color of the rings may be regarded as white light minus those colors that are interfering destructively. The inner circles, called **isochromes**, consist of increasingly lower-order colors (see the Michel–Levy interference color chart in Figure 5.11). A common center for both the black cross and the isochromes is termed the *melatope*, which denotes the origin of the light rays traveling along the optical axis of the crystal.

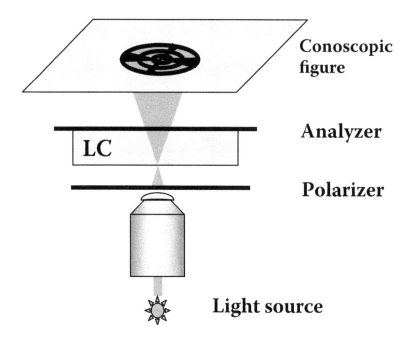

FIGURE 5.12
Schematics of conoscopy (illustration is courtesy of Oleg Pishnyak).

Biaxial materials display two melatopes (Figure 5.13c) and a far more complex pattern of interference rings.[28]

Conoscopy is used to study the field-induced effects in nematic[29] and smectic ferroelectrics.[30] It is most useful when one has a large domain with uniform director orientation without any structure in the lateral direction. For example, with orthoscopy one cannot distinguish between truly and quasi-isotropic (homeotropic) situations. However, with conoscopy one can easily see the difference: in the case of isotropic director structure one sees only

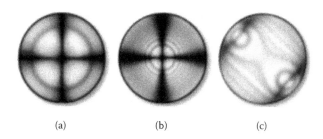

(a)   (b)   (c)

FIGURE 5.13
Conoscopic images under white light illumination. (a and b): Uniaxial material: an interference pattern consisting of two intersecting black bars (termed isogyres) forming Maltese cross-like pattern and circular distributions of interference colors; (c) biaxial material.

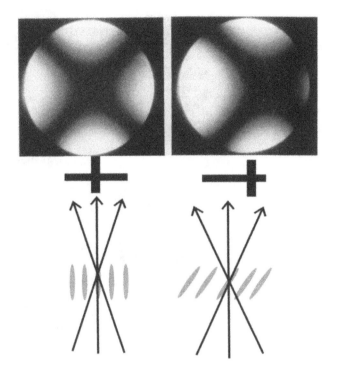

FIGURE 5.14

Illustration of the difference in the conoscopic images for homeotropic and tilted director structures. The images are taken on a free-standing film of a liquid crystal mixture ZKS 2504 in the SmA and SmC phases, respectively.[25]

a uniformly dark picture, whereas for homeotropic uniaxial geometry a dark cross is seen in gray background (see Figure 5.14). The melatope is off centered for a tilted director structure, such as SmC in homeotropic geometry, as illustrated in Figure 5.14.

PM images bear two-dimensional information, integrating the 3D pattern of optical birefringence over the path of light. To obtain 3D director patterns we may use fluorescent confocal polarizing microscopy (FCPM), which is illustrated in Figure 5.15.[31,32]

We employ the property of anisotropic fluids to orient fluorescent dye molecules (for example, N,N′-bis(2,5-di-tert-butylphenyl)-3,4,9,10-perylenedicarboximide), which are dissolved in small quantities in the "host" material.

Light excites dye molecules, which emit fluorescent signal. The amplitude of this fluorescence signal is maximal when the linear polarization of the light is parallel to the director $\vec{n}$ ($\vec{P} \parallel \vec{n}$), and is minimal if $\vec{P}$ is perpendicular to $\vec{n}$. For an intermediate angle α between $\vec{P}$ and $\vec{n}$, the intensity of light is also intermediate, $I \sim \cos^4 \alpha$, provided the efficiency of both absorption and

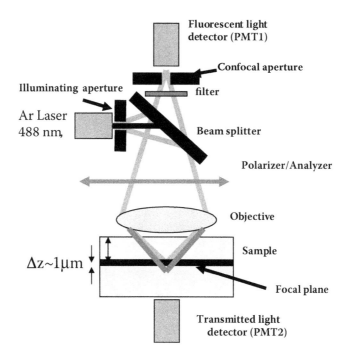

Fluorescent light detector (PMT1)

Confocal aperture

Illuminating aperture

filter

Ar Laser
488 nm,

Beam splitter

Polarizer/Analyzer

Objective

Sample

$\Delta z \sim 1\mu m$

Focal plane

Transmitted light detector (PMT2)

FIGURE 5.15
Illustration of the fluorescent confocal polarizing microscopy (courtesy of O. Lavrentovich).

fluorescence is proportional to the scalar product of $\vec{n}$ and $\vec{P}$. In polarized light, the measured fluorescence signal is determined by the spatial orientation of the molecules; confocal mode of light illumination and fluorescence detection allows optical slicing in both horizontal and vertical planes.

References

1. D. Dunmur, P. Palffy-Muhoray, *J. Phys. Chem.*, 92, 1406 (1988).
2. M. Schadt, W. Helfrich, *Mol. Cryst. Liq. Cryst.*, 17, 355 (1972); H.J. Coles, R.R. Jennings, *Mol. Phys.*, 31, 571 (1976); B.R. Ratna, M.S. Vijava, R. Sashidar, B.K.J. Sadashiva, *Pramana, Suppl.*, 1, 69 (1975).
3. E. Bartholinus, Experimenta crystalli Islandici disdiaclastici (1670), reprinted in *Polarized Light*, Ed. W. Swindell, Dowden, Hutchingson & Ross, Inc., Stroudsburg, PA (distributed by Halsted Press-John Wiley & Sons) (1975).
4. C. Huygens, "Traite de la lumiere" (1690), reprinted in *Polarized Light*, Ed. W. Swindell, Dowden, Hutchingson & Ross, Inc., Stroudsburg, PA (distributed by Halsted Press-John Wiley & Sons) (1975).
5. R.C. Jones, *J. Opt. Soc. A*, 31, 488 (1941).
6. R.C. Jones, *J. Opt. Soc. A*, 31, 488 (1941).

7. S. Diele, P. Göring, *Handbook of Liquid Crystals*, Vol. II(B), p. 887, Ed. D. Demus, J. Goodby, G.W. Gray, H.-W. Spiess, V. Vill (1998).

8. F. Reinitzer, *Monatsh. Chem.*, 9, 421 (1888).

9. P.P. Crooker, in *Chirality in Liquid Crystals*, Ed. H.S. Kitzerow, C. Bahr Springer-Verlag, New York (2001).

10. M.H. Li, H.T. Nguyen, G. Sigaud, *Liq. Cryst.*, 20, 361 (1996); M.H. Li, V. Laux, H.T. Nguyen, G. Sigaud, P. Barois, N. Isaert, *Liq. Cryst.*, 23, 389 (1997).

11. B.A. DiDonna, R.D. Kamien, *Phys. Rev. Lett.*, 89, 215504 (2002).

12. L.G. Fel, *Phys. Rev. E*, 52, 702 (1995).

13. L. Radzihovsky, T.C. Lubensky, *Europhys. Lett.*, 54, 206 (2001).

14. D. Vorländer, *Z. Phys. Chem., Stoechiom. Verwandtschaftsl*, 105, 211 (1923).

15. Y. Matsunaga, S. Miyamoto, *Mol. Cryst. Liq. Cryst.*, 237, 311 (1993); H. Matsuzaki, Y. Matsunaga, *Liq. Cryst.*, 14, 105 (1993).

16. J. Thisayukta, Y. Nakayama, S. Kowauchi, H. Takezoe, J. Watanabe, *J. Am. Chem. Soc.*, 122, 7441 (2000).

17. G. Dantlgraber, A. Eremin, S. Diele, A. Hauser, H. Kresse, G. Pelzl, C. Tschierske, *Angew. Chem.*, 41, 2408 (2002).

18. A. Jákli, Y.M. Huang, K. Fodor-Csorba, A. Vajda, G. Galli, S. Diele, G. Pelzl, *Adv. Mater.*, 15(19), 1606 (2003).

19. D.W. Berreman, T.J. Scheffer, *Phys. Rev. Lett.*, 25, 577 (1970); D.W. Berreman, *J. Opt. Soc. of America*, 62, 502 (1972).

20. C. Maugin, *Bull. Soc. Franc. Miner.*, 34(6), 71 (1911).

21. C.W. Oseen, *Trans. Faraday Soc.*, 29, 883 (1933).

22. H.I. de Vries, *Acta. Cryst.*, 4, 219 (1951).

23. O. Parodi, *J. de Physique*, 36, C1–325 (1975).

24. P.G. de Gennes, *The Physics of Liquid Crystals*, Clarendon Press, Oxford (1974); V.A. Belyakov, *Diffraction Optics of Complex-Structured Periodic Media*, Springer-Verlag, Berlin (1992).

25. A. Dahlgren, Optical investigations of chiral liquid crystals, Ph.D. thesis, Chalmers University of Technology, Göteborg, Sweden (2000).

26. O. Lehmann, *Z. Phys. Chem.*, 4, 462 (1889).

27. M.C. Mauguin, *Bull. Soc. Franc. Miner.*, 34, 71 (1911).

28. L. Yu, A. Saupe, *Phys. Rev. Lett.*, 45, 1000 (1980).

29. P.G. de Gennes, J. Prost, *The Physics of Liquid Crystals* Clarendon Press, Oxford (1993); F. Brochard et al., *Phys. Rev. Lett.*, 28, 1681 (1972).

30. E. Gorecka et al., *Jpn. J. Appl. Phys.*, 29, 131 (1990).

31. O.D. Lavrentovich, *Pramana*, 61(2), 373 (2003).

32. I.I. Smalyukh, S.V. Shiyanovskii, O.D. Lavrentovich, Three-dimensional imaging of orientational order by fluorescence confocal polarizing microscopy, *Chem. Phys. Lett.*, 336, 88 (2001).

6

Defect Structures

In the examples of the previous chapter we assumed that the surfaces impose uniform director alignment in the entire sample. Uniform alignment results in uniform pictures when observing the sample between crossed polarizers. For example, uniaxial samples with uniform homeotropic alignment would appear black between crossed polarizers, whereas those with uniform planar alignment appear bright when the directions of the crossed polarizers do not coincide with the optical axis. This is not the case for chiral materials, where the director has a tendency of forming a helical structure, which competes with the uniform surface alignment and results in defects. In addition, equivalent conditions for different director structures or impurities make uniform alignment exceptional and the director field is inhomogeneous unless special treatment is applied. Because the director field determines the local optical properties, we usually observe a wide variety of visual patterns. These patterns are due almost entirely to the defect structures that occur in the long-range molecular order of the anisotropic fluids. Indeed, historically the underlying structures of the nematic and smectic-A liquid crystal phases were proposed based on the stable structures of the defects that characterize these phases.

Defects are areas where the director structure changes abruptly. They are, in general, present even in thin films where the surface alignments cannot produce uniform director structures, or when they are competing with each other or with alignment given by some bulk fields, or when impurities or dispersions impose defects. The detailed description of defects with their structures, energies and dynamical behavior requires knowledge beyond the scope of this book. In this book we will give only a flavor to this subject, to the extent that it will be needed to understand the macroscopic physical behavior of anisotropic fluids.

Study of defects has importance ranging from nanoscience through display industry and biology, and even to cosmology.[1] For example, cosmological defects are virtually impossible to produce artificially even in the most powerful particle accelerators. However, it is proposed by Zurek and Turok[1] that the same physics governs the formation of defects in cosmology and in nematic liquid crystals, only the scales are different. Consequently we can learn about the cosmological defects if we study the topological defects of nematic liquid crystals. On the other hand, this is one of the reasons people are interested in them — if they can be found today; they will be a unique direct link to the physics of the first moments of the universe.

For those interested in details of this beautiful subject we suggest reading first the book of M. Kleman,[2] and some excellent reviews of earlier[3,4] and more recent developments are found in references.[5,6,7] For detailed discussion of textures of liquid crystals, see books by Demus,[8] Gray,[9] and Dierking.[10]

6.1 Nematic Liquid Crystals

One well-known characteristic feature of nematic liquid crystals is the thread-like texture that can be observed with a polarizing microscope. The name nematic, derived from the Greek word "thread," reflects that feature. By examining the thin and thick thread-like structures in nematic liquid crystals, Otto Lehman[11] and Georges Friedel[12] deduced that this phase involves long-range orientational order. The first step to the interpretation of the threads as disclinations of the director field has been made by Oseen.[13] Later Frank[14] derived Oseen's theory of curvature elasticity on a more general basis and presented it in a simpler form (see Appendix C.1).

As we understand now, the line defects of liquid crystals are due to discontinuities in the "inclination" of the molecules and are referred to as "disclinations." The inclinations of the molecules are described by the director field $\vec{n}$, which is the average direction of the long axes of the rod-like molecules. Because the director may vary from place to place in a sample, we write it as the function of the position: $\vec{n}(\vec{r})$. At the defects the director cannot be defined uniquely. There are two main classes of defects, or singularities: line and point defects. The line defects are commonly discussed in terms of their behavior in the plane normal to the line. A defect is said to be of strength S if, on moving around a closed path in that plane, the director rotates by S multiples of 2π. Because $\vec{n}$ and $-\vec{n}$ are equivalent, half integer values of S are allowed. Frank has called such half integer defects "Möbius defects," because they have the same topology as Möbius strips.

Mathematically, the director field in the XY plane can be described by the angle ϕ as:

$$\vec{n}(x,y) = (\cos\phi, \sin\phi, 0) \qquad (6.1)$$

It can be shown[15] that the director-field around the defects correlates to the electro- and magnetostatic fields in two dimensions, and to the flow of incompressible liquids, so ϕ can be given as:

$$\phi_p = S\alpha_p + \psi_p \qquad (6.2)$$

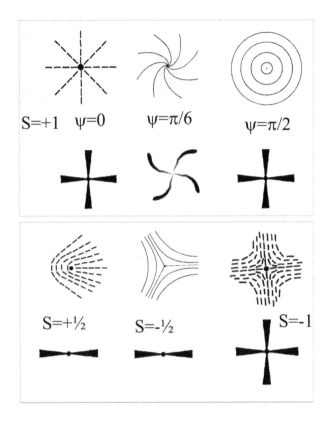

FIGURE 6.1
The director-field of threads normal to the plane of the figure in two-dimensional approximation
(upper rows) and the texture around the defects seen between crossed (vertical and horizontal)
polarizers (lower rows).

where $S = \pm\frac{1}{2}, \pm 1, \pm\frac{3}{2}, ...,$ and $\alpha_p = i \tan^{-1}\frac{y-y_p}{x-x_p}$. The singularity is positioned
at x_p and y_p, and ψ_p is an additive constant. The director-fields around line
singularities in the plane normal to the lines and their appearance between
crossed polarizers are shown in Figure 6.1. Note that the textures around
$S = +1$ and $S = -1$ defects look the same (dark crosses) between crossed
polarizers. They can be distinguished by slightly rotating the analyzer or
polarizer, because the dark stripe rotate in the opposite sense for the defects
of opposite sign.

Vertical line defects can be induced, for example, in samples with not
perfect homeotropic anchoring (see Figure 6.2a). Lines parallel to the film
surfaces are usually present spontaneously in films with planar surface
anchoring when the sample is cooled rapidly from the isotropic phase (see
picture Figure 6.2b).

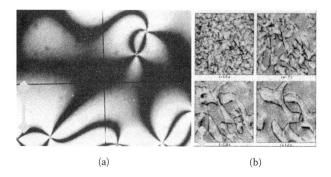

(a)   (b)

FIGURE 6.2

Vertical and horizontal line defects in samples. (a) Schlieren texture of a 70-μm sample of MBBA (N-(p-methoxy-benzylidene)-p-n-butylaniline) at 20°C in homeotropic anchoring. To induce the vertical line defects an electric field of 6 Vrms 300 Hz applied vertically. (b) Horizontal line defects appeared spontaneously in planar cell on cooling from the isotropic phase. The subsequent photos illustrate the evolution of a string network in a liquid crystal. The four snapshots have the same size, but were obtained at different times. Notice the progressive dilution of the string network (picture from http://www.damtp.cam.ac.uk/user/gr/public/cs_phase.html). Bar: 100 μm.

The disclinations interact with each other through forces similar to the magnetic interactions between currents. Specifically, the interaction between two parallel threads p and q at $\vec{r}_p$ and $\vec{r}_q$ is given[16] by:

$$\vec{F}_{p-q} = -2\pi K S_p S_q \frac{\vec{r}_{p-q}}{r_{p-q}^2} \tag{6.3}$$

where $\vec{r}_{p-q} = \vec{r}_p - \vec{r}_q$. Disclinations of opposite signs attract each other. They can merge and disappear or form a new disclination with a characteristic S number equal to the sum of the S of the merging threads.

The structure of the cores of the disclinations has attracted considerable interest and has been carefully studied for the S = +1 case.[16] A three-dimensional structure can be described by a z component of the director field $\cos\theta$ in the core that goes to zero ($\theta = \pi/2$) outside the core reproducing the two-dimensional director structure given in (6.2). It is obtained that:

$$\tan\frac{\theta}{2} = \left(\frac{r}{r_{\pi/2}}\right)^{|S|} \tag{6.4}$$

where $r_{\pi/2}$ denotes the core radius. Optical observations show that the core of the defect lines can be quite large. The reason for this is that defect lines with integer S values can "escape into the third dimension," that is, the director can simply rotate to lie along the line in the central region, obviating the need for core.[17]

Point defects are another class of defects, which can be associated with line defects and usually seen in capillary tubes, where the boundary conditions

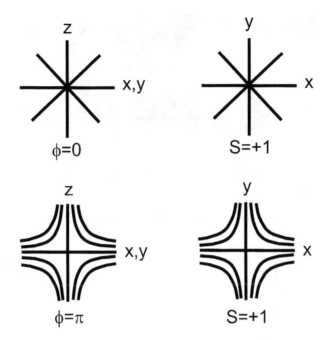

FIGURE 6.3
Director-field around singular points.

are radial at the surface and where the director escapes in opposite directions in different parts of the capillary.

The director structures around singular points associated with $S = +1$ and $S = -1$ are sketched in Figure 6.3.

The polarizing optical texture and the corresponding director structures in a capillary with homeotropic boundary conditions are illustrated in Figure 6.4. Note that the disclination line in the middle along the capillary axis is due to the escape to the third direction. The vertical lines indicate regions where the different escape directions meet. And the crosses are point defects.

Study of defect structures is also an important tool to distinguish between uniaxial and biaxial nematic (N_b) phase structures.[18] Theoretical studies have shown that the defect of the N_b phase has unusual properties related to the non-Abelian topology.[19] In principle, the N_b phase exhibits three types of $1/2$ disclinations (of both signs) corresponding to each director, and one $S = 1$ disclination (see Figure 6.5d). Because rotations of the different axes do not communicate, the $1/2$ disclinations corresponding to different directors cannot cross themselves.[20] Since the textures of the original biaxial nematic phase did not show any integer-numbered singularities,[18] it is often argued[21] that the N_b textures consist of only $1/2$ defects (see texture in Figure 6.6). In any case, it is generally believed that defect structures alone are not sufficient to distinguish between biaxial and nematic phases.

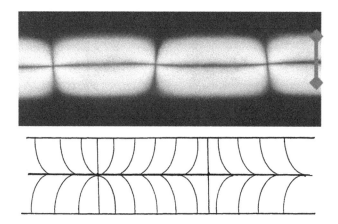

FIGURE 6.4
Point singularities in a capillary. The points are arranged along the center and alternate between type π, 1 and 0, 1 arrangements of Figure 6.3. Bar indicates 150 μm. (Ph.D. thesis of Jiming Liu, Kent, 1991).

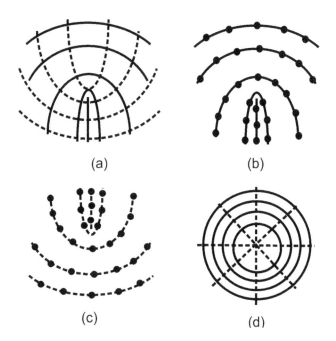

FIGURE 6.5
Representatives of four distinct classes of line defects in biaxial nematics. The solid and dashed lines represent two director fields, n_1 and n_2.[23] (a) a $1/2$ disclination in both n_1 and n_2; (b) a $1/2$ disclination in n_1 and no disclination in n_2; (c) a $1/2$ disclination in n_2 and uniform in n_1; (d) S = 1 disclinations with $\psi = 0$ in n_1 and with $\psi = \frac{\pi}{2}$ in n_2.

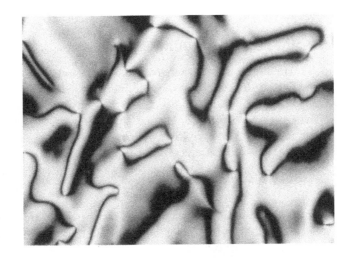

FIGURE 6.6
Schlieren texture of N,N'-bis (2,3,4-trido decyloxy benzylidene)p terphenyl diamine.[22] Note that only two-brush disclinations are seen.

Textures of the N_b phase in lyotropic liquid crystals have been studied in capillaries.[22] An example of the textures and the corresponding director-field is shown in Figure 6.7. In comparing with Figure 6.4, we note that the center defect line disappears, as a consequence of the biaxiality; although n_1 escapes along the capillary axis (and would be dark in uniaxial system between one of the crossed polarizers set parallel to the capillary axis), n_2 will have no extinction there.

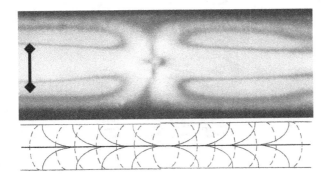

FIGURE 6.7
Texture of KL (potassium laurate) in cylindrical capillary and the corresponding director structure (solid lines and dashed lines correspond to n_1 and n_2, respectively). Bar: 150 μm (Ph.D. thesis of Jiming Liu, Kent, 1991).

6.2 Smectic and Lamellar Liquid Crystals

In contrast to nematic liquid crystals, where the defects can be rigorously classified by their topology,[23] defects of quasi-two-dimensional fluids such as smectic and lamellar structures rather can be described by using geometrical arguments.[24] The reason for this is the layered structure with the requirement of the equidistant layer spacing. In macroscopic defects, the layer spacing must be kept constant. This tells us some geometrical characteristics of the possible defects, because it implies that neither twist ($\vec{n} \cdot \nabla \times \vec{n}$) nor bend [$\vec{n} \times (\nabla \times \vec{n})$] distortions are allowed. Thus, textures in smectic A phase consist of pure splay ($\nabla \cdot \vec{n} \neq 0$).

A defect, or singularity, with pure splay the director should point radially outward (Figure 6.8c). Both a line disclination and a point defect can have this property. It is important to note, however, that disclination of strength 1 does not escape in smectic A structures.[24] Singularities in term of equally spaced surfaces are known as "Cyclides of Dupin." Parts of this family constitute the so-called focal conic domains, investigated by many authors, starting with G. Friedel in 1922[12] and Bouligand.[25] A more recent and detailed description of the focal conic defects is given by Boltenhagen et al.[26] All focal conic singularities are generalizations of the pair of defect lines that one gets

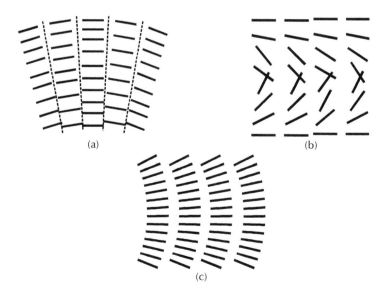

FIGURE 6.8

Illustration of deformation of layered structures. (a) Layer splay (director bend); (b) twist; (c) layer bend (director splay). It can be seen that only the layer bend leaves the layer spacing constant.

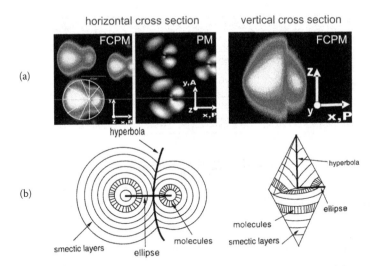

FIGURE 6.9
Three-dimensional Fluorescent Confocal Polarizing Microscopic (FCPM) pictures of focal conic defects of SmA phase with homeotropic alignment (upper row), and their graphical illustration (bottom row). (a) Horizontal cross section (along the substrates); (b) vertical cross section.

by taking equally spaced SmA layers, wrapping them into concentric cylinders, bending the result into a torus, and then adding more and more equally spaced layers, going beyond the stage where the hole of the torus disappears. One of the resulting defect lines is the circular axis of the torus, whereas the other one is the line normal to the plane of the circle, passing through its center. In the focal conic generalization of this texture, the circle becomes an ellipse, and the straight line a hyperbola that shares the focus with the ellipse. (see Figure 6.9).

Textures of SmA samples with focal conic defects are shown in Figure 6.10.

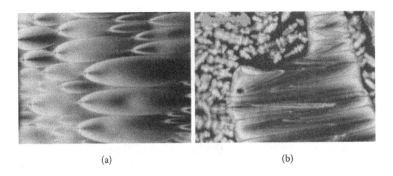

FIGURE 6.10
SmA with focal conics. (a) Photo courtesy of O. Lavrentovich; (b) focal conic texture coexisting with "Batonnets" when SmA forms directly from the isotropic phase. Bar: 100 μm.

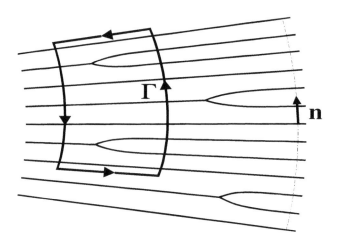

FIGURE 6.11

Illustration of edge dislocations induced by a bend stress, due to constant layer thickness requirement. A closed loop Γ enclosing two elementary edge dislocations in an area of bend in the director (100μm).

In addition to focal conic singularities, smectic liquid crystals exhibit edge dislocations in the layer spacing, i.e., a layer can be added that does not extend throughout the entire material (Figure 6.11).[27] Such edge dislocations are difficult to observe, because the layers are typically 3 nm in thickness. The introduction of edge dislocations allows some bend in a texture if the boundary conditions require it.

Integrating the director field along a loop Γ that contains edge dislocations we get:

$$\oint_\Gamma \vec{n} \cdot d\vec{r} = \oiint_\Omega (\vec{\nabla} \times \vec{n}) \cdot d\vec{\Omega} \neq 0 \qquad (6.5)$$

Here Ω is the area enclosed by the loop, and we utilized the Stokes theorem to see that, in the area of dislocations $\vec{\nabla} \times \vec{n} \neq 0$, i.e., bend ($(\vec{\nabla} \times \vec{n}) \times \vec{n}$) and twist ($(\vec{\nabla} \times \vec{n}) \cdot \vec{n}$) deformations are possible. Parabolic focalconic arrays induced by dilation of layers have importance in lyotropic lamellar systems, too.[28]

Focal conic domains also appear in smectic C phases. Their structure was extensively studied, especially in the ferroelectric SmC* materials,[29,30] where, due to competition between the uniform surface anchoring and the helical structure inside the bulk, periodic arrays of defect lines (so-called unwinding lines) form (Figure 6.12) near the surfaces.[32] The distance between the lines

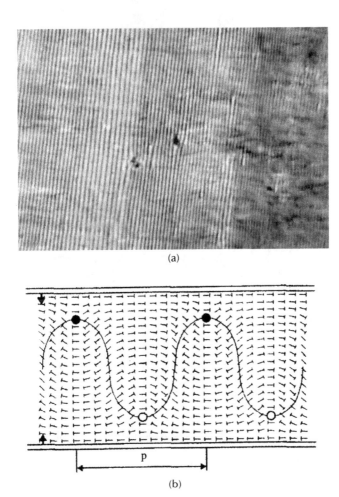

(a)

(b)

FIGURE 6.12

Texture and the proposed director structure[31] in films with planar anchoring. The helical pitch p is smaller than the film thickness L. (a) Texture of a 4-(2'-metilbutyloxy)-fenil-4-alkyloxy-benzoate film at room temperature. p = 5 μm, L = 20 μm (picture from A. Jákli, Ph.D. thesis, Budapest, 1985). (b) Illustration of the cross section of the director structure. The dots illustrate the unwinding lines running about p/2 distances from the surfaces. The nails indicate the director pointing outward from the plane of the drawing. The arrows show the direction of the ferroelectric polarization.

is equal to that of the helical pitch, and they are approximately of half a pitch distance from the substrates.

In case of strong planar and azimuthal anchoring when the director is fixed at the surfaces, typically a so-called chevron texture forms, where the layers have a kink in the middle of cells. The formation of the chevron layer structure leads to defects, where regions of opposite kink directions meet.

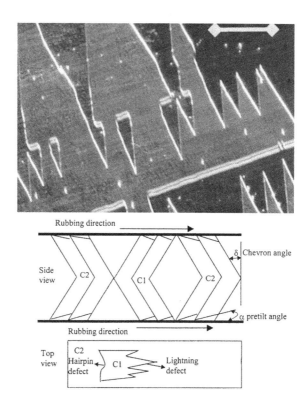

FIGURE 6.13

Typical texture of surface stabilized SmC* film with chevron domains (from http://www.kth.se/fakulteter/tfy/kmf/lcd/lcd~1.htm). Bar length: 25 μm. Illustration of the layer and director structure. (Illustration is adapted from the Ph.D. dissertation of Chenghui Wang, Kent, 2003.)

Typical chevron texture with the widely accepted layer and director structures are illustrated in Figure 6.13.

Especially peculiar and complicated structures can form in liquid crystal of bent-core molecules.

In the tilted polar smectic (SmCP) phase, the texture depends on the overall chirality, i.e., whether the director field is synclinic or anticlinic (see a few examples in Figure 6.14).

Note the presence of fan-shaped domains, with crosses either parallel or obliquely with respect to the polarizers. The oblique crosses indicate tilt of the optical axis with respect to the layer normal by the director tilt angle, θ. This is explained in Figure 6.15. Also note that the synclinic SmA and SmC_s textures have higher birefringence than the anticlinic SmC_a texture. This can be followed on Figure 6.14, where the green areas (4-μm thick cells)

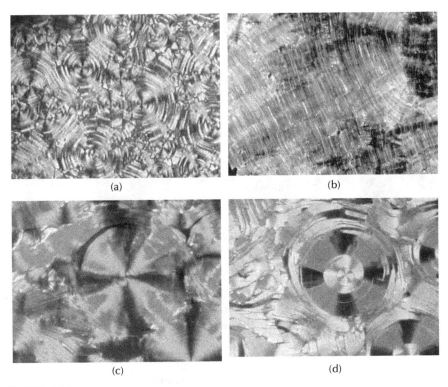

(a)　　　　　　　　　　(b)

(c)　　　　　　　　　　(d)

FIGURE 6.14

Textures in the SmCP phase. (a) 4-Chloro-1,3-phenylene bis[4-4(4-octyloxy phenyliminomethyl) benzoates[32] in the chiral state; (b) the same material in the racemic state (both at 125°C); (c) a sulphur-containing material B-10(S)[33] at 120°C and at 130°C (d). The domains with red color correspond to anticlinic structure, whereas those with green color have synclinic structure. Bar: 100 μm.

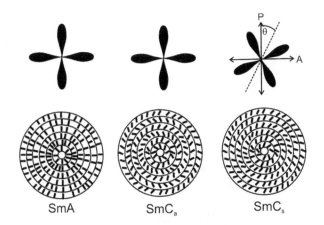

SmA　　　　SmC$_a$　　　　SmC$_s$

FIGURE 6.15

Illustration how the Maltese crosses observed between vertical and horizontal crossed polarizers can be related to different director structures. Although by the direction of the crosses SmA and SmC$_a$ are indistinguishable, their birefringent colors are different.

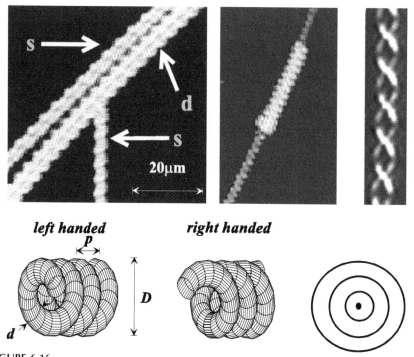

FIGURE 6.16

Textures of helical filaments forming on cooling from the isotropic phase[36] (upper row) and the proposed telephone cord structure consisting of smectic layers.

correspond to the high birefringence, and the red areas have low birefringence due to the anticlinic structure.

Even more astonishing textures appear in those phases (so-called B_7 structures), which form peculiar helical superstructures, including helical filaments.[34] Typical helical filaments with a propsed telephone-wire-type structure, where the smectic layers form concentric cylinders, are shown in Figure 6.16.

Another type of interpretation of the textures was proposed by Nastishin et al.,[36] who suggested that the B_7 phase is a smectic and columnar phase at the same time. The geometry of the helical filaments (ribbons) is that one of the central region of a screw disclination with a giant Burgers vector split into two disclination lines of strength $1/2$, which bound the ribbon.

The spontaneous chirality that results in the telephone wire type textures at the transition persists in the fully formed B_7 phase, resulting in various astonishing helical superstructures,[34] as illustrated in Figure 6.17.

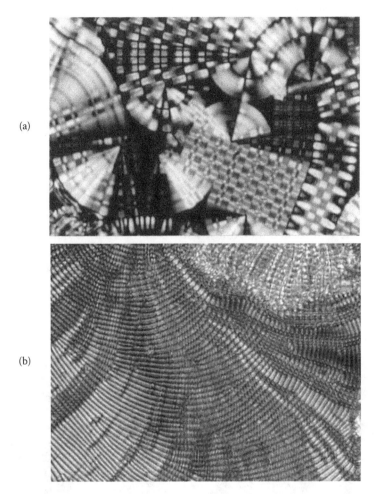

(a)

(b)

FIGURE 6.17
Helical superstructures in the B_7 phases. (a) 2-Nitro-1,3-phenylene bis[4-(4-n-alkyloxyphenylim-inomethyl)benzoates];[34] (b) of a binary mixture.[37]

6.3 Cholesteric Liquid Crystals

The characteristic feature by which cholesterics differ from the nematics is the spontaneous formation of twisted structures, reflecting the existence of a preferred screw sense. For this reason, the defect lines no longer merge and cancel each other as in nematics; instead, complicated stable networks of disclination lines may form. The "streaks" in planar cholesteric films that often form a crackle consist of bundles of thin individual lines (Figure 6.18). A single line itself may show a number of complicated features.[38]

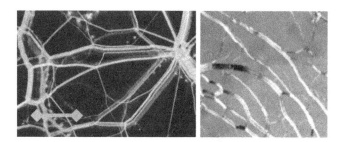

FIGURE 6.18
Oily streak structure of cholesteric liquid crystals. Bar: 100 μm.

A useful structural concept introduced by Kleman and Friedel[39] postulates a quasi-layered structure and explicitly takes into account the natural twist of the system. Concerning defects, we may think of cholesteric liquid crystals as a smectic with an in-plane nematic behavior, similar to the smectic C* phase. Instead of using the concept of a layered structure to account for the twist, we may also consider the field of twist axis $\vec{t}$ in addition to the director field $\vec{n}$. The two concepts are essentially equivalent, with the twist field being identical with the layer normal. The twist field accordingly suffices the condition $\vec{t} \cdot curl\vec{t} = 0$, which means that in this twist field no twist deformation is allowed. The concept of "layers" or twist-field is an approximation, which may not be valid in the core of the defects. We assume that the core structures of cholesterics (especially those with weak chirality) are similar to that of nematics.

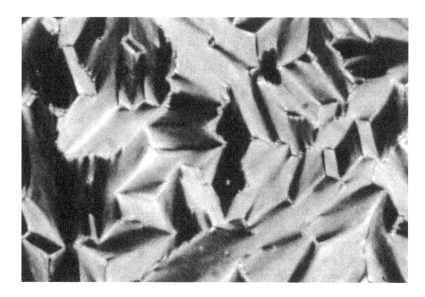

FIGURE 6.19
"Fan" texture of MBBA + 20% chiral dopant between glass plates at 20°C. Crossed polarizers, Bar: 100 μm.[4]

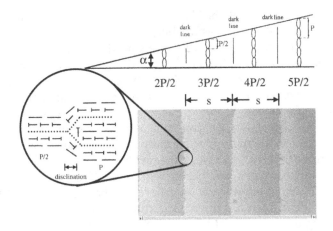

FIGURE 6.20
A cholesteric liquid crystal with pitch of about 0.3 μm in wedge geometry showing Grandjean steps (Grandjean–Cano wedge cell).

In an ideal planar texture of chiral nematic phase, the twist is uniform and the twist axis is normal to the plates. The director field is thus given by (6.1) with $\phi = 2\pi z/p$, where p is the pitch of the helical structure and $z \parallel \vec{t}$. The simplest disclinations are lines running parallel to surfaces of constant $\vec{n}$ in planar textures. They are formed in wedge-shaped samples when the surface orientation is fixed, and are known as Grandjean steps.[39] Figure 6.20 shows the Grandjean lines as proposed by Kleman and Friedel.[40] The lines can be interpreted as combinations of a pair +1/2 and –1/2 lines in the field of the twist axis (indicated by circles). The Grandjean lines separate areas which differ in the number of helical turns of $\vec{n}$. The distance s between the subsequent Grandjean lines in a wedge cell of uniform wedge angle α (Grandjean–Cano cell) thus will give us information about the helical pitch as $p = 2\alpha \cdot s$.

An excellent recent review about the cholesteric defects was written by Lavrentovich and Kleman.[40]

6.4 Defect Phases

The introduction of the disclinations into a liquid crystal structure usually leads to an increase of the free energy of the system. Therefore, defects are not stable, and one generally is able to decrease the number of disclinations by annealing (see Figure 6.2b), or other ways. There are, however, a few cases where disclinations are energetically favorable, and the ordered array of disclinations or defect walls are necessary conditions for the existence of the phase. The existence of these defect phases reflects the fine balance between competing factors.

The simplest example of the defect phases is the twist grain boundary smectic A phase (TGBA*). It was predicted theoretically by Renn and Lubensky,[41] based on the analogy between the smectic and the superconducting materials introduced by de Gennes.[42] According to this analogy, the TGBA* phase is equivalent to the Abrikosov flux lattice phase of the type II superconductor in an external magnetic field. The first experimental observations of the TGBA* phase were reported by Goodby et al.[43] In case of TGBA,* the smectic A phase is formed by chiral molecules. Similar to the chiral nematic (cholesteric) phase, the molecular chirality tends to result in a helical structure even in the smectic A phase. However, neither bend nor twist is compatible with the constant layer spacing requirement, so the material becomes frustrated. This frustration leads, in general, to the suppression of the SmA phase range, or in case of very strong chirality, to the formation of twist grains (see Figure 6.21). The phase is composed of regions of smectic A liquid crystal blocks separated from each other by an array of equally spaced grain boundaries. As we cross each grain boundary, the director rotates by a small angle, so a twist in the director structure forms. The phase is stable, because the increase in the free energy due to the introduction of the grain boundaries is less than the decrease of free energy due to the introduction of twist. The pitch p of the TGB phase is determined by the distance between grains, l, and the rotation of the director in crossing a grain boundary, α, as $p = 2\pi l/\alpha$. For example, if $l = 25$ nm and $\alpha = \pi/9$, p = 450 nm, i.e., such structure should show selective reflection exactly as observed in cholesteric materials with similar half pitch values.

Renn and Lubensy have also proposed the structure of tilted twist grain boundary (TGBC and TGBC*) phases.[44] The existence of the TGBC phase was demonstrated by Nguyen et al.[45]

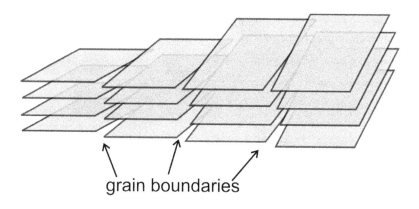

grain boundaries

FIGURE 6.21
Structure of the twist grain boundary (TGB) phase.

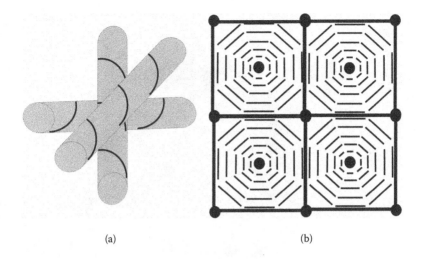

(a)   (b)

FIGURE 6.22
The first model of the blue phases.[51] (a) The three-dimensional structure with the director orientation in different points; (b) a cross section parallel to one of the orthogonal axes. (Dots represent points where the director is undefined).

Other examples for the frustrated defect phases include the so-called blue phases (BP) that appear between the isotropic and cholesteric or smectic phase. They often exhibit a submicron-length twist and show selective reflection in the visible range. The first examples observed by Reinitzer,[46] Lehman,[47] Friedel,[48] then by Gray,[49] showed blue color (that is why the name), but later materials reflecting light in other visible and UV ranges were also observed. In addition to the color, it was also shown that the Bp phase has no birefringence, but is optically active, indicating chiral, deformable structure of high symmetry. Based on this, without many experimental details, it was already suggested in 1969[50] that the blue phase has a three-dimensional body-centered cubic structure formed by cylindrically symmetric local director configuration in which the preferred orientation rotates along all directions perpendicular to the symmetry axis.

The local configuration corresponds to a double-twist cylinder, where at the edge of the cylinders the director is twisted with respect to some angle. In the original model shown in Figure 6.22, it was 90°, in more recent models it is 45°[52] or 54.7°, depending on the 3D structure.

Later it was found that there are three distinct type of blue phases.[53,54] The lower-temperature modifications, BPI and BPII, possess cubic lattice of different symmetries (O^8 and O^2, respectively). The high-temperature modification, BPIII, appears on cooling from the isotropic phase as an amorphous "blue fog." Freeze fracture electron micrographs of BPIII appear to show a disordered packing of filamentary objects.[54] Textures of the different blue phases are shown in Figure 6.23.

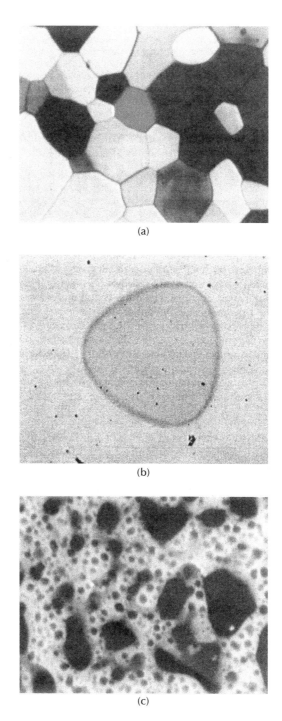

(a)

(b)

(c)

FIGURE 6.23
Optical micrographs of the three different blue phases. (a) BPI; (b) BPII; (c) BPIII (pictures reproduced from Hauser et al.[55]). Pictures indicate 80 μm × 50 μm areas.

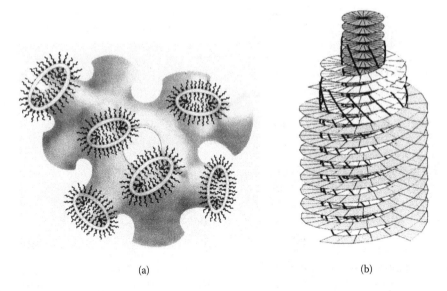

(a)   (b)

FIGURE 6.24
(a) Sketch of the direct sponge phase, where the surfactant bilayer of thickness d of multiply connected topology separates two polar solvent domains. [figure reproduced from A.C. Tromba, A.M. Figueiredo Neto, Lyiotropic liquid crystals, Oxford University Press, Oxford (2005)]; (b) The structure of the SmBlue phase. The smectic layers form double-twist cylinders that can be packed together to form hexagonal or cubic lattice.[60]

The structure of the BPIII phase actually resembles the L_3 (so-called sponge) phase of lyotropic liquid crystals,[56] and the smectic blue phases[58] observed and studied recently. Both the sponge and smectic blue phases are optically isotropic, and the smectic blue are optically active as well. Theoretical arguments show that the reason for the defect structure is the negative value of the saddle-splay elastic constant, K_{24},[58] which makes the defects energetically favorable. A sketch of the sponge phase and of the smectic blue phase is shown in Figure 6.24.

We note that bent-core smectic liquid crystals are also good candidates to form optically isotropic SmBlue type phases due to their bent-shape, which may favor negative saddle-splay elastic constant. In fact, there are a number of experimental observations of optically isotropic and optically active smectic phases.[60]

6.5   Defect-Mediated Phase Transitions

As we have seen in Chapter 3, the usual approach to describe a phase transition is starting in the higher symmetric phase, considering excitations and finding which ones prevail at low temperature.

Furnished with the basics of defects, one can also consider the opposite approach (increasing temperature). For example, describing the melting from the smectic to nematic phase, the topological excitations in smectics are dislocations (Figure 6.11), which are characterized by a line tension γ. At increasing temperatures, the line tension decreases as $\gamma = \gamma_o - k_B T x^{-1}$, where x is the length of the disclination. When T is high enough that γ becomes negative, a spontaneous formation of dislocations appears, leading to the nematic phase.[61] This approach was first used by Kosterlitz and Thouless,[66] who described the 2D melting of two-dimensional crystals, where the important defects are the lattice dislocations.

Halperin and Nelson recognized[61] that the defect-mediated-melting theory implies an additional result. Dislocations disrupt the local positional order, but have only negligible effect on the bond orientational order, which therefore remains of long-range order. Accordingly, the two-dimensional crystal will first melt into an intermediate phase with a quasi-long-range bond order, then — at a higher temperature — into a two-dimensional liquid. The intermediate phase has been called "hexatic" phase. Additional theoretical predictions indicate that the hexatic order is reduced by out-of-plane fluctuations, which are usually larger in the inside layers than at the surface.[63] This argument can be applied also to smectic materials, predicting that with the quenching of the fluctuations at the surface, the top layer of a smectic film crystallizes sooner than the bulk.[64] Examples can be found in many liquid crystals in materials with nematic SmA transition[65] or in the *SmA–SmC* transition, where the free surface usually induces a tilt.[66] Such surface effects are due to the symmetry breaking (missing neighbor effect) and are mainly responsible for the observations that the phase transitions in thin films occur usually at higher temperatures than in bulk.

References

1. W.H. Zurek, *Nature,* 317, 505 (1985); N. Turok, *Phys. Rev. Lett.,* 63, 2625 (1989).
2. M. Kleman, Points, lines and walls, in *Liquid Crystals, Magnetic Systems and Various Ordered Media,* John Wiley & Sons, New York (1983).
3. W.F. Brinkman, P.E. Cladis, Defects in liquid crystals, *Phys. Today,* 48 (May 1982).
4. A. Saupe, *Mol. Cryst. Liq. Cryst.,* 21, 211 (1973).
5. M. Kleman, O.D. Lavrentovich, Y. A. Nastishin, Dislocations and disclinations in mesomorphic phases, in *Dislocations in Solids,* Ch. 66, Ed. F.R.N. Nabarro, J.P. Hirth, Elsevier, Amsterdam (2004).
6. H.R. Trebin, *Liq. Cryst.,* 24, 127 (1998); O.D. Lavrentovich, *Liq. Cryst.,* 24, 117 (1998).
7. R. Repnik, L. Mathelitsch, M. Svetec, S. Kralj, *Eur. J. Phys.,* 24, 481 (2003).
8. D. Demus and L. Richter, *Textures of Liquid Crystals,* ISBN 0–89573–015–4, Verlag Chemie, New York, or ISBN 3–527–25796–9, Weinham (1978).
9. G.W. Gray, J.W. Goodby, *Smectic Liquid Crystals,* ISBN 0–249–44168–3, Leonard Hill Publishing (1984).

10. I. Dierking, *Textures of Liquid Crystals*, Wiley-VCH, Weinheim (2003).

11. O. Lehman, *Flüssige Crystalle*, Engelmann, Leipzig (1904).

12. G. Friedel, *Ann. Phys.*, 18, 273 (1922).

13. C.W. Oseen, *Arkiv Matematik Astron. Fysik A*, 19, 1 (1925); ibid., 22, 1 (1930); *Fortschr. Chem. Phys. U. Phys. Chem.*, 20, 1 (1929); *Trans. Faraday Soc.*, 29, 883 (1933).

14. F.C. Frank, *Disc. Faraday Soc.*, 25, 19 (1958).

15. C.M. Dafermos, *Quart. J. Mech. and Appl. Math.*, 23, 49 (1970).

16. C. Williams, P. Pieranski, P.E. Cladis, *Phys. Rev. Lett.*, 29, 90 (1972).

17. P.E. Cladis, M. Kleman, *J. Phys. (Paris)*, 33, 591 (1972); R.B. Meyer, *Philos. Mag.*, 27, 405 (1973).

18. L.J. Yu, A. Saupe, Z. *Phys. Rev. Lett.*, 45, 1000 (1980).

19. N.D. Mermin, *Rev. Mod. Phys.*, 51, 591 (1979).

20. G. Toulouse, *J. Physique Lett.*, 38, 67 (1977).

21. S. Chandrasekhar, G.G. Nair, K. Pfraefcke, D. Singer, *Mol. Cryst. Liq. Cryst.*, 288, 7 (1996).

22. J. Liu, Ph.D. Thesis, Kent State University (1991).

23. V.P. Mineyev, G.E. Volovik, *Phys. Rev. B*. 18, 3197 (1978).

24. P.E. Cladis, *Phil. Mag.*, 29, 641 (1974).

25. Y. Bouligand, *J. Phys. (Paris)*, 33, 525 (1972).

26. P. Boltenhagen, M. Kleman, O.D. Lavrentovich, Focal conic domains in smectics, in *Soft Order in Physical Systems*, Ed. Y. Rabin and R. Bruinsma, Plenum Press, New York (1994).

27. S.T. Lagerwall, R.B. Meyer, B. Stebler, *Ann. Phys.*, 3, 249 (1972).

28. P. Pershan, *Phys. Today*, 34, May (1982).

29. J.W. Godby, *Ferroelectric Liquid Crystals. Principles, Properties and Applications*, Gordon and Breach, Philadelphia (1991).

30. S.T. Lagerwall, *Ferroelectric and Antiferroelectric Liquid Crystals*, Wiley-VCH, Weinheim (1999).

31. M. Glogarova, L. Lejcek, J. Pavel, V. Janovec, J. Fousek, *Mol. Cryst. Liq. Cryst.*, 91, 309 (1983).

32. G. Pelzl et al., *Liq. Cryst.*, 26, 401 (1999).

33. G. Heppke, D.D. Parghi, H. Sawade, *Liq. Cryst.*, 27, 313 (2000).

34. G. Pelzl, S. Diele, A. Jákli, C.H. Lischka, I. Wirth, W. Weissflog, *Liq. Cryst.*, 26, 135 (1999)

35. A. Jákli, C.H. Lischka, W. Weissflog, G. Pelzl, A. Saupe, *Liq. Cryst.*, 27, 1405 (2000).

36. Yu.A. Nastishin, M.F. Achard, H.T. Nguyen, M. Kleman, *Eur. Phys. J. E*, 12, 581 (2003).

37. A. Jákli, D. Krüerke, H. Sawade, G. Heppke, *Phys. Rev. Lett.*, 86 (25), 5715 (2001).

38. J. Rault, *Solid State Commun.*, 9, 1965 (1971).

39. M. Kleman, J. Friedel, *J. Physique*, 30, C4 (1969).

40. O.D. Lavrentovich, M. Kleman, Cholesteric liquid crystals: Defects and topology, in *Chirality in Liquid Crystals*, Ch. 5, Ed. H.S. Kitzerow, C. Bahr, Springer-Verlag, Heidelberg (2001).

41. S.R. Renn, T.C. Lubensky, *Phys. Rev. A*, 38, 132 (1988).

42. P.G. de Gennes, *Solid State Commun.*, 10, 753 (1972).

43. J.W. Goodby, M.A. Waugh, S.M. Stein, E. Chin, R. Pindak, J.S. Patel, *Nature*, 337, 449 (1989)

44. S.R. Renn, T.C. Lubensky, *Mol. Cryst. Liq. Cryst.*, 209, 349 (1991).

45. H.T. Nguyen, A. Bouchta, L. Navailles, P. Barois, N. Isaert, R.J. Twieg, A. Maaroufi, C. Destrated, *J. Phys. II (France)*, 2, 1889 (1992).

46. F. Reinitzer, *Monatsch. Chem.*, 9, 421 (1988).
47. O. Lehmann, *Z. Phys. Chem.*, 56, 750 (1906).
48. G. Friedel, *Ann. Phys. (Paris)*, 18, 273 (1922).
49. G.W. Gray, *J. Chem. Soc.*, 3733 (1956).
50. A. Saupe, *Mol. Cryst. Liq. Cryst.*, 7, 59 (1969).
51. R.M. Hornreich, S. Shtrikman, *Phys. Rev. A*, 24, 635 (1981); S. Maiboom, J.P. Sethna, P.W. Anderson, W.F. Brinkman, *Phys. Rev. Lett.*, 46, 1216 (1981).
52. K. Bergmann, H. Stegemeyer, *Z. Naturforsch.*, 34A, 351 (1979).
53. H. Stegemeyer, K. Bergmann, in *Liquid Crystals of One and Two Dimensional Order*, Ed. W. Helfrich, G. Heppke, p. 161 Springer, Berlin (1980).
54. J.A. Zasadinsky, S. Meiboom, M.J. Sammon, D.W. Berreman, *Phys. Rev. Lett.*, 57, 364 (1986).
55. A. Hauser, M. Thieme, A. Saupe, G. Heppke, D. Kruerke, *J. Mater. Chem.*, 7, 2223 (1997).
56. D. Roux, M.E. Cates, U. Olsson, R.C. Ball, F. Nallet, A.M. Bellocq, *Europhys. Lett.*, 11, 229 (1990).
57. E. Grelet, B. Pansu, M-H. Li, H.T. Nguyen, *Phys. Rev. Lett.*, 86, 3791 (2001).
58. B. di Donna, R.D. Kamien, *Phys. Rev. Lett.*, 89, 215504 (2002).
59. E. Grelet, B. Pansu, M.-H. Li, H.T. Nguyen, *Phys. Rev. E* , 64, 0101703(R) (2001).
60. A. Jákli et al., *Adv. Mater.*, 15(19), 1606 (2003); J. Ortega et al., *Phys. Rev. E*, 68, 011707 (2003); G. Pelzl, A. Eremin, S. Diele, H. Kresse, W. Weissflog, *J. Mater. Chem.*, 12, 2591 (2002).
61. D.R. Nelson, *Defects and Geometry in Condensed Matter Physics*, Cambridge University Press, London (2002).
62. J.V. Selinger, *J. Phys. (Paris)*, 49, 1387 (1988).
63. W. de Jeu, A. Fera, O. Konovalov, B.I. Ostrovski, *Phys. Rev. E*, 67, 020701 (R) (2003).
64. J. Als-Nielsen, F. Christense, P.S. Pershan, *Phys. Rev. Lett.*, 48, 1107 (1982).
65. S. Heinekamp, R.A. Pelcovits, E. Fontes, E.Y. Chen, R. Pindak, R.B. Meyer, *Phys. Rev. Lett.*, 52, 1017 (1984).
66. J.M. Costerlitz, D.J. Thouless, *J. Phys. C*, 6, 1181 (1973).

7

Magnetic Properties

7.1 Isotropic Materials

7.1.1 Diamagnetism

Diamagnetism is the consequence of Lenz's law applied to electronic currents: "When a current loop is exposed to a magnetic field the current flows so to oppose the change in flux through the loop." For atomic units the currents persist, because there is no dissipation.

Although diamagnetism is a quantum phenomenon, to illustrate the origin of diamagnetism we consider the classical picture of an electron orbiting in a circular path with radius r around a nucleus. The position of the electron is given by the unit vector $\vec{b}$ along the radius (Figure 7.1).

Without external magnetic field, the acceleration of the orbiting electron can be calculated as:

$$m\vec{a} = -k\frac{Ze^2}{r^2}\vec{b} \tag{7.1}$$

where $k = \frac{1}{4\pi\varepsilon_0}$ and $\varepsilon_0 = 8.85 \cdot 10^{-12} \frac{AS}{Vm}$ is the vacuum permittivity. With $\vec{a} = -\omega_0^2 r \cdot \vec{b}$, from Eq. (7.1) the orbital frequency becomes:

$$\omega_0 = \pm\left(\frac{kZe^2}{mr^3}\right)^{1/2} \tag{7.2}$$

Applying magnetic induction $\vec{B}$ (units: Tesla = Weber/m²) along the normal of the orbiting plane, we have to add the Lorentz force, $\vec{F} = -e\vec{v} \times \vec{B} = -er\omega B\vec{b}$, to Eq. (7.1). Accordingly,

$$-m\omega^2 r = \frac{kZe^2}{r^2} - er\omega B = -mr\omega_0^2 - er\omega B \tag{7.3}$$

or $e\omega B = m(\omega - \omega_0)(\omega + \omega_0)$. With $\omega = \omega_0 + \Delta\omega$, where $\Delta\omega \ll \omega_0$, we get that $e\omega_0 B \approx 2m\omega_0\Delta\omega$. This means that the orbital angular frequency ω_0 will be

FIGURE 7.1
Classical picture of an orbiting electron.

changed due to the magnetic field (Lorentz force) by $\Delta\omega = eB/2m$ (Larmor frequency). This change in rotational frequency gives rise to a current per electron: $I = -e\frac{\Delta\omega}{2\pi} = -\frac{e^2B}{4\pi m}$. The magnetic moment p_m (unit: Am²) of the loop with area A becomes: $\vec{p}_m = I \cdot \vec{A} = -\frac{e^2}{4\pi m}\pi r^2\vec{B} = -\frac{e^2 r^2}{4m}\vec{B}$, which gives for the magnetic susceptibility:

$$\chi_{dia}^a \equiv \frac{\vec{p}_m}{\vec{H}} = -\mu_o \frac{e^2 r^2}{4m} \tag{7.4}$$

where $\mu_o = 4\pi \times 10^{-7}$ Vs/Am is the permeability of the vacuum, and $\vec{H} = \vec{B}/\mu_o$ is the magnetic field in a nonferromagnetic material (unit: A/m).[†]

For Z electrons and N identical atoms per unit volume, the diamagnetic susceptibility is dimensionless and can be expressed as[††]:

$$\chi_{dia} = -\mu_o N \frac{Ze^2 r^2}{4m} \tag{7.5}$$

It is important to note that the diamagnetic susceptibility is temperature independent, except the density change, which is usually small! It is also important to point out that the susceptibility is negative, which means that the induction of magnetic moment is in opposition to an applied magnetic induction, so the contribution of magnetic induction to the free energy W is positive.

$$\Delta W = W_{mag} - W_o = -\int B_\alpha dM_\alpha = -\mu_o^{-1}\int B_\alpha \chi_{\alpha\beta} dB_\beta = -\frac{1}{2}\mu_o^{-1}\chi_{\alpha\beta}B_\alpha B_\beta > 0 \tag{7.6}$$

Concerning macroscopic responses, it is important to emphasize that the magnetic interaction is the weakest, in the sense that the sum of individual molecular interactions (averaged over orientational distribution function)

[†] In SI units, one has to multiply $\Delta\chi$ by μ_o.
In CGS system: $\vec{B} = \mu\vec{H} = \vec{H} + 4\pi\vec{M} = \vec{H} + 4\pi\chi\vec{H}$, consequently: $\mu = 1 + 4\pi\chi$). In SI units: $\vec{B} = \mu_o\mu\vec{H} = \mu_o\vec{H} + \vec{M} = \mu_o\vec{H} + \mu_o\chi\vec{H}$, consequently: $\mu = \mu_o(1 + \chi)$.
[††] The exact quantum expression is different, and reads as: $\chi_{dia} = -\mu_o N \frac{Ze^2}{6m}\sum_i \langle r_i^2 \rangle$.

gives the macroscopic response, i.e., the internal fields and molecular interactions can be neglected.

7.1.2 Paramagnetism

Pramagnetism is due to unpaired electron spins, which have magnetic moment, which aligns toward the field. Unpaired spins introduced by metal centers result in magnetization along the field direction. The paramagnetic susceptibility χ_{para} is given as:

$$\chi_{para} = \frac{N\mu_0 g_e^2 \mu_B^2 J(J+1)}{3k_B T} \tag{7.7}$$

Here J is the total electron spin quantum number; μ_B is the so-called Bohr magneton; and g_e is the electronic g-value (g-value ~ 2 is a kind of cavity factor that describes the modification of the magnetic field experienced by the electron spin arising from the electron distribution in the molecules).

The most important difference between dia - and paramagnetism is that χ_{para} is positive.

7.1.3 Ferromagnetism

Ferromagnetism may occur as the result of strong spin–spin interactions. This requires unpaired spins and high order. The ferromagnetic susceptibility χ_{ferro} can be given as:

$$\chi_{ferro} = \frac{N\mu_0 g_e^2 \mu_B^2 J(J-1)}{3k_B(T-\theta)} \gg 1 \tag{7.8}$$

In this expression θ is the Curie temperature: $\theta > 0$ corresponds to ferromagnetic order; $\theta < 0$ means antiferromagnetic order, and is called Néel temperature.

7.2 Magnetic Properties of Liquid Crystals

Just as many other macroscopic properties, magnetic susceptibility of liquid crystal is also anisotropic, i.e., it can be described by a tensor, which, in a uniaxial system fixed to the director, can be expressed as:

$$\chi_{\alpha\beta} = \begin{bmatrix} \chi_\perp & 0 & 0 \\ 0 & \chi_\perp & 0 \\ 0 & 0 & \chi_\parallel \end{bmatrix} \tag{7.9}$$

In this expression $\chi_{\|}$ and $\chi_{\perp}$ are the susceptibility along the director and normal to it. They relate to the average molecular susceptibility $\bar{\chi}_{mol}$ (unit: m^3/kg), the mass-density ρ, and the uniaxial order parameter S as:

$$\chi_{\|} = \rho\bar{\chi}_{mol} + \frac{2}{3}\rho\Delta\chi_{mol}S \tag{7.10}$$

$$\chi_{\perp} = \rho\bar{\chi}_{mol} - \frac{1}{3}\rho\Delta\chi_{mol}S \tag{7.11}$$

The magnetic susceptibility tensor is a very useful measure of the macroscopic symmetry; it gives the direction of the uni-axis with large precision. The weighted average of the $\chi_{\|}$ and $\chi_{\perp}$ give the average susceptibility density:

$$\frac{\chi_{\|} + 2\chi_{\perp}}{3} = \rho\bar{\chi}_{mol} \tag{7.12}$$

From (7.10) and (7.11) the magnetic anisotropy can be given as:

$$\Delta\chi = \chi_{\|} - \chi_{\perp} = \rho\Delta\chi_{mol}S \tag{7.13}$$

Sometimes it is also useful to determine the relative anisotropy as:

$$\frac{\chi_{\|} - \chi_{\perp}}{\chi_{\|} + 2\chi_{\perp}} = \frac{\Delta\chi_{mol}}{3\bar{\chi}_{mol}}S \tag{7.14}$$

These expressions show that the magnetic anisotropy $\Delta\chi_{mol}$ is proportional to the order parameter, and its temperature dependence gives the temperature dependence of the order parameter.[†]

An example for the temperature dependence of $\Delta\chi_{mol}$ for 5CB measured by different techniques: diamagnetic anisotropy,[2] depolarized Raman scattering,[3] nuclear magnetic resonance[4] and birefringence,[5] and compared also to the mean-field theory, is shown in Figure 7.2. It can be seen that the optical and magnetic data as well as the mean-field behavior do not well correspond to each other for this material. However, we note that for other nematogens, the correspondences are much better.[6]

Most liquid crystals are diamagnetic, and the susceptibility is mainly arising from the delocalized charges distributed in plane perpendicular to $\vec{B}$, i.e., the aromatic groups give large contribution to the diamagnetic susceptibilities (see Figure 7.3).

For most calamitic liquid crystals $|\chi_{\perp}| > |\chi_{\|}|$, i.e., the diamagnetic anisotropy is positive, which means that the magnetic field aligns the molecules along

---

[†] We note that for biaxial materials, $\Delta\chi_{mol} = a \cdot S + b \cdot D$, where D is the biaxiality order parameter.

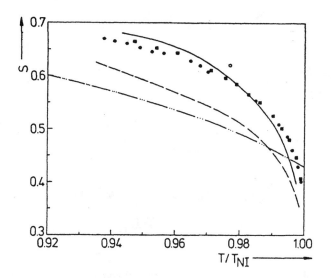

FIGURE 7.2
Order parameters for 5CB (figure reproduced from Figure 5 of Ref. 1 Solid dots and squares: magnetic susceptibility measurements,[1] open circle: nuclear magnetic resosnance,[3] solid line: depolarized Raman scattering; dotted line: birefringence,[4] and _...._: mean-field theory.[17]

the long axis. In other words, the Lenz effect is minimized when $\vec{B} \| \vec{n}$ (this contributes to the smallest extent to the free energy).

A few examples of molecular diamagnetic anisotropies are shown in Figure 7.4. Since the densities of the materials are close to $1 \text{g}/\text{cm}^3$, in practice the values in the graph would correspond to the same dimensionless values with 10^{-7} in CGS units (e.g., $\Delta \chi_{mol} = 10^{-10} \text{ m}^3/\text{kg} \rightarrow \Delta \chi = 10^{-7}$ (CGS)) or $4\pi \times 10^{-7}$ (SI) = $1.256 \cdot 10^{-6}$ (SI).

It can be seen that materials with benzene rings in their rigid linear core have relatively large positive diamagnetic anisotropies, whereas cyclohexanes promote negative diamagnetic anisotropies.

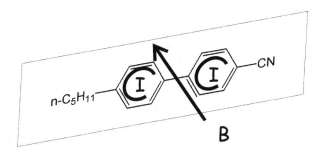

FIGURE 7.3
Representation of the origin of diamagnetism of liquid crystals with benzene rings.

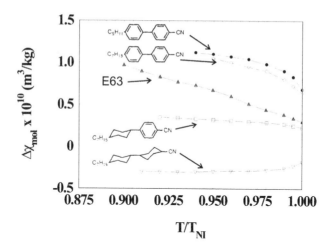

FIGURE 7.4

Average molecular diamagnetic anisotropies of several liquid crystals in their nematic phase at the function of reduced temperature T/T_{NI}. E63 is a five-component mixture of different cyanobiphenyls. (Graph reproduced using Figure 2 and 3 of Ref. [1] and Figure 3 of Ref. [22].)

Paramagnetic liquid crystals are found in metal-containing mesogens,[6] e.g., salicylaldimine complexes of copper.[7] These materials have both dia- and paramagnetic anisotropies, which may have different signs. Since negative magnetic anisotropies (both diamagnetic and paramagnetic) align the molecules perpendicular to the magnetic fields, materials with larger magnitude of negative paramagnetic anisotropies than of positive diamagnetic anisotropies will be aligned perpendicular to the magnetic fields, whereas those with larger positive diamagnetic anisotropies than negative paramagnetic anisotropies will align along the fields. In the literature, there are examples for both situations.[8]

Since ferromagnetism requires long-range spin–spin interactions, for pure liquid crystals the ferromagnetic order is very unlikely. Although (there are some organic ferromagnets,[9] they have very low Curie temperature.[10] The only organic materials that may have a chance[10] for ferromagnetism are the metals containing liquid crystalline polymers.

For achieving ferromagnetism in liquid crystals, a different approach was proposed by Brochard and de Gennes[11] by means of dispersion of ferromagnetic particles, e.g., γ-Fe_2O_3 (ferrite) or Fe_3O_4 (magnetite) in liquid crystals. Depending on the phase of the liquid crystals, the dispersions are called ferronematics or ferrosmectics. To obtain single magnetic domains, sufficiently small dispersed particles are needed.

The coupling to the liquid crystal is imagined either via magnetic anisotropy of liquid crystal and magnetic moment of particles, or via elastic interactions if the particle shape is anisotropic.

It is important to note that special surfactants are needed to stabilize the magnetic suspension against aggregation. These surfactants are more compatible with aqueous lyotropic systems, explaining that experimental success is realized mainly only in lyotropics with $0.04\,\mu m < d < 0.35\,\mu m$ particle diameters.[12]

7.3 Magnetic Field–Induced Director Deformation

The free energy density of a liquid crystal in the presence of external bulk interaction is:

$$F(\theta,\theta') = \frac{1}{2}K\theta'^2 + F_e(\theta) \tag{7.15}$$

where K is the relevant curvature elastic constant, and $F_e(\theta)$ represents the interaction energy with external bulk force. In case of magnetic field H acting on a liquid crystal with diamagnetic anisotropy, $\Delta\chi$, the magnetic contribution to the free energy is:[†]

$$F_M = -\frac{1}{2}\chi H^2 - \frac{1}{2}\Delta\chi(\vec{n}\cdot\vec{H})^2 \tag{7.16}$$

Assuming that the magnetic field makes an angle θ with the director and causes a deformation along the z axis, the Euler–Lagrange equation in the bulk (see Chapter 4) reads as:

$$K\frac{d^2\theta}{dz^2} + \Delta\chi H^2 \sin\theta\cos\theta = 0 \tag{7.17}$$

This last equation can also be considered as the balance of the elastic and magnetic torque densities, where the magnetic torque density is defined as:

$$\vec{\Gamma}_M = \vec{M}\times\vec{H} = \left(\hat{\chi}\cdot\vec{H}\right)\times\vec{H} \tag{7.18}$$

For uniaxial materials characterized by a director $\vec{n}$, $\vec{\Gamma}_M$, can be expressed[14] with the diamagnetic anisotropy χ_a, and the angle between the director θ and the amplitude of magnetic field $\vec{H}$ as:

$$\vec{\Gamma}_M = \chi_a(\vec{n}\cdot\vec{H})\vec{n}\times\vec{H} = \chi_a H^2 \sin\theta\cdot\cos\theta \tag{7.19}$$

[†] In SI units one has to multiply $\Delta\chi$ by μ_0.

We define the magnetic coherence length as:

$$\xi(H) = (K/\Delta\chi)^{1/2}/H \tag{7.20}$$

With this definition (7.17) looks like:

$$\xi^2 \frac{d^2\theta}{dz^2} + \sin\theta\cos\theta = 0 \tag{7.21}$$

Multiplying (7.21) by $d\theta/dz$ we get that,

$$\xi^2 \frac{d}{dz}\left\{\frac{1}{2}\left(\frac{d\theta}{dz}\right)^2\right\} + \frac{d}{dz}\left(-\frac{1}{2}\cos^2\theta\right) = 0 \tag{7.22}$$

This can be integrated and gives:

$$\xi^2\left(\frac{d\theta}{dz}\right)^2 = \cos^2\theta + const. \tag{7.23}$$

As an example, let us consider a simple twist with strong anchoring as sketched in Figure 7.5a.

Far from the wall ($z \to \infty$) we expect $\theta = \pi/2$ and $d\theta/dz = 0$. Thus, the integration constant must vanish, and

$$\xi_2 \frac{d\theta}{dz} = \pm\cos\theta \tag{7.24}$$

$+/-$ signs correspond to "right" and "left" handed twist. For example, for $+$ sign:

$$\frac{dz}{\xi_2} = \frac{d\theta}{\cos\theta} = -\frac{du}{\sin u} \tag{7.25}$$

where $u = \pi/2 - \theta$. With $t = \tan(u/2)$ we get that $\sin u = \dfrac{2t}{1+t^2}$, $du = \dfrac{2dt}{1+t^2}$.

Therefore $\frac{dz}{\xi_2} = -\frac{dt}{t}$ and $t = \exp(-z/\xi_2)$.

The integration constant ensures that for $z = 0$ (at the wall), $\theta = 0$ and $u = \pi/2$, as required.

From the definition of the magnetic coherence length with $\Delta\chi = 10^{-7}$, $K_{22} = 10^{-6} dyne$, $H = 10^4 Oersted$ (CGS units), we get that $\xi_2 \sim 3\mu m$.

Similar to the twist case, the magnetic coherence lengths can be defined for bend and splay deformations as:

$$\xi_1 = \left(\frac{K_{11}}{\Delta\chi}\right)^{1/2} \bullet \frac{1}{H} \quad and \quad \xi_3 = \left(\frac{K_{33}}{\Delta\chi}\right)^{1/2} \frac{1}{H} \tag{7.26}$$

The magnetic field-induced director deformations involving bend or splay are illustrated in Figure 7.5b.

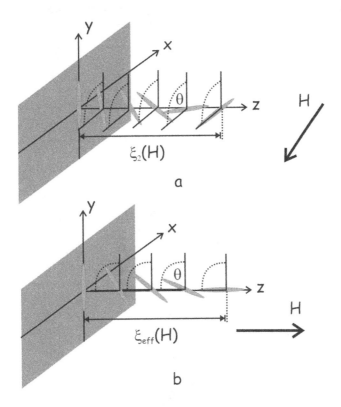

FIGURE 7.5
Competition of wall alignment and magnetic field alignment in case of pure twist (a) and in geometry involving a combination of bend and splay (b).

Realignments of the director structure in films with initially uniform planar and homeotropic alignments were first studied by Freedericks[14] on the geometries illustrated in Figure 7.6. A magnetic field acting on a material with positive diamagnetic anisotropy tries to rotate the director parallel to the field. Since the field does not determine the direction of the rotation, the deformation starts only above a threshold field (H_c). The deformations near the threshold involve splay, twist and bend distortions in the geometries corresponding to (a), (b) and (c), respectively.

Based on the experimental results, Freedericks concluded that the critical magnetic field H_c is inversely proportional to the film thickness L, i.e., $H_c \cdot L = const.$

Theoretical verification of these experimental results was provided by Zöcher,[15] using the following arguments.

- Although the transition has to break symmetry, for calculating the critical fields we assume that the transition is nearly of second order, i.e., the director deformation $\delta \vec{n}(\vec{r}) \ll 1$ at the threshold. In this case, the director structure of the deformed state can be written as $\vec{n} = \vec{n}_o + \delta \vec{n}(\vec{r})$, where $\delta\vec{n} \perp \vec{n}_o$ (since $n^2 = 1$). With this assumption, the relevant part of the magnetic free energy density F_m (see [7.16]) can be written as:

$$F_M \sim -\frac{1}{2}\Delta\chi\,(\delta n)^2 H^2 \tag{7.27}$$

- Assuming strong anchoring, i.e., $\delta\vec{n}(0, L) = 0$, we can expand δn in a Fourier series, as $\delta n = \sum_q \delta n_q \sin qz$, where $q = v\frac{\pi}{L}$ (v is positive integer). The energy per unit area then can be written as:

$$\Phi = L\cdot(F_d + F_M) = \frac{L}{4}\sum \delta n_q^2 (Kq^2 - \Delta\chi H^2) \tag{7.28}$$

where K is the relevant elastic constant (K_{11}, K_{22} and K_{33} for the geometries corresponding to Figure 7.6a, b, and c, respectively).

The unperturbed state is stable if $\Phi > 0$ for all δn_q, i.e., if $\Delta\chi H^2 < K\,q^2$ and becomes unstable when $\phi = 0$.

The smallest value of q is π/L (distortion of half thickness), which means that:

$$\Delta\chi H_c^2 = K\left(\frac{\pi}{L}\right)^2 \tag{7.29}$$

Accordingly, the threshold field is:

$$H_c = \sqrt{\frac{K}{\Delta\chi}}\cdot\frac{\pi}{L} \tag{7.30}$$

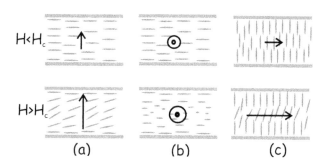

(a)   (b)   (c)

FIGURE 7.6

The Freedericks transition for nematic films in magnetic fields H. At low fields the molecules are parallel to the easy-axis of the wall. For fields higher than the threshold the molecules (at least those far from the surfaces) rotate parallel to the magnetic field. (a) splay, (b) twist and (c) bend.

This agrees with the sample thickness dependence as observed by Freedericks. From the definition of the magnetic coherence length (see (7.20)) it is seen that, at the threshold for the Freedericks transition, the magnetic coherence length is $\xi_i(H_{c,i}) = \frac{L}{\pi}$.

7.4 Magnetic Effects in Liquid Crystals

Applying magnetic field in isotropic phase, an anisotropy can be induced. This is called the Cotton–Mouton effect. The magnetic field–induced anisotropy is mainly detected by measuring the optical birefringence. The magnetic field–induced birefringence in the isotropic phase is proportional to the square of the magnetic field.

$$\Delta n = CH^2 \tag{7.31}$$

where C is the Cotton–Mouton constant. The coefficient C is typically two orders of magnitudes larger in liquid crystalline materials[16] than in simple organic liquids, such as nitrobenzene. The Cotton–Mouton coefficients are expected to diverge near the nematic phase as:

$$C \propto (T - T_c)^{-\gamma} \tag{7.32}$$

Here T_c is the critical temperature (slightly lower than T_{N-I}); γ is the critical constant (~1 in mean-field model[17]).

Another interesting magnetic effect that can be observed in transparent substances is the Faraday effect.[18] This effect causes a rotation of a plane of polarization of light propagated along the magnetic field as the beam transverses through the material when it is placed in a magnetic field (see Figure 7.7).

It is found that the specific rotation $\rho \equiv \frac{\alpha}{d}$ (the rotation per unit length) is linearly related to the magnetic induction, i.e.,

$$\rho = Ve \cdot B \tag{7.33}$$

where *Ve* is the Verdet constant. Some examples of the Verdet constants are listed in Table 7.1.

The Faraday effect originates from the influence of static magnetic field on the motion of electrons. The electric field of the light polarizes the material (electrons are displaced from the equilibrium position). This induces a motion with velocity $\vec{v}$ along the field. In the presence of a magnetic displacement $\vec{B}$, a Lorentz force $\vec{F} = e \cdot \vec{v} \times \vec{B}$ is induced, resulting in a dipole moment component along $\vec{F}$. Since $\vec{v} \parallel \vec{E}$, the dipole moment induced by $\vec{F}$ is along $\vec{B} \times \vec{E}$, and the material relation becomes:

$$\vec{D} = \hat{\varepsilon}\vec{E} + i\varepsilon_o\gamma\vec{B} \times \vec{E} \tag{7.34}$$

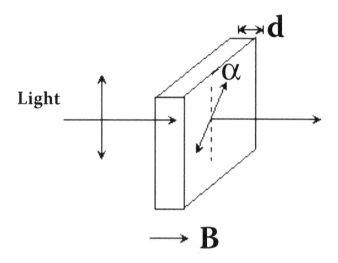

FIGURE 7.7
Illustration of the Faraday rotation.

We note here that the imaginary factor i is due to the $\pi/2$ phase leg between $\vec{v}$ and $\vec{E}$. Such a relation represents a chiral material, which can be characterized by an asymmetric and complex dielectric tensor:

$$\varepsilon_{chiral} = \begin{pmatrix} \varepsilon_1 & -i\gamma & -i\gamma \\ i\gamma & \varepsilon_2 & -i\gamma \\ i\gamma & i\gamma & \varepsilon_3 \end{pmatrix} \tag{7.35}$$

where γ measures the optical activity. We emphasize, however, that in the present case the material itself is not optically active, but it is induced by the magnetic field only. This induced optical activity explains the rotation of the light polarization.

Magnetic fields have primary importance in nuclear magnetic resonance (NMR) measurements of liquid crystals. NMR measurements give information about the order parameter of different flexible segments of the molecules.

TABLE 7.1

Verdet Constants at λ = 5893 Å at Room Temperature

Substance	Ve [10^{-5} degrees/(G·mm)]
Water	2.2
Diamond	2.0
Glass (flint)	5.3
Phosphorus	22
MBBA	6.7

Note: G is the acronym for Gauss (10^4G = 1T).

They also can reveal local biaxiality. NMR is routinely used by chemists to check the synthetic products.

Magnetic fields are also very important for aligning anisotropic materials, especially relatively thick samples (d > 20 µm), when the surface alignment methods are not effective.

7.5 Measurements of the Magnetic Susceptibility[19]

In all these methods, one measures only one component of the magnetic susceptibility (the largest one for paramagnetic and the smallest for diamagnetic samples). The analysis is based on measuring the force on a sample placed in inhomogeneous magnetic fields:

$$\vec{F} = \frac{1}{2}\mu_o^{-1}(\chi - \chi_1)\int_V \frac{\partial B^2}{\partial z}dV = \frac{1}{2}\mu_o^{-1}(\chi - \chi_1)V\left\langle\frac{\partial B^2}{\partial z}\right\rangle \tag{7.36}$$

where χ_1 is susceptibility of gas, for example air, N_2 or He. We note that air is not the most suitable for the measurements, because air is paramagnetic due to O_2.

If the material is placed in tubes with cylindrical cross section of area $A = r^2\pi$, (7.36) simplifies to:

$$\vec{F} = \frac{1}{2}\mu_o^{-1}(\chi - \chi_1)\int \frac{\partial B^2}{\partial z}Adz = \frac{1}{2}\mu_o^{-1}(\chi - \chi_1)A(B_2^2 - B_1^2) \tag{7.37}$$

where B_1 (B_2) is the magnetic induction at the bottom (top) of the sample.

Experimentally, mainly two variants are used, based on Gouy's and Faraday's methods (Figure 7.8).

In the microbalance using Gouy's method, the sample is partially placed in uniform magnetic field (Figure 7.8a). In this case:

$$F = -\frac{1}{2}\mu_o^{-1}(\chi - \chi_1)AB_1^2 \tag{7.38}$$

In the Faraday balance, the entire tube is placed in an inhomogeneous magnetic field, which can be characterized by a constant gradient, $\vec{\nabla}B$, where $B_1 \geq B_2 = B_1 - \Delta B$ and $B_2^2 - B_1^2 = -2B_1\Delta B$. In this case, the force reads as:

$$F = -\mu_o^{-1}(\chi - \chi_1)AB_1\Delta B = -\mu_o^{-1}(\chi - \chi_1)VB_1\frac{\Delta B}{\Delta z} \tag{7.39}$$

For a diamagnetic gas $\chi_1 \ll \chi$, so $F \sim -\mu_o^{-1}\chi VB_1\frac{\Delta B}{\Delta z} = -\mu_o^{-1}m\chi_m B\frac{\Delta B}{\Delta z}$. To estimate the actual forces, we consider the example of *PAA* with $\chi_\parallel^m \approx 6\cdot10^{-8} m^3/kg$. Taking $B = 1T$, $\frac{\partial B}{\partial z} = 1T/m$ and $m = 10^{-3}kg$, we have $|\vec{F}| = \frac{1}{4\pi\times10^{-7}}\times6\times10^{-8}\times1\times1\times10^{-3}$

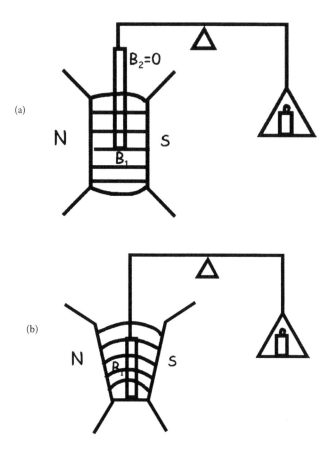

FIGURE 7.8
Schematics of the microbalances used to measure magnetic susceptibilities. (a) Gouy balance; (b) Faraday balance.[19]

$N \sim 5 \times 10^{-5}\,N$. This corresponds to an effective mass of $m_{eff} = \dfrac{|\vec{F}|}{g} \sim 5 \cdot 10^{-6}\,kg = 5mg$. To measure such small forces we need microbalance, or much larger gradient of magnetic induction.

7.6   Measurements of the Anisotropy of the Magnetic Susceptibility

The anisotropy of the magnetic susceptibility can be measured by several methods.

1. One possibility is to do the previous measurements both in the liquid crystalline and in the isotropic phases. Since, disregarding the density

change, χ is basically temperature independent, one gets that:

$$\Delta\chi = \frac{3}{2}(\chi_\| - \bar{\chi}) \quad and \quad \Delta\chi = 3(\bar{\chi} - \chi_\perp) \tag{7.40}$$

2. Another way of determining $\Delta\chi$ is based on the measurement of the magnetic Freedericks transition. In principle, any measurement that probes some anisotropic property, e.g., dielectric constant, capacitance or electric conductivity (probe: electric field), thermal conductivity (probe: temperature gradient), would detect the Freedericks transition. For geometries *a* and *c* in Figure 7.6, the probe fields will be along the film thickness, but for the arrangement corresponding to Figure 7.6b they should be parallel to the substrates.

   It is important to note that measurement of the transmitted light intensity is not sensitive to the threshold, because changes are observable only at much higher fields: when $\xi(H) \sim \lambda$ (where λ is the wavelength of the light). Among optical observations the conoscopic measurements are considered to be the most sensitive to detect the threshold.

3. A third method for measuring the anisotropy of the diamagnetic susceptibility is termed oscillating tube method.[21] In this case, a cylindrical tube with sample volume V is hung between the plates of a magnet via a thin torsion wire (typically tungsten). The sample is slightly rotated by a motor, then stopped and allowed to rotate back and forth. The angular frequency of small oscillations is: $\omega_o^2 = D/I$, where I is the moment of inertia of the filled tube about the center axis, and D is the torsion constant of the thin wire. If a magnetic field is applied, a magnetic torque acts during the rotation of the director and will tend to keep the director parallel to the magnetic field. The balance between the magnetic and elastic torque will result in an angle θ between the field and director, and the torque per unit volume is given as: $\bar{N} = \frac{1}{2}\Delta\chi B^2 \sin 2\theta$. If $\Delta\chi_{dia} > 0$, the elastic torque is transferred to the tube and acts as a restoring torque, which alters the angular frequency to ω. If the time τ for realignment is large compared to the oscillation period, $\Delta\chi$ can be calculated with the measured ω and ω_o values as:

$$\Delta\chi = \frac{\mu_o D}{VB^2}\left(\frac{\omega^2}{\omega_o^2} - 1\right) \tag{7.41}$$

This method is best suited for lyotropics, which have small diamagnetic anisotropy, $\Delta\chi \sim 10^{-9}$. For thermotropic liquid crystals the oscillation is much more complicated,[22] and the measurement is less sensitive.

References

1. A. Buka, W.H. de Jeu, *J. Physique*, 43, 361 (1982).
2. K. Miyamo, *J. Chem. Phys.*, 69, 4807 (1978).
3. J.W. Emsley, G.R. Luckhurst, G.W. Gray, A. Mosley, *Mol. Phys.*, 35, 1499 (1978).
4. R.G. Horn, *J. Physique*, 39, 4807 (1978).
5. W.H. de Jeu, W.A.P. Claasen, *J. Chem. Phys.*, 68, 102 (1978); H. Gasparoux, J.R. Lalanne, B. Martin, *Mol. Cryst. Liq. Cryst.*, 51, 221 (1979).
6. D.W. Bruce, *J. Chem. Soc. Dalton Trans.*, 2983 (1993); S.A. Hudson, P.M. Mailtis. *Chem. Rev.*, 93, 861 (1993).
7. J.L. Serrano, P. Romero, M. Marcos, P.J. Alonso, *J. Chem. Soc. Chem. Commun.*, 859 (1990); I. Bikhanteev, Yu. Galyametdinov, A. Prosvirin, K. Kriesar, E.A. Soto-Bustamante, W. Haase, *Liq. Cryst.*, 18, 231 (1995).
8. D. Dunmur, K. Toriyama, Magnetic properties of liquid crystals, in *Physical Properties of Liquid Crystals*, Vol. 2, Ch. 4, Ed. D. Demus, J. Goodby, G.W. Gray, H.-W. Spiess, V. Vill, Wiley-VCH, Weinheim, (1999).
9. D. Gatteschi, *Europhys. News*, 25, 50 (1994).
10. S. Takahashi, Y. Takai, H. Morimoto, K. Sonogashira, *J. Chem. Soc. Chem. Commun.*, 3 (1984).
11. F. Brochard, P.G. de Gennes, *J. de Phys.*, 31, 691 (1970).
12. P. Fabre, C. Casagrande, M. Veyssie, V. Cabuil, R. Massart, *Phys. Rev. Lett.*, 64, 539 (1990); C.Y. Matuo, F.A. Tourinho, A.M. Figueiredo-Neto, *J. Mag. Mater.*, 122, 53 (1993).
13. P.G. de Gennes, J. Prost, *The Physics of Liquid Crystals*, 2nd ed., Clarendon Press, Oxford, (1993).
14. V. Freedericks, V. Zolina, *Zh. Russ. Fiz. Khim. Oschch.*, 62, 457 (1930); *Trans. Faraday Soc.*, 29, 9199 (1933); V. Freedericks, V. Zvetkoff, *Sov. Phys.*, 6, 490 (1934).
15. H. Zöcher, *Trans. Faraday Soc.*, 29, 945 (1933).
16. J.D. Litster, in *Critical Phenomena*, Ed. R.E. Mills, p. 393, McGraw-Hill, New York (1971).
17. W. Maier, A. Saupe, *Z. Naturforsch.*, A13, 564 (1958); ibid., A14, 882 (1959); ibid., A15, 287 (1960).
18. G.R. Fowles, *Introduction to Modern Optics*, Ch. 6, Holt, Rinehart and Winston, Inc., New York (1975)
19. L.E. Bates, *Modern Magnetism*, Cambridge University Press, London (1961).
20. W.H. de Jeu, A.P. Glassen, *J. Chem. Phys.*, 68, 102 (1978).
21. S. Plumley, Y.K. Zhu, Y.W. Hui, A. Saupe, *Mol. Cryst. Liq. Cryst.*, 182B, 215 (1990).
22. A. Jákli, D-R. Kim, M.R. Kuzma, A. Saupe, *Mol. Cryst. Liq. Cryst.*, 198, 331 (1991).

8

Electrical Properties

8.1 Dielectrics

Dielectric fluids such as liquid crystals are leaky insulators, i.e., they have low electrical conductivity and polarize in the presence of an electric field $\vec{E}$. This means that the electric field induces internal charge reorganization, or distortion such as a net electric dipole moment per unit volume P appears. This is the polarization with units C/m^2. The sign of the polarization is defined as $P > 0$ if the direction is pointing from negative to positive charges.

The polarization per unit electric field of an anisotropic material is described by the electric susceptibility tensor $\hat{\chi}^e$ as:

$$P_\alpha = \varepsilon_0 \chi^e_{\alpha\beta} E_\beta \tag{8.1}$$

where $\alpha, \beta = x, y, z$, and $\varepsilon_0 = 8.85 \times 10^{-12} \ C^2/Nm^2$ (or C/Vm) is the vacuum permittivity.

It is usual to define the dielectric tensor $\hat{\varepsilon}$ as:

$$\hat{\varepsilon} = \hat{I} + \hat{\chi}^e_0 \tag{8.2}$$

The reason for introducing $\hat{\varepsilon}$ is that it gives the electric displacement $\vec{D}$ (its magnitude is the free surface charge per unit area).

$$\vec{D} = \vec{P} + \varepsilon_0 \vec{E} = (1 + \chi)\vec{E} = \varepsilon_0 \hat{\varepsilon} \vec{E} \tag{8.3}$$

The dielectric permittivity is a macroscopic quantity; it relates the external electric field $\vec{E}$ to the macroscopic polarization.

In the special case of isotropic materials, the tensor becomes a scalar, i.e., $\hat{\varepsilon} - \hat{I} \rightarrow \varepsilon - 1$.

For uniaxial materials, like ordinary calamitic (rod-shape) nematics, SmA, SmB, the dielectric tensor is symmetric and has a traceless form in a coordinate system fixed to the director:

$$\hat{\varepsilon} = \begin{pmatrix} \varepsilon_\perp & 0 & 0 \\ 0 & \varepsilon_\perp & 0 \\ 0 & 0 & \varepsilon_\| \end{pmatrix} \tag{8.4}$$

where $\varepsilon_\perp$ and $\varepsilon_\|$ are the components normal and along the director. The eigenvalues of the dielectric tensor are related to the order parameter as:

$$\varepsilon_\| = \bar{\varepsilon} + \frac{2}{3}\varepsilon_a S \tag{8.5}$$

and

$$\varepsilon_\perp = \bar{\varepsilon} - \frac{1}{3}\varepsilon_a S \tag{8.6}$$

where $\varepsilon_a = \varepsilon_\| - \varepsilon_\perp$ is the dielectric anisotropy, and $\bar{\varepsilon} = \frac{1}{3}(\varepsilon_\| + 2\varepsilon_\perp)$ is the isotropic part of the dielectric constant, thus independent of the orientational order. Apart from density changes, $\bar{\varepsilon}$ is typically continuous through all uniaxial liquid crystal phase transitions. A uniaxial material is said to be positive when $\varepsilon_a > 0$ and said to be negative when $\varepsilon_a < 0$.

Typical temperature dependences of the dielectric constants $\varepsilon_\perp$ and $\varepsilon_\|$ are shown in Figure 8.1a for a nematic material with positive dielectric anisotropy.

For materials with smectic order below the nematic phase, usually the dielectric anisotropy is negative in the smectic phase, and the dielectric anisotropy may change sign in the nematic or smectic phase, due to the presence of smectic clusters (so-called cybotactic groups) in the nematic phase. Some examples for this behavior are shown in Figure 8.1b and 8.1c, where alkylaoxybenzene materials (-N = N- bond in the middle of the molecules) are with positive dielectric anisotropies in the nematic phase and negative ones in the underlying smectic A phase. The decrease in $\varepsilon_\|$ and increase of $\varepsilon_\perp$ is due to the correlation in the arrangement of neighbor dipoles in the smectic phase. The longitudinal component of the dipole in the azoxy group prefers antiparallel orientation, while the transverse component is parallel.

Some phases, like the SmC are spontaneously biaxial due to the director tilt. In this case, one needs to determine three different dielectric constants, which require measurements in three different geometries, such as shown in Figure 8.2.

In the local molecular frame, the dielectric tensor has three different eigenvalues in its diagonal form as:

$$\hat{\varepsilon} = \begin{pmatrix} \varepsilon_1 & 0 & 0 \\ 0 & \varepsilon_2 & 0 \\ 0 & 0 & \varepsilon_3 \end{pmatrix} \tag{8.7}$$

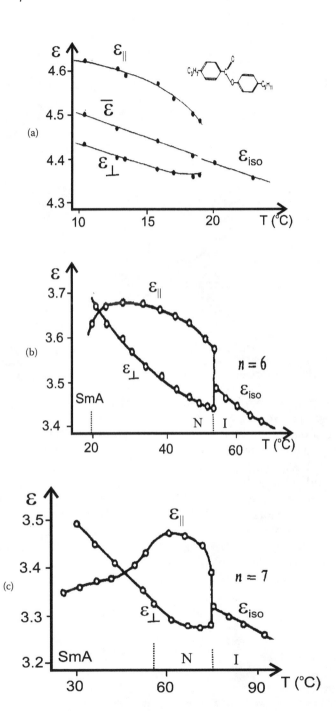

FIGURE 8.1

Temperature dependences of the static dielectric (f = 1 kHz) permittivities of 4-pentylphenyl-4-propoylbenzoate (ME 35)[15] (a) and 4,4′-n-alkylaoxybenzene having nematic and SmA phases;[1] (b) n = 6; (c) n = 7.

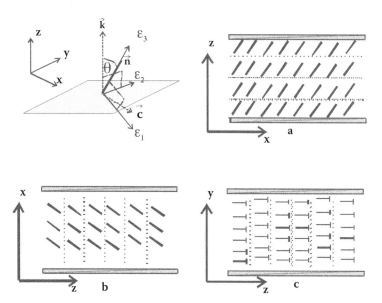

FIGURE 8.2
The definitions of the dielectric tensor components in the local molecular frame, and main alignment structures sandwiched SmC materials in thin films. For testing the dielectric constant, small electric fields, not enough to cause any reorientation, are applied across the plates.

where ε_3 is parallel to the director, ε_2 is parallel to $\vec{c} \times \vec{k}$ and ε_1 is normal to ε_3 and ε_2. (see Figure 8.2). Here we must note that, due to the different order of the core and aliphatic chains, the tilt angle is not a well-defined parameter. One may get a few degrees difference when determining it by optical method or by X-ray data. During the following discussion, we will disregard this difficulty and assume that the tilt angle is well defined.

In a laboratory frame, the dielectric tensor can be expressed with the help of the rotation tensors. The local frame in the SmC phase can be reached from the laboratory system by rotating the director about the two-fold symmetry axis $\vec{c} \times \vec{k}$ by an angle θ, then by rotating the c-director around the layer normal $\vec{k}$ by the azimuth angle ϕ.[†] These transformations are expressed by the matrices:

$$T_\theta = \begin{pmatrix} \cos\theta & 0 & -\sin\theta \\ 0 & 1 & 0 \\ \sin\theta & 0 & \cos\theta \end{pmatrix} \qquad (8.8)$$

[†] We note here that these two angles are sufficient if the system is uniaxial, i.e., the molecules can be regarded cylindrical around the director. Since the individual molecules do not have exact cylindrical shape, it is true if the molecules can freely rotate around their long axes. Due to the tilt with respect to the layer normal, the rotation around the long axis is slightly biased, but we disregard this here.

and

$$T_\phi = \begin{pmatrix} \cos\phi & \sin\phi & 0 \\ -\sin\phi & \cos\phi & 0 \\ 0 & 0 & 1 \end{pmatrix} \qquad (8.9)$$

With these, the dielectric tensor can be expressed as $\hat{\varepsilon}(\theta,\phi) = \hat{T}_\phi \hat{T}_\theta \hat{\varepsilon} \hat{T}_\theta^{-1} \hat{T}_\phi^{-1}$, which explicitly gives:

$$\hat{\varepsilon}(\theta,\phi) = \begin{pmatrix} \varepsilon_\perp \cos^2\phi + \varepsilon_2 \sin^2\phi & \varepsilon_b \sin\phi\cos\phi & -\varepsilon_a \sin\theta\cos\theta\cos\phi \\ \varepsilon_b \sin\phi\cos\phi & \varepsilon_\perp \sin^2\phi + \varepsilon_2 \cos^2\phi & \varepsilon_a \sin\theta\cos\theta\sin\phi \\ -\varepsilon_a \sin\theta\cos\theta\cos\phi & \varepsilon_a \sin\theta\cos\theta\cos\phi & \varepsilon_\parallel \end{pmatrix} \qquad (8.10)$$

where $\varepsilon_a = \varepsilon_3 - \varepsilon_1$ is the dielectric anisotropy, and $\varepsilon_b = \varepsilon_2 - \varepsilon_1$ is the dielectric biaxiality. In addition, we defined the effective dielectric constant within the smectic layers as $\varepsilon_\perp = \varepsilon_1 \cos^2\theta + \varepsilon_3 \sin^2\theta$ and along the smectic layer normal as $\varepsilon_\parallel = \varepsilon_1 \sin^2\theta + \varepsilon_3 \cos^2\theta$. Eq. (8.10) can be used to calculate the effective dielectric constant in any structures. For example, the effective dielectric constant measured in the geometries shown in Figure 8.2a through Figure 8.2c can be calculated by equating $\phi = 0$ and multiplying the dielectric tensor with the corresponding electric field vector along the z, x and y axes, respectively. Since the dielectric constant is calculated from the induced polarization along the applied field, the effective dielectric constants become $\varepsilon_\parallel$, $\varepsilon_\perp$ and ε_2., in the geometries of Figure 8.2a, b and c, respectively.

Experiments on SmC and SmC* material indicate that the dielectric biaxiality is not necessarily negligible with respect to the dielectric anisotropy. An example for the temperature dependence of the dielectric constants for a commercially available ferroelectric SmC* material is shown in Figure 8.3.

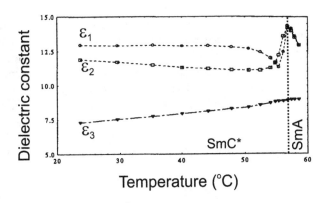

FIGURE 8.3
Temperature dependence of the dielectric tensor components[2] of a ferroelectric liquid crystal mixture FLC 6430 from Hoffmann La Roche[2]. d = 6 µm; p = 0.4–0.5 µm; f = 100 kHz.

It can be seen that $\varepsilon_b = 0$ in the SmA phase, and it is about -1 far from the SmA-SmC* transition.

8.2 Relations between Microscopic and Macroscopic Parameters

The macroscopic electric polarization has several microscopic origins.

$$\vec{P} = \vec{P}_{dis} + \vec{P}_o \tag{8.11}$$

$\vec{P}_{dis}$ is the charge displacement (distortion) polarization. If the field is small enough, $\vec{P}_{dis}$ is proportional to the field, and $\vec{P}_{dis} = \alpha \cdot \vec{E}_{loc}$, where α is the polarizability and $\vec{E}_{loc}$ is the local field. It is customary to separate it to the electronic and ionic polarizability, $\vec{P}_e + \vec{P}_i$. $\vec{P}_o$ is the orientational polarization, which exists in anisotropic materials of polar molecules. Although thermal motion (if macroscopic symmetry allows) averages out the molecular dipole moments, external fields, opposing the thermal disorder, result in a macroscopic polarization $\vec{P}_o$. $\vec{P}_o$ is a linear function of weak electric fields but saturates as the field becomes very strong; once a dipole is fully aligned with the field, its average dipole moment cannot increase any more.

The dielectric permittivity, which is a macroscopic parameter, can be related to microscopic parameters such as the polarizability α, molecular dipole moment μ, and the number density N, as:[3]

$$\varepsilon - 1 = \frac{NFh}{\varepsilon_0} \left[\alpha + \frac{\mu^2 F}{3k_B T} \right] \tag{8.12}$$

In this expression, F and h are reaction field and cavity field factors, respectively. They account for the field-dependent interaction between the molecules and the environment.

Although it looks as the molecular contributions to the permittivity are approximately additive, since $\varepsilon - 1$ is proportional to N, just as in the case of diamagnetic susceptibility, here F and h are also density dependent.

The distortion polarizability contributes only to the refractive index, and the dielectric constant can be related to the molecular dipole moment through the Kirkwood–Fröhlich equation, which in isotropic liquids reads[4] as:

$$(\varepsilon - n^2) \frac{2\varepsilon + n^2}{\varepsilon(n^2 + 2)} = \frac{N}{9\varepsilon_0 k_B T} \mu^2 \tag{8.13}$$

This means that, knowing n, ε, and N, we can get μ^2, i.e., the mean-square effective molecular dipole.

For anisotropic materials, assuming that the isotropic cavity factor is still valid the Kirkwood–Froehlich equation can be easily generalized when written in the form.[5]

$$\left(\varepsilon_i - n_i^2\right)\frac{2\varepsilon_i + n_i^2}{\varepsilon_i(n_i^2 + 2)} = \frac{N}{9\varepsilon_0 k_B T}\mu_i^2 \qquad (8.14)$$

where for uniaxial materials $i = \parallel, \perp$, denoting the coordinates parallel and perpendicular to the symmetry axis.

It is useful to note that the dielectric anisotropy of anisotropic liquid crystals can be related to the angle β that the molecular dipole moments make with the long axis.

$$\varepsilon_a = \frac{NhFS}{\varepsilon_0}\left\{\Delta\alpha + \frac{F\mu^2}{2k_B T}(3\cos^2\beta - 1)\right\} \qquad (8.15)$$

If the anisotropy of the distortion polarization is small compared to the dielectric anisotropy, one gets that $\varepsilon_a > 0$ if $\beta < 54.7°$, and $\varepsilon_a < 0$ for $\beta > 54.7°$. Although this rule is not always correct, generally we can say that a material with small β has positive dielectric anisotropy, whereas those with net dipole normal to the molecular long axis have negative anisotropy. This is illustrated in Figure 8.4 with the structures of the molecules that have very large positive and negative dielectric anisotropies.

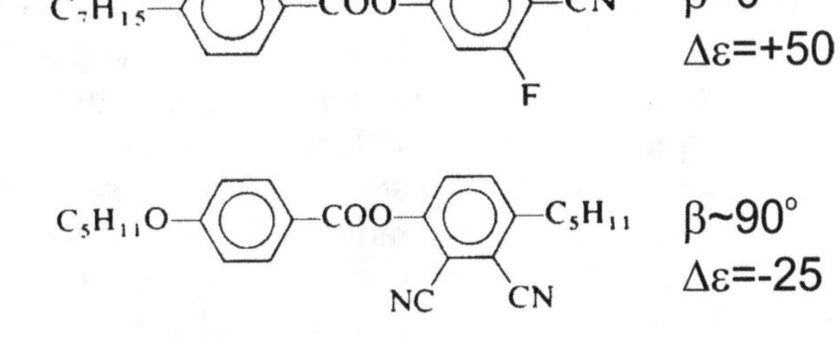

FIGURE 8.4
Schematic representation of a mesogen with an off-axis dipole moment making an angle β with the long axis, and the structure of molecules with very high positive and negative dielectric anisotropies.

8.3 Dielectric Spectroscopy

The dielectric spectroscopy of anisotropic fluids started in the 1970s[6,7] by the extension of the Debye model from isotropic media[8] (described in Appendix D) to uniaxial systems based on statistical mechanical Kubo formalism,[9] but no quantitative estimates about the critical frequencies or the susceptibilities were obtained. Quantitative estimates were given first on molecules with dipole moments along the long axis,[10] then for general dipole directions using the rotational Brownian picture in Maier–Saupe mean-field potential.[11] This theory was subsequently refined in the 1990s.[12]

The rotational diffusion coefficients for ellipsoids characterized by half axes of revolutions a, b, c, with respect to each axes can be given as:[13]

$$D_j = \frac{k_B T}{\xi_j}, \quad j = a, b, c \tag{8.16}$$

where $\xi_j = 8\pi \eta j^3$, $j = a,b,c$. For uniaxial ellipsoids, where b = c, and p = b/a is the aspect ratio, it is useful to determine the equivalent sphere with radius $a_{eq} = ap^{2/3} = (ab^2)^{1/3}$ and the corresponding diffusion coefficient $D_{eq} = k_B T / 8\pi \eta a_{eq}^3$. With these, it can be derived for very long rods (p << 1) that:

$$\frac{D_b}{D_{eq}} \approx \frac{3}{2}[1 + p^2 \ln p] \tag{8.17}$$

$$\frac{D_b}{D_{eq}} \approx -3p^2 \ln p \tag{8.18}$$

Substituting parameters typical to uniaxial nematic liquid crystals, such as p ~ 0.1, a = 2 nm, $\eta = 0.1$ Pa·s, we get D_b ~ 2×10^6 Hz, D_a ~ 10^8 Hz, which correspond to half of the relaxation frequencies.

Based on these estimates, the dielectric spectra were calculated at the function of the angle β between the resulting dipole moment and molecular long axis. The results of the calculations[11] were found to be in reasonable agreement with the experiments and are illustrated in the dielectric loss spectra as shown in Figure 8.5 in the parallel and perpendicular alignments, respectively. It can be seen that there are two well-separable modes when the director is parallel to the electric field (typically this is the homeotropic alignment): the lower frequency mode is typically in the 10-MHz range and is due to the flip-flop motion around the short axis of the rod-shape molecules; the higher frequency modes are in the gigahertz frequency range and can be associated with the rotation around the long axis. They become more important at increasing angles β between the resulting dipole moment and the long molecular axis (Figure 8.5a). In case of the planar alignment (electric field is perpendicular to the director), experimentally only one broad dispersion can be observed in the 100 MHz–1GHz range (see Figure 8.5b), which actually are the superpositions of several different rotation mechanisms.

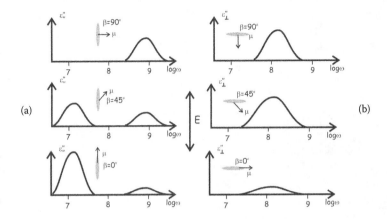

FIGURE 8.5
Calculated imaginary part of the dielectric permittivity when the electric field is parallel (a) and perpendicular (b) to the director.[11]

Experimentally, it is indeed observed that the complex dielectric permittivity of a nematic liquid crystal has two dispersion regions: one when the measuring external electric field is parallel and one when it is perpendicular to the director.[14] A typical experimental dielectric spectra of a nematic liquid crystal 4-pentylphenyl-4-propoylbenzoate is shown in Figure 8.6.

An experimental Cole–Cole plot of 6CB is shown in Figure 8.7.

The rotational diffusion constant and relaxation frequency depend on temperature mainly through the temperature dependence of the viscosity, which

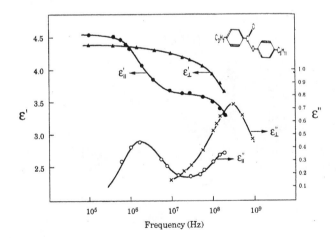

FIGURE 8.6
Complex permittivity components for 4-pentylphenyl-4-propoylbenzoate.[15] Note: it is the same material as of Figure 8.1a.

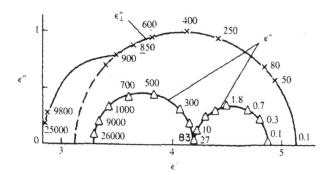

FIGURE 8.7

Cole–Cole plot of permittivity for the nematic phase of 4-heptyl-4-cyanophenyl (6CB). The numbers adjacent to the data points indicate the frequency in Hz.

usually follows the Arrhenius behavior: $f_r \sim D \propto e^{-\frac{E_a}{k_B T}}$, where E_a is the activation energy. It is customary to plot the logarithmic of f_r versus $1/T$, which therefore gives straight lines in each phase with different slopes, corresponding to the different activation energies in the different phases. A measured result is shown in Figure 8.8 for a material with a number of different phases.

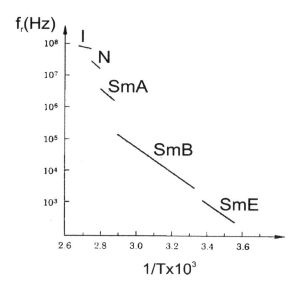

FIGURE 8.8

Relaxation frequencies plotted on a logarithmic scale as a function of inverse temperature for 4-pentylphenyl-4′-carboxylate[16].

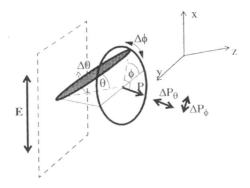

FIGURE 8.9
Schematic representation of the Goldstone mode and soft modes involving variation of the azimuth angle ϕ and tilt angle θ, respectively.

8.4 Dielectric Torque

The contribution of the electric field to the free energy of a uniaxial dielectric material is:

$$F_{el} = -\int \vec{D}d\vec{E} = -\frac{\varepsilon_o}{2}\left(\varepsilon_\perp E^2 + \varepsilon_a(\vec{n}\cdot\vec{E})^2\right) \tag{8.21}$$

For $\varepsilon_a > 0$ the free energy is smallest when the director is parallel to the electric field (homeotropic alignment). Similarly, for $\varepsilon_a < 0$ the free energy is smallest when the director is perpendicular to the electric field (planar alignment in thin films). If initially the director were in the "wrong" direction not corresponding to the lowest free energy, a dielectric torque would arise, which would try to orient the director to the proper direction. If other torques would compete with the electric torque, the equilibrium alignment would be determined by the balance of the different torques. Practically the most important situation is when the surface alignment promotes a director orientation, which is disfavored by external electric fields. In this case, the director becomes realigned when electric field is applied, and it relaxes back to the original orientation when the field is removed. This is the principle of the electro-optic displays that we will summarize in Chapter 8.

Similar to the magnetic torque, an electric torque can be given as:

$$\vec{\Gamma}_E = -\vec{D}\times\vec{E} = -(\hat{\varepsilon}\cdot\vec{E} + \vec{P}_o)\times\vec{E} \tag{8.22}$$

Here P_o is the ferroelectric polarization of the material, which will be discussed later. If the director (the direction corresponding to the largest eigenvalue of the dielectric tensor) makes an angle θ, and the ferroelectric

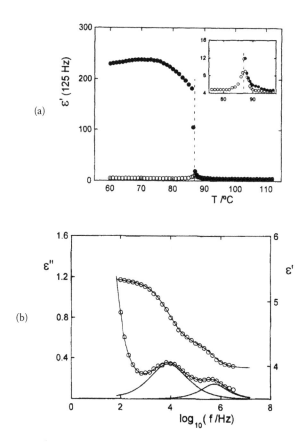

FIGURE 8.10
(a) Real part of the dielectric permittivity at 125 Hz vs. temperature of a ferroelectric liquid crystal: 6-(2S-2-chloro-3-methylbutanyloxy)-2-naphtyl 4-decyloxybenzoate;[18] (•) bias field 0; (o): bias field 0.4 V/μm. (b) Complex dielectric permittivity in the SmC* phase of the same material at 0.4 V/μm bias field. The higher frequency relaxation is due to the soft mode; the lower frequency one is the remaining of the Goldstone mode.

polarization makes an angle ϕ with respect to the electric field, the total electric torque can be written as:

$$\vec{\Gamma}_E = -\varepsilon_a(\vec{n}\cdot\vec{E})\vec{n}\times\vec{E} - \vec{P}_o\times\vec{E} = -\varepsilon_a E^2\sin\theta\cdot\cos\theta - P_oE\sin\phi \qquad (8.23)$$

The first term on the right hand side is the dielectric torque, and the second term is the ferroelectric torque. In case of dielectric materials, such as nematic liquid crystals, the ferroelectric torque is zero, and the balance between the electric and elastic torques becomes analogous to that between the diamagnetic and elastic torques. Accordingly, all the calculations described for magnetic fields between Eq. (7.17) and (7.30) can be repeated by simply substituting $\chi_a H^2$ with $\varepsilon_o\varepsilon_a E^2$. For example, similar to the magnetic coherence length defined in Eq. (7.20), one also can define the electric coherence length as:

$$\xi(E) = \frac{1}{E}\sqrt{\frac{K}{\varepsilon_o \varepsilon_a}} \tag{8.24}$$

where K is a relevant elastic constant describing the director deformation.

It is also easy to see that Freedericks transition can be induced by electric field, too. In analogy with (7.30) one can get the threshold for the transition as:

$$E_{ci} = \frac{\pi}{L}\sqrt{\frac{K_{ii}}{\varepsilon_o |\varepsilon_a|}} \tag{8.25}$$

where $K_{ii} = K_{11}$ or K_{33} for the splay (S) or for the bend (B) Freedericks transitions, respectively. In this expression L is the film thickness, which means that the threshold voltage $V_c = E_c \cdot L$ is independent of the film thickness. The director deformations induced by the electric fields are also similar to those by the magnetic fields. The situations in the most common case when the electric field is applied across the film, are shown in Figure 8.11.

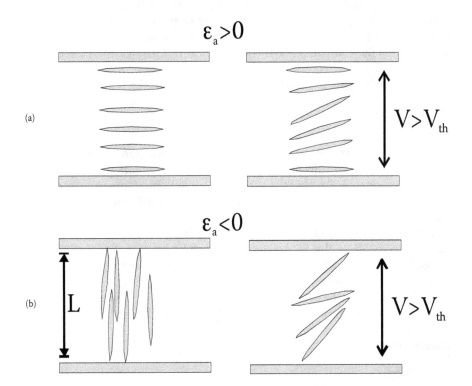

FIGURE 8.11
Splay and bend electric Freedericks transitions in (a) homogeneous and (b) homeotropic nematic cells with (a) positive and (b) negative dielectric anisotropies.

Comparing electric and magnetic interactions, we can utilize the balance of these two interactions as:

$$\mu_o^{-1}\Delta\chi B^2 = \varepsilon_0\Delta\varepsilon E^2 \tag{8.26}$$

Inserting parameters typical for liquid crystals ($\Delta\chi \sim 10^{-7}$ (SI), $\Delta\varepsilon \sim 10$), we get that comparable contribution requires $B/E = \frac{1}{c}\sqrt{\frac{\Delta\varepsilon}{\Delta\chi}} \sim 3\cdot10^{-5}$. This means that that even as much as B ~ 10 Tesla of magnetic induction can be balanced by as small as E ~ 0.3 V/μm electric fields. Note, that E > 1V/μm electric fields are easily accessible for thin films (d < 10 μm), whereas a magnetic induction of B = 10 Tesla can be achieved only by powerful superconducting magnets. It can be seen, therefore, that for thin films the electric field is much more effective than the magnetic to induce reorientation.

However, we have to emphasize that the magnetic interactions are much more clean, because they do not cause electrohydrodynamic or other kinds of instabilities, which are very usual for electric fields. A recent review of electrohydrodynamic instabilities can be found in several more specialized textbooks.[19,20]

If the director is already parallel to the electric field, the free energy decreases as ε_a increases. This means an ordering of the anisotropic state, which in the isotropic phase is equivalent to a field-induced transition to the nematic liquid crystal phase. This is similar to the case when the pressure induces a transition to the lower symmetric phase. Such a situation is described by the Clausius–Clapeiron equation that relates the increment of the phase transition temperature, ΔT_{NI}, to the pressure. In our case this is equivalent to the field-induced increase of the isotropic–nematic transition as:

$$\Delta T_{NI} \approx \frac{(\varepsilon_{||} - \varepsilon_{is})E^2}{2\Delta H} \tag{8.27}$$

where ΔH is the enthalpy for the *N–I* transition.[21] Below the isotropic–nematic transition this effect results in an increased uniaxial orientational order parameter by suppressing director fluctuations. For the chiral smectic C* phase a linear term (P_sE) due to the spontaneous polarization P_s must be taken into account, too.

Electric fields can also give rise to changes in the symmetry and, for example, may induce a tilt in the *SmA* phase. Another example is the field-induced biaxiality observed in nematic materials[22] with $\varepsilon_a < 0$ and $\vec{E}\perp\vec{n}$. In this case, the rotation around the long axis becomes hindered, and $\varepsilon_\perp$ splits into two values. The degree of biaxiality is defined as $\varepsilon_b = \varepsilon_{\perp1} - \varepsilon_{\perp2}$. Usually the induced biaxiality is orders of magnitude smaller than of the dielectric anisotropy.

8.5   Electric Conductivity

Although organic dielectric fluids may be very weakly conducting ($\sigma \sim 10^{-17}$ ohm^{-1}m^{-1}), in practice they are not perfectly insulating, and their conductivity typically ranges between $\sigma \sim 10^{-6}$ and 10^{-12} ohm^{-1}m^{-1}. For display application,

$\sigma < 10^{-10}$ $(\Omega m)^{-1}$ is desired. (Note: often the conductivity is given in Siemens/ m units, where $1\ S = 1\ \Omega^{-1}$.) The contaminations can be ionic, mainly introduced by reaction byproducts, but neutral impurities may also dissociate into anions A^- and cations B^+, with subsequent recombination according to $AB \Leftrightarrow A^- + B^+$.

We consider low rate of ionization, i.e., the concentration of neutral molecules is much larger than that of the ions. In such systems, the equilibrium ion concentration can be determined by the following differential equation:[18]

$$\frac{dn(t)}{dt} = \beta N - \alpha n^2(t) \tag{8.28}$$

where μ is the mobility of the ions (could be different for the positive and negative ions), N is the concentration of the neutral impurity molecules, n is the concentration of the positive or negative ions per unit volume, α is the recombination coefficient (it is multiplied by n^2, because two ions have to be simultaneously present to achieve recombination), and β is the ionization coefficient. This equation gives that the concentration of the impurities can be given as:[24]

$$n_o = \sqrt{\beta N/\alpha} \tag{8.29}$$

In a weak uniform electric field does not change this concentration, but only results in a drift of the impurities, so the drift current is given by:

$$J = qn_o(\mu_+ + \mu_-)E \tag{8.30}$$

Here q is the charge of an ion, and μ_+ (μ_-) are the mobility for the positive (negative) charges.

The conductivity $(\sigma = J/E)$ of a weakly ionized material (just like most liquid crystals) can be written as:

$$\sigma = q(\mu_+ + \mu_-)\left(\frac{\beta N}{\alpha}\right)^{1/2} \tag{8.31}$$

where $\mu_+ \sim \mu_- = \mu \sim 10^{-6}cm^2s^{-1}V^{-1}$, and typically $\beta/\alpha \sim 10^{14}cm^{-3}$.

Using the argument that an ion, moving under a concentration gradient, experiences the same resistance from the solvent as when it migrates in a potential gradient, we conclude that the friction coefficient ξ for electric conduction is the same as for diffusion.[25] Accordingly, for ions that are large enough, the charge mobility is related to the diffusion (Nernst–Einstein relation[26]) as:

$$\mu = \frac{qD}{k_BT} \tag{8.32}$$

With this, we can estimate the number of impurities in the weakly conducting materials as:

$$n_o = \frac{\sigma k_B T}{q^2 D} \tag{8.33}$$

If we take the example of cyanobiphenyls with $\sigma \sim 10^{-9}\ 1/(\Omega m)$ *and* $D \sim 10^{-11}$ *m^2/s, we get that:*

$$n_o = \frac{10^{-9} \times 10^{-23} \times 300}{3 \times 10^{-38} \times 10^{-11}} \approx 10^{19}\, m^{-3}$$

Taking $\sim 200\ cm^3 = 2 \times 10^{-4}\ m^3$ per mole, we get that $n_o \sim 5 \times 10^{15}/mole$. As one mole contains 6×10^{23} molecules, this number means that $n_o \sim 10^{-7}/molecule$. This is a very small value, i.e., it is hard to eliminate.

Conductivity is a thermally activated process, and the temperature dependence follows the Arrhenius behavior, i.e.:

$$\sigma = \sigma_o \exp(-E_a/k_B T) \tag{8.34}$$

Here E_a is the activation energy of the charge carriers. It depends on phase structure; therefore, $\sigma(T)$ can be used to monitor the phase transitions.

In case of anisotropic materials, the conductivity is also anisotropic, such as the dielectric and diamagnetic coefficients; however, its anisotropy is related to the dopants, and usually $\sigma_\parallel / \sigma_\perp \neq \varepsilon_\parallel / \varepsilon_\perp$. For acceptor–donor type impurities, typically $\sigma_\parallel/\sigma_\perp \sim 1.05 - 1.3$ (increasing slightly with concentration), whereas for ionic impurities the anisotropy is nearly 2.

Defining the anisotropy of conductivity as

$$\Delta\sigma = \sigma_\parallel - \sigma_\perp = qn(\mu_\parallel - \mu_\perp) \tag{8.35}$$

with the Nernst–Einstein relation that relates the mobility to diffusion coefficient of the charge carriers, we get for anisotropic media:

$$\sigma_\parallel/\sigma_\perp = \mu_\parallel/\mu_\perp = D_\parallel/D_\perp \tag{8.36}$$

The diffusion anisotropy depends on the shape on the molecules and on the structure of the phase; for example, for columnar liquid crystals the ratio can be as large as 10^6! In nematic phase of rod-shape molecules, the anisotropy is positive and typically increases at lower temperatures in proportion to the increase of the order parameter. However, the presence of a smectic phase below the nematic may modify this picture, because the conductivity anisotropy is usually negative in the smectics due to the reduced diffusion constant normal to the smectic layers in connection with the density modulation in this direction.

The electric conductivity has important effects on electric measurements. To illustrate this, we consider a sandwich cell, when the material is placed between two conducting plates. Initially the distribution of the ions is

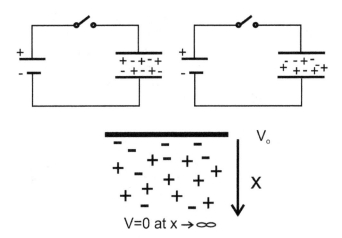

FIGURE 8.12
Distribution of the free charges in a weakly conducting dielectric fluid in sandwich cell geometry when an electric potential difference is applied between the plates.

random, but in a potential gradient the medium will be polarized, as shown in Figure 8.12.

To estimate how thick is the charged layer, we utilize that at any point at distance x from the charged plate the charge density can be written as:

$$\rho(x) = n_+(x) \cdot q - n_-(x) \cdot q \tag{8.37}$$

where n_+ (n_-) is the concentration of the positive and negative ionic carriers. Since the charge carriers are not interacting with each other at low concentrations, their spatial distribution follows the Boltzmann distribution,[27] i.e.:

$$\rho(x) = n_o q \left[\exp\left(-\frac{qV(x)}{k_B T}\right) - \exp\left(\frac{qV(x)}{k_B T}\right) \right] \tag{8.38}$$

For small applied voltages this gives:

$$\rho(x) \approx -2 n_o q^2 V(x)/k_B T \tag{8.39}$$

Utilizing that $\vec{\nabla}\vec{D} = \rho$, $\vec{D} = \varepsilon\vec{E}$, and $E = -\frac{\partial V}{\partial x}$, we obtain the Poisson–Boltzmann equation that reads,

$$\frac{\partial^2 V}{\partial x^2} = \frac{2 n_o q^2}{\varepsilon_o \varepsilon k_B T} V(x)$$

This provides that:

$$V(x) = V_o \times \exp(-\kappa x) \tag{8.40}$$

where $\kappa^{-1} = \sqrt{\frac{\varepsilon_0 \varepsilon k_B T}{2 n_0 q^2}}$ is the Debye screening length. For example, taking $n_0 = 10^{20}$ m^{-3}, $\varepsilon \sim 10$, T $\sim$ 300K, we get $\kappa^{-1} \sim 3 \times 10^{-7}$ m = 300 nm. This means that the potential is screened outside a thickness of about 0.3 µm near the electrodes.

Such a situation can be modeled as the response of a capacitor $C_{el} = \frac{\varepsilon_0 \varepsilon A}{\kappa^{-1}}$ with thickness κ^{-1} connected in series to the cell containing the fluid material characterized by a parallel resistance R and capacitance $C = \frac{\varepsilon \cdot \varepsilon_0 A}{d}$. In general, $d >> \kappa^{-1}$, so $C_{el}/C \sim \kappa d >> 1$.

The impedance of the corresponding cell Z_c then can be given as $Z_c = Z + \frac{1}{i\omega C_{el}}$, where $Z = \frac{R}{1 + i\omega RC}$. At low frequencies (i.e., when $\omega RC << 1$), $Z = R$, i.e., $Z_c = R + \frac{1}{i\omega C_{el}}$. This means that the voltage drop on the electrode capacitance is: $V_{el} = I \cdot Z_{el} = V_0 \frac{Z_{el}}{Z_c} = V_0 \frac{1}{1 + i\omega RC_{el}}$, which at very low frequencies is equal to the externally applied voltage, i.e., the voltage drops completely on the electrode area, and the studied material does not experience any electric field.

We note that this statement is true only in case of small fields, when the charge on the electrodes can be compensated by the free charges. If the charges on the electrodes are larger than the total free charges, some DC field will be experienced by the material. The minimum electric field needed to successfully apply DC field on the material can be calculated by keeping in mind that $Q = C \cdot V = \varepsilon_0 \varepsilon \frac{A}{d} V = \varepsilon_0 \varepsilon A E$, whereas the total free charges are given as $Q_{ion} = q \cdot n_0 \cdot A \cdot d$. At the threshold these two charges are equal, which gives: $E_{th} = \frac{q n_0 d}{\varepsilon_0 \varepsilon}$. With q = 1.6×10^{-19} C, $n_0 = 10^{19}$ m^{-3}, d = 10^{-5} m and $\varepsilon_0 \varepsilon = 10^{-10}$ C/Vm, we get $E_{th} = 10^5$ V/m. This means that if we apply more than 1 V on 10 µm cell, we can switch the director even with DC fields.

Defining the critical frequency f_c from the condition that there the voltage drop is equally shared between the electrode and the material, i.e., $2\pi f_c RC_{el} = 1$, we obtain that:

$$f_c = \frac{\sigma}{2\pi \kappa d \varepsilon_0 \varepsilon} \tag{8.41}$$

For example, with $\sigma_{LC} \sim 10^{-7} (\Omega m)^{-1}$; $\kappa^{-1} \sim 0.5$ µm; d $\sim$ 10 µm; $\varepsilon \sim 10$ we get: $f_c \sim$ 10Hz.

Mathematically, the conductivity can also be treated as a complex number (or tensor for anisotropic materials), i.e., it has both real and imaginary components ($\sigma = \sigma' + i\sigma''$). It is real much below f_c, and its imaginary component has maximum at f_c. Accordingly, it contributes both to the real and imaginary component of the complex dielectric constant as: $\varepsilon_\sigma'' = \sigma'/\varepsilon_0 \omega$ and $\varepsilon_\sigma' = \sigma''/\varepsilon_0 \omega$.

The effect of conductivity to the dielectric response is especially important in colloids, i.e., when nonconducting particles are suspended in weakly conducting fluids (for example in liquid crystals). In this case, a charge cloud with a thickness of Debye screening length will be present around each colloidal particle. In a steady field, the diffuse layer has a dipole character with a charge separation distance $l \sim a + 2\kappa^{-1}$, where a is the radius of the suspended particle and κ^{-1} is the Debye screening length (see (8.40)).[28] If the sense of the field is reversed, the dipole must revert by diffusion and electro migration (inertial effects can be neglected up to 10^7 Hz even for as large particles as 0.2 μm diameter), which requires a time of $\tau \sim l^2/D$ (D is the ion diffusivity). Thus temporal forcing of the charge cloud at a frequency ω will be on the same time scale as ionic transport when

$$f_c \approx \frac{D\kappa^2}{(2+a\kappa)^2} \tag{8.42}$$

For 0.2 μm particles with $D = 10^{-10}$ m^2/s and $\kappa^{-1} \sim 0.1$ μm, this would mean about 1 kHz characteristic frequency.

Just as the real (DC) part of the conductivity contributes to the dielectric spectra (see Eq. 0.25), so the dielectric loss, i.e., the imaginary part of ε^*, contributes to the electric current even in a purely insulating medium, as $\sigma = \omega\varepsilon_o\varepsilon''$. Accordingly, the apparent conductivity can be expressed as:

$$\sigma(\omega) \sim \sigma_{dc} + \frac{\varepsilon_o[\varepsilon(0) - \varepsilon(\infty)]\omega^2\tau}{(1+\omega^2\tau^2)} \tag{8.43}$$

In case of anisotropic materials, the anisotropy of the electric conductivity is not determined by the anisotropy of the ionic mobility, but by the ratio of the Debye relaxation times.

$$\frac{\sigma_{\parallel}}{\sigma_{\perp}} = \frac{\varepsilon_{\parallel}(0) - \varepsilon_{\parallel}(\infty)}{\varepsilon_{\perp}(0) - \varepsilon_{\perp}(\infty)} \cdot \frac{\tau_{\parallel}}{\tau_{\perp}} \tag{8.44}$$

since in general $\tau_{//} \gg \tau_{\perp}$ (see Figure 8.6). In this range the ratio of the electric conductivity can be as high as 10^4, whereas the ratio of the DC conductivities is less than 2.

8.6 Piezoelectricity

Piezoelectricity is one of the basic properties of crystals, polymers and liquid crystals. The prefix "piezo-" is derived from the Greek word "press," reflecting the nature of the first piezoelectric effects observed in crystals.

It has already been mentioned by Coulomb that electricity might be produced by pressure. Then Haüy (the "father of crystallography") and later Antoine Becquerel[29] performed experiments on certain crystals showing electrical effects when compressed; however, it turned out later that what they observed was mainly contact electricity.[30] In 1880 Pierre and Jacques Curie[31] detected positive and negative charges on some crystals compressed in particular directions, these charges being proportional to the pressure and disappearing when the pressure was withdrawn. This effect was not due to contact electricity, so the Curie brothers can be regarded as discoverers of piezoelectricity. Their approach to the phenomenon was largely influenced by Pierre Curie's previous work on pyroelectricity and crystal symmetry, which showed that polar electricity can be developed only in particular directions depending on the symmetry properties of the crystals. Their experiments were carried out on crystal slabs of blend, sodium chlorate, boracite, tourmaline, quartz, calamine, topaz, tartaric acid, sugar and Rochelle salt, using thin metal foil electrodes and a Thomson electrometer. One year later Lippmann[32] proposed, on the basis of thermodynamic principles, that the inverse effect (electrically induced pressure) must exist too. The Curie brothers were also those who experimentally verified this converse piezoelectric effect. In 1890 W. Voight published the first complete and rigorous formulation of piezoelectricity;[33] however, it was regarded as a curiosity for a long time, until the first practical application of this effect was mentioned by Langevin in 1918.[34] He applied the converse and direct piezoelectric effects to emit and detect underwater sound waves by means of large quartz plates, and thus opened the field of ultrasonics and hydroacoustics. Soon after this, Cady[35] demonstrated that the frequency can be stabilized if a resonating quartz crystal is associated with an electrical oscillator. Such quartz crystal oscillators were first used by the U.S. National Bureau of Standards as frequency standards, and later they became very useful in the development of broadcasting.

Discoveries of new piezoelectric materials started a new boom of research in the 1940s when A.V. Shubnikov predicted that piezoelectric properties would be found in amorphous and polycrystalline materials. His predictions were confirmed soon by observing that ferroelectric ceramics are strongly piezoelectric.[36] The existence of piezoelectricity for certain synthetic and biological polymers has also been known for a long time.[37] In particular, piezoelectricity in bone and tendon has been extensively studied.[38]

The piezoelectric effects in the polymers are generally small, but can be increased in synthetic polymer films when they are subjected to a strong DC electric field at elevated temperatures.[39] Due to their flexibility and the possibility to prepare films of large area, these materials opened up new applications and device concepts, which could not have been realized with conventional crystalline piezoelectric substances. The best-known and commercially most attractive example is the polyvynilidefluoride (PVDF), which has been utilized as the active element in many applications ranging from infrared detector technology to loudspeakers. Today piezoelectric devices are found in television sets, radios, wristwatches, small computer games, automobiles, etc. Many communication and navigation systems use piezoelectric resonators for frequency control, generation and selection. In spite of the more than century-long history, the study of piezoelectricity is still in an accelerating stage.

The definition of piezoelectricity extended considerably in time. Originally it was used only for crystals in connection with compressions, but later it was generalized to polymers and other materials for any strains and stresses, including shear. For historical reasons, in most cases direct and converse piezoelectric effects are distinguished. When electric polarization is produced by mechanical stress we speak about *direct piezoelectric effect*. Mathematically this can be expressed by the equation:

$$P_i = P_{oi} + \sum_{jk} d_{i,jk} T_{jk} \tag{8.45}$$

where P_i is the polarization, P_{oi} is the spontaneous polarization (if there is any), and T_{jk} is the stress tensor. The third rank tensor coefficients $d_{i,jk}$ are called piezoelectric coefficients.

In the *converse* (sometimes called "reciprocal" or "inverse") *effect*, the material becomes strained when electric field is applied. Mathematically, this is expressed as:

$$S_{jk} = S_{ojk} + \sum_{i} d_{i,jk} E_i \tag{8.46}$$

Here S_{jk} is the strain tensor ($S_{jk} = \partial s_i / \partial x_k$), where s_i is the displacement of a volume element with respect to the equilibrium position. S_{ojk} is called spontaneous strain, and E_i is the external electric field.

Both effects are manifestations of the same fundamental property of the substance: linear coupling between a second-rank tensor (like strain or stress) and a first-rank tensor (such as electric field vector). This is emphasized by the fact that the same coefficients d_{ijk} enter both in Eq. (8.45) and (8.46).

We note that, sometimes the symmetric stress and strain tensors with six components are represented as six element vectors T_λ and S_λ, where $\lambda = 1/2(k + j)\delta_{jk} + [9 - (i + j)](1 - \delta_{jk})$, i.e., the 11➔1; 22➔2; 33➔3; 23 = 32➔4; 13 = 31➔5; 12 = 21➔6 transformations are used.[40] In this case, the piezoelectric constants are formally expressed as 3×6 element second-rank tensors. These notations are simpler, but much less transparent than the third-rank tensor notation, so in the following, we will keep the mathematically more transparent notation.

The nonvanishing piezoelectric coefficients are determined by the symmetry of the material and can be determined by the Curie principle, which states that the tensor coefficients characterizing material properties should be invariant under the symmetry transformations of the substance.[41]

From the definitions of (8.45) and (8.46), it follows that d_{ijk} will transform as the product of x_i, x_j, x_k. Accordingly, in a system with inversion symmetry, transformations like x −x, y −y, and z −z, would require that $d_{-i,-j-k} = (-1)^3 d_{i,jk}$ for any i,j,k, which is equivalent to the statement that systems with inversion symmetry cannot be piezoelectric. At the other side of the symmetry range, the materials with C_1 symmetry allow the presence of all piezoelectric constants. Since $d_{i,jk} = d_{i,kj}$, it means $3 \times 6 = 18$ constants. The number of the possibly nonzero components can be determined for each crystallographic groups (see Appendix C.2). All the piezoelectric materials have no inversion symmetry; 20 out of the 32 crystallographic groups[42] and 3 of the continuous point groups belong to this category.

Phase	Molecular shape	Symmetry	Non-zero piezoelectric constants
N, SmA, L_α	Cylinder	$D_{\infty h}$	N/O
N*, SmA*	Chiral cylinder	D_∞	$d_{1,23} = -d_{2,13}$
SmAP,	Bent shape	C_{2v}	$d_{3,11}$; $d_{3,22}$; $d_{3,33}$; $d_{1,13}$; $d_{2,23}$
SmC	Cylinder	C_{2h}	N/O
SmC*	Chiral cylinder	C_2	$d_{3,11}$; $d_{3,22}$; $d_{3,33}$, $d_{1,13}$; $d_{2,23}$; $d_{1,23}$; $d_{2,13}$;
SmCP	Bent shape	C_2	$d_{3,12}$
SmC$_G$	Bent shape	C_1	$d_{3,11}$; $d_{3,22}$; $d_{3,33}$; $d_{1,13}$; $d_{2,23}$; $d_{1,23}$; $d_{2,13}$; $d_{3,12}$ all $d_{i,jk}$
Col$_h$; H$_1$,	Cylinder	D_{6h}	N/A
H$_2$	Chiral cylinder	C_2	$d_{3,11}$; $d_{3,22}$; $d_{3,33}$, $d_{1,13}$; $d_{2,23}$; $d_{1,23}$; $d_{2,13}$; $d_{3,12}$
Col*$_{tilt}$	Bowl shape	$C_{\infty v}$	$d_{3,33}$; $d_{3,11}=d_{3,22}$
P$_h$	Chiral bowl	C_∞	$d_{3,33}$; $d_{3,11}$; $d_{3,22}$
P*$_h$	Tilted bowl	C_2	$d_{3,33}$; $d_{3,11}$; $d_{3,22}$
P$_{tilt}$			$d_{3,11}$; $d_{3,22}$; $d_{3,33}$; $d_{1,13}$; $d_{2,23}$; $d_{1,23}$; $d_{2,13}$; $d_{3,12}$

FIGURE 8.13
List of most important liquid crystal phases with their symmetry and nonvanishing piezoelectric coupling constants.

By symmetry, not only solids can represent linear coupling between electric (polarization, electric field) and mechanical (stress, strain) quantities, but liquid crystals and other organized fluids can do so, too. Structured fluids have low symmetry, and a rich variety of piezoelectric type coupling constants may exist. Lack of inversion symmetry can be due to chirality or special molecular shapes and asymmetric packing. Chiral liquid crystal phases are the cholesterics, chiral smectics and chiral columnar phases.

Figure 8.13 shows that a number of liquid crystal phases can have linear coupling between electric field and mechanical strain. However, liquid crystals are liquid at least in some directions, so they cannot typically maintain elastic restoring forces. Accordingly, without generalization of the definition of strain, linear electromechanical effects can be regarded piezoelectric only in directions where the material is of elastic type (e.g., in the direction normal to the layers in smectics, and in the plane normal to the columns in columnar liquid crystals). The borderline between elastic and viscous behavior, however, is washed out in "soft materials" such as liquid crystals. For example, we can ask if hexatic phases (SmB, SmI, etc.) with quasi-long-range order are elastic or not. It definitely would be very important to generalize the definition of piezoelectricity to fluid or partially fluid directions. The mathematical definition given for solids can formally maintained in the fluid directions by generalizing the definition of strain tensor $\vec{S} = \nabla \cdot \vec{s}$ to hold even for displacements caused by viscous flows. This obviously can be done only for periodic motions, when $\vec{s}$ can be regarded as the amplitude of the vibration. For static cases, the displacement and the piezoelectric coupling constant determined by this method would increase in time, and even for periodic situations the piezoelectric constant could show abnormal frequency behavior due to the viscous nature of the stress. For these reasons, other expressions, such as "linear electromechanical effect," may be preferable

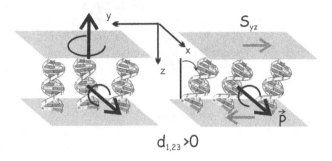

FIGURE 8.14
Explanation of the piezoelectric response in SmA* materials.

for fluid-like motions. One may observe the "real" piezoelectric behavior of liquid crystals, for example when their structure is frozen in a glassy state, or the molecules are crosslinked, like in elastomers. Keeping in mind, however, that the structure and symmetry of a frozen in state is the same as of the partially fluid one, the word "piezoelectricity" may be used even in the fluid directions.

From Figure 8.13 we see that the SmA* and cholesteric phases may posses a piezoelectric response in a geometry where electric polarization is induced normal to the shear plane. This is illustrated in Figure 8.14. The smectic A* materials are composed of chiral molecules (illustrated as helix). Because of the molecular chirality, they do not have mirror symmetry (mirror would invert the handedness), but they have two 2-fold symmetry axes: one is in plane of the layers (x) the other is parallel to the layer normal (z). When a shear in the yz plane is applied, a small tilt of the molecules is induced. When this happens, the material will not be symmetric with respect to 180° rotation along the layer normal. It means that a vector along the remaining symmetry axis along x would not be averaged out. This means, for example, that the x component of the molecular dipole moments will not be averaged out, but would result in a net polarization. The polarization is sensitive to the sign of the strain and changes sign when the shear is reversed. This, therefore, is a piezoelectric effect given by the coupling constant $d_{x,yz} = P_x/S_{yz}$. The inverse of the effect, i.e., an electric field along x would result in a shear strain, which would also lead to a tilt of the director. The optical consequence of this effect, rotation of the optic axis (the director) proportional to the electric field, is known as electro-clinic effect.[43,44] It is especially strong near to a transition to the spontaneously tilted SmC* phase and provides very fast (~1–10 μs) threshold-less switching.

In addition to the above-mentioned converse piezoelectric effect, there is an additional quadratic electromechanical effect associated with the electro-clinic. This is due to the decrease of the layer spacing upon the director tilt. Since the number of layers cannot vary quickly, the change of the layer spacing will result in a variation of the film thickness. This results in a force normal to the cover plate. This effect is strongest when the polarization is parallel to the electric fields. Although the induced tilt angle changes sign with the electric field, the resulting layer contraction is the same independent of the sign of the field. Because the force acting on the substrates is independent of the sign of the electric field (quadratic electromechanical effect), this latter effect is not a piezoelectric phenomenon.

We emphasize that in the piezoelectricity of the SmA* the chirality and the tilt are needed for the electric polarization. This concept has enormous importance in explaining piezoelectric and ferroelectric phenomena in structured fluids. The converse effect, i.e., the electroclinic, is another representation of the chirality-tilt-polarity triangle. In that case the molecular chirality and the field-induced polarity lead to the tilt.

Note that the tilt does not necessarily require a layer structure, and the same concept can be employed to explain the piezoelectricity of the cholesteric liquid crystal. In this case, the shear along the helical axis leads to a tilt of the director toward the helical axis and a polarization normal to the shear plane. This "shear electricity" was predicted by Prost[45] and experimentally was verified in case of cholesteric elastomers, where the molecules are weakly cross-linked and thus can sustain elastic strains.[46]

Let us note here that a kind of linear electromechanical effect (electric field-induced director rotation) observed in cholesterics,[47] does not fall into the category of piezoelectricity, since it does not involve a collective displacement of the center of mass.

We see from Figure 8.13 that a number of liquid crystal phases, SmC* of chiral rod-shape and the tilted columnar phase of chiral disc shape molecules, as well as the *SmCP* of achiral bent-core, and the tilted bowl-shape molecules all have C_2 symmetry with eight independent piezoelectric coefficients. The direct[48,49] and converse[51,52] piezoelectric effects have been mostly studied in the fluid SmC* liquid crystals.

The direct piezoelectric (or mechano-electrical) effect is due to the shear polarization[48] arising from the distortion of the helix as illustrated in Figure 8.15.

Direct piezoelectric type signals were also detected in thermotropic and lyotropic SmC* substances in geometries where the flow was induced by oscillating air pressure.[52] Piezoelectric response of SmC* elastomers[53,54] and of glasses[55] have also been observed. In the latter case, a fluid liquid crystal was cooled to a glassy state in the presence of electric field, so the glass became poled. It is remarkable that the magnitudes of the piezoelectric constants (50pC/N) were found to be comparable to the poled piezoelectric polymer PVDF.[56]

The converse piezoelectric effect, when electric field induces mechanical deformation, has special importance because of its influence on the electro-optical responses; they are mainly unwanted since they result in misalignment, but with clever design can be used to re-heal alignment.[57] It is interesting to note that in the audio frequency ranges, the vibrations result in audible acoustic effects, implying their possible use in electromechanical transducers.[58]

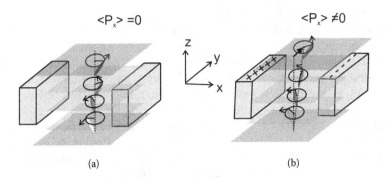

(a)   (b)

FIGURE 8.15
Illustration of the shear deformation of the director structure and the induction of net polarization in SmC* phase. The smectic layers are parallel to the xy plane. The tilt plane rotates from one layer to the other. The spontaneous polarization $\vec{P}$ is normal to the tilt plane (see detailed explanation later). (a) Undisturbed helical structure and the macroscopic polarization is averaged out. (b) Due to the action of the shear along y, the axial symmetry of the configuration is broken and a nonzero polarization component appears along x.[48]

A systematic study[59] with accurate control of the alignment revealed that the vibrations parallel to the smectic layers and the film surface are generally the strongest, especially if the polarization is also parallel to the plates. Transversal vibrations involving variation of sample thickness are the strongest for alignments with polarization normal to the plates. The vertical vibration is due to the electroclinic effect and the mechanical coupling between the director tilt and smectic layer spacing as illustrated in Figure 8.16.

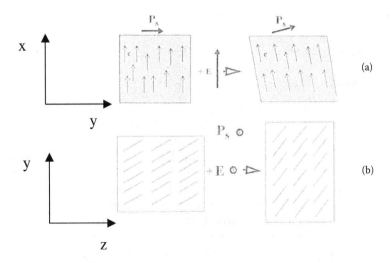

FIGURE 8.16
Illustration of the linear electromechanical effect of SmC* materials due to the backflow (a) and the electroclinic mode (b).

The vibration along the film surfaces (shear strain) (see Figure 8.16a) is due to the coupling between the director rotation about the layer normal (Goldstone mode) and viscous flow (backflow).[59] This mode is excited by the electric field via the torque $ExP_o = EP_o sin\phi$. The magnitudes of small oscillations are accordingly proportional to $sin\phi$ and are largest when the polarization is parallel to the plates. Using a nematic-like description for the flow processes inside the smectic layers and taking into account the relevant constraints (the director rotates around a cone, the electric field couples linearly to the polarization, which is always perpendicular to the director), the mechanical stress induced by the electric field was found to be parallel to the film surface. For unwound bookshelf textures:

$$\sigma_{xy} = \frac{\gamma_2 \cos 2\phi - \gamma_1}{2\gamma_1} EP_o \sin \phi \qquad (8.47)$$

where we have followed the notations described in Chapter 4, i.e., γ_1 is the rotational viscosity and $\gamma_2 = \eta_b - \eta_c$, where η_b and η_c are the shear viscosities with the director in the direction of the flow gradient and parallel to the flow, respectively.

The vibration normal to the film substrates (see Figure 8.16b) is due to the variation of the layer spacing. In contrast to the case of the SmA* phase, in a tilted phase the resulting displacement is proportional to the field.

Piezoelectric effects were also studied[60] in ferroelectric columnar liquid crystals, which have the same C_2 symmetries as of the SmC* materials.[61] Piezoelectrical effects were observed also on various biological systems,[62] in lyotropic liquid crystals and in membranes.[63]

8.7   Flexoelectricity

Closely related phenomenon to the piezoelectricity in liquid crystals is the flexoelectricity introduced by R.B. Meyer.[64] Flexoelectricity means a linear coupling between the distortion of the director and the electric polarization. The constituent molecules of the nematic liquid crystals are rotating around their axes, and in absence of electric fields they are nonpolar. However, polar axes can arise in a liquid crystal made up from polar pear- or banana-shape molecules when they are subjected to splay or bend deformations, respectively. In these cases, the polar structures correspond to more efficient packing of the molecules (see Figure 8.17).

It was shown that, for weak deformations corresponding to the continuum limit, the induced flexoelectric polarization $\vec{P}_f$ is proportional to first-order space derivatives of $\vec{n}$. Higher-order derivatives are negligible in case of small a/l (a: molecular dimension, l: periodicity of deformation). Taking into account

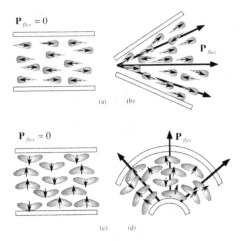

FIGURE 8.17
Illustration of the flexoelectricity assuming polar noncentrosymmetric molecules. Upper row: pear-shape molecules; Bottom row: banana-shape molecules. (After Meyer.[64])

the D_h symmetry of the system and the head-tail symmetry ($\vec{n} \Leftrightarrow -\vec{n}$), the flexoelectric polarization can be given by two coupling constants:

$$\vec{P}_f = e_1 \vec{n}(div\,\vec{n}) + e_3(curl\,\vec{n}) \times \vec{n} \qquad (8.48)$$

where e_1 (e_3) is the splay (bend) flexoelectric coefficient. From the equation we see that the dimension of the flexoelectric coefficients is the electric potential.

The molecular statistical approach to calculate the flexoelectric coefficients was developed independently by Helfrich[65] and Derzhanski and Petrov.[66] The calculation is based on the requirement to ensure maximum packing condition. The excess number ($\Delta N = N_+ - N_-$) of the molecules with dipole moment μ determines the electric polarization: $P = \Delta N \mu$. Dividing this by the distortion we get the flexoelectric coefficient.

Without going into details, we just give the results of Helfrich's calculation. Accordingly:

- Pear-shape molecules:

$$e_1 = \frac{2\mu_{\|} K_{11}}{k_B T} \theta_o \left(\frac{a}{b}\right)^{1/3} N^{1/3} \qquad (8.49)$$

With the estimates: $\mu_{\|} = 4 \cdot 10^{-30} Cm (\approx 1D)$, $K_{11} \sim 10^{-11}$ N, $k_B T \sim 5 \times 10^{-21}$ J, $\theta_o \sim 0.1$, $a/b = 1$, $N = 10^{21}$ $1/m^{-3}$, we get that $e_1 = 2 \times 10^{-12}$ C/m, which is close to the experimentally observed values.

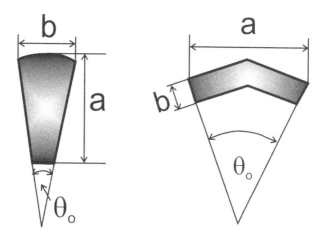

FIGURE 8.18
Molecular models for (a) pear-shape molecules; (b) banana-shape molecules, used for calculation of flexoelectric coefficients.

- Banana-shape molecules:

$$e_3 = \frac{\mu_\perp K_{33}}{2k_B T}\, \theta_0 \left(\frac{b}{a}\right)^{2/3} N^{1/3} \tag{8.50}$$

By dimensional considerations, an upper limit for molecules with very asymmetric shape can be placed as e_1, $e_3 \leq \frac{\mu_e}{a^2}$, where the molecular dipole μ_e is in the range of 1–5 Debye (1 D = 3.336 × 10^{-30} cm), and a ~ 2–4 × 10^{-9} m the typical molecular dimensions for low molecular weight liquid crystals. This gives that $e_3 < 10^{-12}$ C/m^{-1}. Measurements of the distortion due to electric field in a dielectrically stable configuration, e.g., homeotropic cell with $\varepsilon_a < 0$ and with weak surface anchoring,[67] gave that $e_3 \sim 1.2 \times 10^{-12}$ C/m.[67] This is as big as the upper limit of the theoretical value, which indicates that not only the dipolar mechanism contribute to flexoelectricity. Indeed, it turned out that flexoelectricity can exist even in materials with nonpolar and symmetric molecules. In this case, it is caused by the interaction between the electric field gradient and the molecular quadrupolar moments. (see Figure 8.18(c)).[68]

We need to note that mathematically the splay and bending deformations are vectors, i.e., the flexoelectric coupling constant is a second-rank tensor. Accordingly, the flexoelectricity should not be confused with piezoelectricity, which is described by a third-rank tensor coupling constant. Piezoelectricity requires lack of inversion symmetry of the phase, whereas flexoelectricity can exist in materials with inversion symmetry, such as in nematics with macroscopic D_h symmetry.

We also do not consider to be piezoelectric any effect which is characterized solely by director deformations. Piezoelectricity should always involve displacement of the center of a volume element of the substance.

8.8 Ferroelectricity

Ferroelectric materials are polar, in which the direction of polarization can be altered by electric fields.

The history of ferroelectricity traces back to 1665, when Seignette isolated the Rochelle salt ($NaKC_4H_4O_6\cdot4H_2O$) for use as laxatives. Physical studies of this material are not known until 1824, when Brewster[69] investigated its pyroelectric properties. In 1880 the Curie brothers mentioned it as the strongest piezoelectric material at that time.[70] The linear electro-optical properties of the Rochelle salt were first described in 1906 by Pockels.[71] The Ratz–Badecker precise investigations[72] led Debye[73] in 1912 to assume the presence of a permanent electric dipole moment. The same year, Schrödinger[74] introduced the name "ferroelectricity" based on the analogy with the ferromagnetism. The first rigorous description of ferroelectricity was given by Joseph Valasek,[75] who studied the hysteresis of the Rochelle salt.

In ferroelectric crystals the primary order parameter is the spontaneous polarization, $P_s \sim 10 \text{ mC}/\text{m}^2$, which can be switched as fast as 10 ns, though the difference in the birefringence between the states of opposite polarization directions is only 10^{-3}.

Ferroelectricity in biological systems is usually referred as bioferroelectricity.[76] It has been widely observed in biological materials[77] and may be common in biological cell components. Fröclich[78] has analyzed the electric field effects on biological membranes on the dipolar properties of the proteins dissolved in them. A relation between ferroelectricity, liquid crystals, nervous and muscular impulses was predicted by von Hippel.[79] Brain memory has been postulated to be based on a ferroelectric mechanism.[80] Beresnev and coworkers noted the close similarity between biomembranes and ferroelectric liquid crystals,[81] particularly the presence of a layered structure with tilted lipid and protein molecules and chiral molecules of cholesterol. The possibility of involvement of ferroelectric phenomena in membrane function was also suggested by several authors.[82]

Ferroelectricity in liquid crystals was first considered by Max Born in 1916,[83] who proposed a model to explain the isotropic–nematic transition by dipole–dipole interactions. He showed that polar rod-shape molecules of dipoles larger than about 5 Debye would align parallel to each other. The concept turned out to be false, since apolar nematic liquid crystals exist, too. In fact there is no evidence for polar nematic phases yet.

Although the first ferroelectric liquid crystalline substances were synthesized in Halle as early as 1909,[84] their ferroelectricity was not appreciated at that time (which is not a surprise, since the concept of ferroelectricity did not really exist). The concept of ferroelectricity in smectics was discussed first by Saupe in 1969 in a case of orthogonal nonchiral smectic in which all molecular dipoles are pointing in the same direction.[85] In the same paper, a possible antiferroelectric and helical smectic C arrangement in enantiomeric molecules was studied, too. However, it was R.B. Meyer[86] who realized that chiral tilted smectic liquid crystals may have spontaneous polarization along the smectic layers normal to the c-director. His simple symmetry argument utilized the chirality + tilt = polarity concept, which we already have explained in connection with the piezoelectricity and electroclinic of the SmA* materials. Accordingly, SmC* locally has only a two-fold symmetry axis along $\vec{k} \times \vec{c}$. Consequently, any vector in that direction (such as molecular dipole) would not average out, since the inversion of this vector is not allowed due to the absence of the mirror symmetry in the tilt plane. The comparison to nonchiral SmC materials of rod-shape molecules is illustrated in Figure 8.19.

8.8.1 SmC* Liquid Crystals

Ferroelectricity in liquid crystals is mostly studied in the SmC* liquid crystals. These studies resulted in over five thousand patents and thousands of papers. Here we just mention those properties that are in the line of our focus. Much more detailed overviews are given in a number of books and reviews.[87,88,89,90,92]

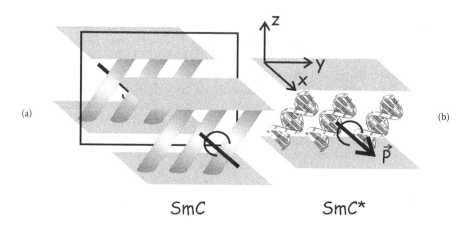

FIGURE 8.19

Illustration of the local symmetry of the chiral and nonchiral SmC phases of rod-shape molecules. Both SmC and SmC* have a two-fold symmetry axis normal to the c-director. In SmC (a) the yz plane is a mirror plane, because the molecules have no handedness; in SmC* (b) the mirror image would have opposite handedness, so it is not a symmetry operation of the material. For this reason the x component of the molecular dipoles do not average out, allowing the presence of a permanent electric polarization of the material. Note: the double-twist representation of the molecules serves only to indicate the chirality, and we do not assume such internal molecular structure.

The first verification of the ferroelectricity of the SmC* liquid crystal[86] was actually carried out on cells where the layers were parallel to the film surface and the electric field was applied along the smectic layers. For the detection of the action of the electric field on the SmC* phase, a conoscopic technique was used. For a sample much thicker than the helix pitch, the conoscopic figure is very similar to the one of the SmA phase: a series of concentric rings centered around the helix axis z. The conoscopic figure is shifted as a whole in the direction perpendicular to the field direction. The shift reversed when the direction of the electric field was reversed The obvious conclusion of this experiment was that the effect of the electric field is linear, so that there must be a spontaneous polarization in the SmC* phase. A more detailed interpretation of the experiment involves the deformation of the SmC* helix by torques exerted by the electric field on the polarization in smectic layers. Such a helical structure is illustrated in Figure 8.20b. This is similar to the helical structure of the cholesteric liquid crystals and is due to the transfer of chirality from the molecular to mesoscopic scales. Due to this helix the macroscopic polarization averages out in the scale over the pitch, which can vary from a few tenths of micrometer over to few hundreds of micrometers. Such a behavior is called heli-electricity[91] and is typically observed in films where the pitch of the material is much smaller than that of the film thickness. Applying electric fields, this structure can easily be unwound, resulting in a uniform polar structure (see Figure 8.20a and 8.20c). At small fields, the net polarization is proportional to the electric field; then at a threshold, typically in the range of 1 V/μm fields, the effective polarization jumps to a value corresponding to the spontaneous polarization. In case of the homeotropic alignment, the film is viewed normal to the smectic layers; however in practice, the transparent boundary plates are normal to the smectic layers (bookshelf alignment), and the field is applied across the plates as shown in Figure 8.20.

In the SmC* liquid crystals, the primary order parameter is the director tilt and not the spontaneous polarization. The polarization, therefore, is often called secondary-order parameter, and the SmC* materials are called improper ferroelectrics. The main reason for this is the weakness of the dipole–dipole interactions in the molecules.[92] The polar order can be estimated by the ratio of the actual spontaneous polarization P_s and P_o, which is the value that would appear when complete polar order is assumed. From its definition, the polarization is the density of the molecular dipoles. Assuming molecular dipoles of 3 Debye ($\sim 10^{-29}$ Cm) and typical molecular weight of 300 g, density of about 1 g/cm^3 = 10^3 kg/m^3, we get that in 1 m^3 we have $\frac{10^3}{0.3} \cdot 6 \cdot 10^{23}$ molecules, which would give $P_o \sim 2 \cdot 10^{-2} C/m^2 = 2000 nC/cm^2$. For a SmC* with $P_s < 100$ nC/cm^2, this means that less than 5% of the dipoles are ordered in one direction. For this reason, in the first approximation the polarization is proportional to the tilt angle. This relation, indeed, is found to be true for materials with moderate or low polarization.[95] However, for materials with large polarization, like $P_s \sim 500$ nC/cm^2, the dipole–dipole interaction becomes considerable, and the proportionality is not true. The deviation is more pronounced at lower temperatures, when the dipole–dipole

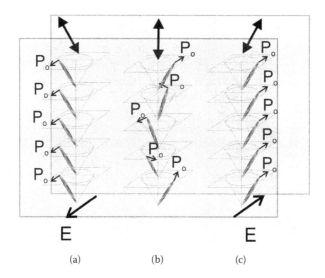

FIGURE 8.20
Illustration of the helical ground state (b) and the field-induced unwound state (a and c). The boundary plates are shown so that they are corresponding to the bookshelf alignment.

interactions are not washed out by thermal fluctuations. An example is shown in Figure 8.21. Materials with high polarizations therefore are not completely improper ferroelectrics.

In addition to the dielectric relaxation modes of anisotropic 3D fluids, just as nematic liquid crystals, helical SmC* liquid crystals (see Chapter 1, Figure 1.11) have interesting low-frequency dielectric modes that are related

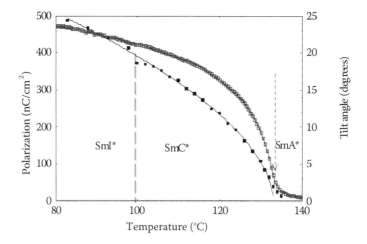

FIGURE 8.21
Temperature dependence of the spontaneous polarization and of the tilt angle of a material RO46–4912 from ROLIC Research LTD.[55, 93] □ tilt angle, •: spontaneous polarization.

to distortion of the spontaneous polarization. In helical chiral smectic C liquid crystals, the polarization is perpendicular to the c-director, and forms a helix, so is averaged out in bulk. External electric field then distorts this structure, leading to a nonzero polarization. This appears as a dielectric response at frequencies below typically a kilohertz. Cells filled with helical smectic C* may have very high dielectric constant[17] with an extra contribution $\Delta\varepsilon_G$ due to the distortion of the helix.

$$\Delta\varepsilon_G = \frac{1}{8\pi\varepsilon_o K}\left(\frac{pP}{\theta}\right)^2 \tag{8.19}$$

here $K \sim 10^{-11}$ N is the elastic constant associated with the deformation of the helical structure, P is the ferroelectric polarization, $\theta \sim 0.5$ *rad* is the director tilt angle and p is the pitch of the helix. With easily achievable polarization of 10^{-3} C/m^2 and pitch of $p \sim 1$ μm, $\Delta\varepsilon_G \sim 10^4$, ie., a few hundred times larger than in the nematic liquid crystals. This mode involves the change of the azimuth angle ϕ and is called Goldstone mode. It relaxes in the kilohertz frequency range above

$$f_G = \frac{2\pi K}{\gamma_\phi p^2}, \tag{8.20}$$

where $\gamma\phi$ describes the viscosity of rotation around the director cone.

Another contribution to the dielectric constant is related to the induction of the polarization when changing the tilt angle, since the polarization is related to the tilt angle. This mode involves variation of the layer spacing (hard deformation) and is strongly suppressed far from the SmA–SmC* transition. Near the phase transition, however, the equilibrium tilt angle is almost zero, and the layer spacing variation is negligible in first order. In this range, therefore, the mode softens, explaining why it is usually referred as "soft-mode." It is, however, still harder than the Goldtone mode, and its contribution is usually measured only under DC bias fields, which unwind the helix and therefore suppress the Goldstone mode. The soft mode relaxes typically in the 100-kHz range. The director fluctuations corresponding to the Goldstone and soft modes are represented in Figure 8.9.

Experimental results for the Goldstone and soft modes are presented in Figure 8.10.

In thin film, which is the most usual geometry in physical studies and in applications, the helical structures become unwound, often resulting in macroscopic polarization in the entire films. Such a macroscopic polarization may be switched between two stable states[94] and the effective polarization shows a hysteresis curve characteristic to ferroelectric crystals. This switching involves a rotation of the director around a cone determined by the tilt angle θ. The inversion of the direction corresponds to the change of the azimuth angle by 180°. In bookshelf geometries, when the layers are perpendicular to the substrates, such a rotation corresponds to the change of the optical axis by 2θ (see Figure 8.22).

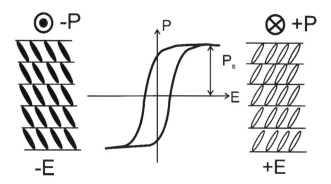

FIGURE 8.22

The hysteresis curve and the schematic director structures in the negative and positive fields as viewed normal to the substrates.

In such geometry, neglecting the dielectric interaction (i.e., if $\varepsilon_o\varepsilon_a E^2 \ll P\cdot E$) and the deformation of the layer structure, the dynamics of the rotation of the polarization can be described by the torque balance equation:

$$\gamma_1\frac{\partial\varphi}{\partial t} = K\cdot\sin^2\theta\cdot\frac{\partial^2\varphi}{\partial t^2} - E\cdot P\sin\varphi - \frac{2w_p}{L}\cdot\sin 2\varphi \tag{8.51}$$

where K is an effective elastic contsant, and w_p is the polar surface anchoring term describing the interaction between the material and the film surfaces.[95] Neglecting the surface and elastic terms (this can be valid for strong fields and weak surfgace anchorings), the torque balance equation gives that:

$$\varphi(t) = 2\arctan(\tan\left(\frac{\varphi_o}{2}\right)e^{-t/\tau}) \tag{8.52}$$

where

$$\tau \approx \frac{\gamma_1}{P\cdot E} \tag{8.53}$$

is the switching time. Since rotational viscosity is comparable to that of nematics ($\gamma \sim 0.1$ Pas), $P \sim 10^{-3}$ C/m^2 with $E \sim 10^6$ V/m $\tau \sim 1$ μs. Such a switching time is much shorter than is possible with nematic materials and bears a great hope for novel electro-optical displays, which will be reviewed in Chapter 9.

From Eq. (8.53), we see that the decrease of the switching time can be achieved by increasing the magnitude of the spontaneous polarization. During this search for higher polarization, a number of groups reported unusual behavior of polarization response, which were assigned to antiferroelectric order,[92,96] however, usually the credit for the unambiguous realization of antiferroelectric order is given to two independent studies by Galerne et al.[97]

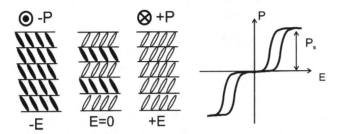

FIGURE 8.23
The typical hysteresis curve and the proposed director structure of the antiferroelectric chiral smectic C (SmC*$_A$) phase.

and Chandani et al.[98] The hysteresis curve and the structure of the director are depicted in Figure 8.23.

Very soon it was recognized that antiferroelectric liquid crystal (AFLC) materials exhibit not only the anticlinic antiferroelectric phase, but also three other subphases.[99,100] One of them, called SmC*$_\gamma$ phase, clearly has so-called "ferrielectric" characteristics, which structure is shown in Figure 8.24.

The natures of the different subphases and their main characteristics are still the subject of active research. Some models state that in the subsequent layers the polarization is always either parallel or antiparallel, and the macroscopic behavior depends on the number of repeating units, which should be more than of two layers range. For example, in the so-called *SmC$_\beta$* phase, the zero field polarization is 2/3 of the saturated vale, which in frame of the orthogonal model would mean interaction of six layers: in five consecutive layers in one direction, and in the sixth the opposite $((5 - 1)/(5 + 1) = 2/3)$. This long-range interaction seems to be very unlikely. For this reason, the so-called "clock model" was suggested. This assumes only interactions of two layers, but does not require that the polarization directions in the neighbor layers be either parallel or antiparallel. This seems to be supported by

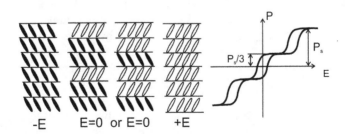

FIGURE 8.24
Characteristics of the ferrielectric liquid crystal SmC*$_\gamma$.

resonant X-ray measurements.[101] Flavors like symmetric clock and distorted clock models also evolved.[102]

Instead of going into details of this active debate, we only consider a few basic questions, which relate to our main focus: the reason for 2D ordering and its relation to molecular structures, to polarity, tilt and chirality. It seems that the antiferroelectric phase never appears directly below the N* phase. The lack of the N* phase sequence of AFLCs is a strong indication that the tendency to form layers in these materials is much stronger than in FLCs, in which the N* phase appears much more often. As we have seen in Chapter 1, in materials with N-Sm phase sequence, the periodicity is quite washed out with diffuse layer boundaries. Such a situation is favorable entropically, because it allows interlayer fluctuations. To maintain this fluctuation in the tilted smectic phase, we have to assure that the directors in the subsequent layers tilt in the same direction (synclinic situation). This indeed happens for materials with relatively low polarization values ($P_s < 10^{-3}$ $C/m^2 = 100$ nC/cm^2), where the dipolar interactions are less important than of the entropic. It seems however, that if the spontaneous polarization of the material is large enough, the dipole–dipole interactions begin to play an increasing role, overruling the entropy gain by the interlayer fluctuations. Indeed, experiments show rather sharp interlayer interfaces (i.e., high-smectic order parameters), i.e., weak interlayer fluctuations.

Another important observation is that in liquid crystals the antiferroelectric phase always seem to occur only from the SmC* phase and not from the SmA* phase. This means that the antiferroelectric phases appear at lower temperatures than of the ferroelectric ones. This is the opposite of the situations in solid states, where the ferroelectric state is the lower temperature phase. This clearly indicates that the cause of the ferroelectric and antiferroelectric orders is different in the fluid states than in the solid materials.

8.8.2   Chiral Tilted Columnar Phase

In 1981 J. Prost suggested[103] that chiral disc-shape molecules that form tilted columnar structures may also be ferroelectric, since they have the same C_2 symmetry as of the SmC* liquid crystals. However, it was not evident whether a polar structure can be switched in such materials. The theoretical suggestion was first verified by Bock and Helfrich in 1992.[104] Similar linear electro-optical effects were then observed on other compounds by Scherowsky and Chen[105] and by Heppke et al.[106] The structure and their switching in electric fields are illustrated in Figure 8.25. It was observed that the switching requires much higher fields than in the SmC* phase, and the switching is slower, typically in the milliseconds range.

Actually, it turned out that the material where the first ferroelectric switching was observed has a ferroelectric structure, where the neighbor polarization directors make a 60° angle with each other.[107] The columns have elliptical cross sections, and they form quasi-hexagonal lattices (each column has six neighbors). Under the electric field the polarization directions have to rotate

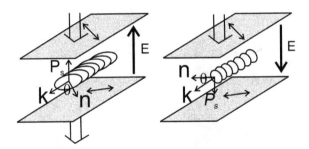

FIGURE 8.25
Structure of tilted columnar liquid crystals of chiral disc-shape molecules, and their behavior in electric fields in planar geometry.

(Goldstone mode), or the magnitude should vary (electroclinic mode). As illustrated in Figure 8.26, these lead to deformation of the lattice structure, explaining the high threshold for switching.

8.8.3 Bowl-Shape Materials

Columnar (or pyramidal) mesophases of bowl-shape,[108] conical,[109] and badminton- shuttlecock[110-] shaped molecules may form polar columns, since

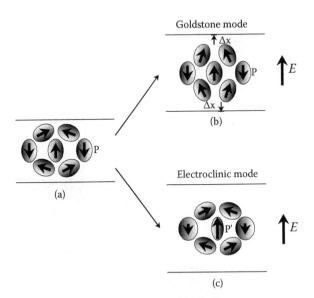

FIGURE 8.26
Illustration of the structure of the ferroelectric type arrangements and their deformation under electric fields of a tilted chiral columnar liquid crystal 1,2,5,6,8,9,12,13-octakis-((S)-2-heptyloxy) dibenzo[e,1]pyrene that has chiral therminal chains and form tilted ferroelectric columnar liquid crystal (corresponding to the molecule of Figure 1.8c).

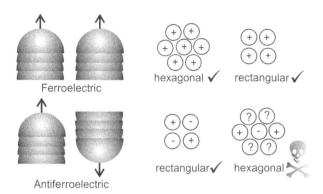

FIGURE 8.27

Illustration of the packing of bowl-shape molecules. The rectangular lattice structure allows both the antiferroelectric and ferroelectric arrangements; however, the hexagonal packing is compatible only with the ferroelectric order.

in the individual columns the molecules have to stack in the same direction to fill the space effectively. The situation is represented in Figure 8.27.

Experiments on the material shown in Figure 1.8b show that it is possible to switch the polarization of the material, strongly indicating the ferroelectricity

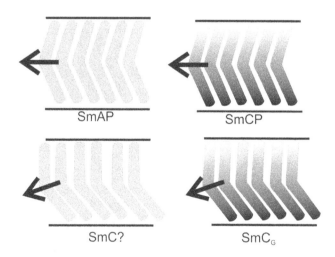

FIGURE 8.28

Possible structures of smectic phases of bent molecules. Vectors indicate the layer polarizations. In SmAP the long axis (a line along the line connecting the end points) is normal to the smectic layers. In SmCP the long axis and the molecular plane are tilted with respect to the layer normal (darker part of the molecule indicates tilt away from the observer). The situation, when the long axis is tilted, but the molecular plane is along the layer normal has not observed yet and there is no widely accepted name for this. The SmC$_G$ phase has a double-tilted structure (G stand for general), i.e., both the molecular plane and the long axis are tilted.

of the pyramidal phase.[111] The measured polarization is in the range of 10^{-2} C/m^2, which indicates strong dipolar interactions. The switching time is slow, in the range of a few seconds. Very recently it was found that achiral urea molecules may also form columnar structures, showing ferroelectric polarization switching in less than 0.1s.[112]

8.8.4 Achiral Bent-Core Molecules

As described in Chapter 1, smectic liquid crystal phases of bent-core molecules can form polar strcutures.

Considering only monolayer structures, four types of fluid director arrangements are possible (see Figure 8.28).[114,115]

1. The orthogonal arrangement, where both the molecular plane and the long axis (the virtual line connecting the end points) are parallel to the layer normal. It is a transversely polarized achiral SmA-type structure with C_{2v} symmetry. This corresponds to the structure proposed originally by Niori et al.[115] but observed only later.[116]

2. The single-tilted arrangement when only the molecular plane is tilted. In this case the smectic layers have monoclinic chiral symmetry C_2. It is denoted as $SmCP$[117] phase (P stands for "polar"). Most of the experimental results concern this phase, and will be summarized below.

3. The other single-tilted structure, when only the long axis is tilted with respect to the smectic layers, has an achiral monoclinic symmetry C_s. To our knowledge, so far there is no experimental evidence of this phase.

4. The double-tilted structure, which has a triclinic configuration with chiral C_1 symmetry. It corresponds to the SmC_G[116] phase proposed by de Gennes[118] (G stands for "general"). Experimental evidences of this phase were found recently both in free-standing films[119] and in bulk samples.[120]

Assuming that the director arrangement is the same in every layer, one would realize ferroelectric ferroelectric *(FE)* ground states just as observed in a few cases.[121,122,123] However, it turned out that the ground states are mainly antiferroelectric *(AFE)*,[119,124] so one needs to consider at least two-layer periodicity. In this case, in the adjacent layers the sign of the chirality can alternate or can remain unchanged over micrometer ranges. Accordingly, two types of domains are possible: *"racemic,"* if the chirality in the adjacent layers alternates, and *"chiral,"* if the adjacent layers have the same handedness. The *AFE racemic SmCP* phase is synclinic, i.e., the molecules in the adjacent layers tilt in the same direction. In the *chiral domains* the *AFE* state is anticlinic, and the optical axis is parallel to the layer normal regardless of the handedness of the domains. The *AFE* arrangement can be easily switched to ferroelectric *(FE)* by applying an external electric field. It is assumed that

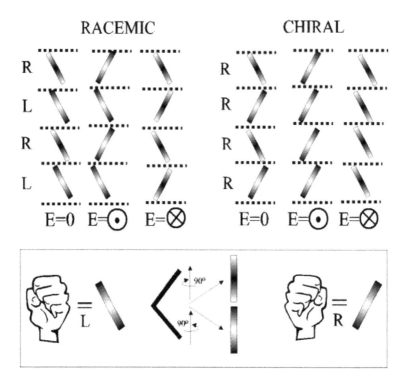

FIGURE 8.29

Sketch of the racemic and chiral structures of the SmCP phase of achiral banana-shape molecules in antiferroelectric (at $E = 0$) and ferroelectric (at $E > E_{th} \sim 5$ V/μm) states. Left column: Racemic structure, in which the chirality alternates in the adjacent layers. The AFE domains are synclinic with coexisting opposite tilt directions. Right column: chiral structure, in which the adjacent layers have the same handedness, but domains of different chirality coexist. The shading illustrates the bent or tilted shape of the molecules (brighter parts are closer to the reader). R (L) is the chirality descriptor corresponding to right (left)-handed layer conformations.

the chirality is conserved during the *AFE*→ *FE* transitions, so the racemic *FE* state is anticlinic, and the chiral *FE* state is synclinic. In the racemic *FE* state, the optical axis is parallel to the layer normal, independent of the sign of the external AC field; therefore, no electro-optical switching can be observed for square wave fields. In the chiral *FE* state, the optical axis makes an angle θ with the layer normal, and there is an electro-optical switching under square wave fields (see Figure 8.29.). This difference makes it possible to distinguish racemic and chiral domains by textural observations.

Although in first approximation the individual layers keep their chirality during switching between *AFE* and *FE* states,[119] extended application of electric fields can alter racemic domains to chiral or vice versa.[125] Racemic domains can be rendered chiral by surface interactions, too.[126]

The $B_{7\text{-}II}$ materials listed in Table 1.2, and that form telephone wire type structures (see Figure 6.16) when cooling from the isotropic phase, can be switched under high $(E > 10\ V/\mu m)$ electric fields that decrease with decreasing temperatures. $B_{7\text{-}II}$ materials can be both ferroelectric[127] and antiferroelectric,[128] and based on polarization current and electro-optical observations, it has been suggested that they have out-of-plane polarization components, i.e., they correspond to the double-tilted SmC_G phase.

References

1. W.H. de Jeu, JW.J.A. Goossens, P. Bordewijk, *J. Chem. Phys.*, 61, 1985 (1974).
2. S. Markscheffel, A. Jákli, A. Saupe, *Ferroelectrics*, 180, 59 (1996).
3. Detailed description, see C.J.F. Bottcher, P. Bordewijk, *Theory of Electric Polarization*, Elsevier, Amsterdam (1978).
4. H. Froehlich, *Theory of Dielectrics* 2nd edition, Clarendon Press, Oxford (1968).
5. P. Bordewijk, *Physica*, 75, 146 (1974).
6. A. Martin, G. Meier, A. Saupe, *Symp. Faraday Soc.*, 5, 119 (1971).
7. P. Nordio, G. Rigatti, U. Serge, *Mol. Phys.*, 25, 129 (1973); G.R. Luckhurst, J. Zannoni, *Proc. R. Soc. London*, A343, 389 (1975).
8. P. Debye, *Polar Molecules*, Chemical Catalogue Co., New York (1927); also available as a Dover Reprint, New York (1945).
9. R. Kubo, *J. Phys. Soc. Jpn.*, 12, 570 (1957).
10. A. Buka, F. Leyvraz, *Phys. Stat. Sol. B*, 112, 289 (1982).
11. A. Jakli, A. Buka, Rotational Brownian Motion and dielectric permittivity in nematic liquid crystals, Reprint KFKI-1984–23 (HU ISSN 0368 5330, ISBN 963 372 183 0) (1984).
12. Yu.P. Kalmykov, *Krystallografya*, 36, 1075 (1991); W.T. Coffey, Yu.P. Kalmykov, *Liq. Cryst.*, 14, 1227 (1993); Yu. P. Kalmykov, W.T. Coffey, *Liq. Cryst.*, 25, 329 (1998).
13. J. Perrin, *J. Phys. Radium* (VII) 5, 33 (1934); ibid., 7, 1 (1936).
14. W. Maier, G. Meier, *Z. Naturforsch.*, 16A, 262 (1961); W.H. de Jeu, Th.W. Lathouwers, *Z. Naturforsch.*, 30A, 79 (1975); A. Buka, B.G. Owen, A.H. Price, *Mol. Cryst. Liq. Cryst.*, 51, 273 (1979).
15. M.F. Bone, A.H. Price, M.G. Clark, D.G. MacDonnell, *Liquid Crystals and Ordered Fluids*, Vol. 4, p. 799, Ed. A.C. Griffin, J.P. Johnsson, Plenum, New York (1984).
16. D. Dunmur, K. Toriyama, Dielectric properties, in *Physical Properties of Liquid Crystals*, Ch. 4, Ed. D. Demus et al., Wiley-VCH, Weinheim (1999).
17. B. Zeks, R. Blinc, *Ferroelectric Liquid Crystals*, Gordon and Breach, (1991).
18. S. Merino, F. de Daran, M.R. de la Fuente, M.A. Perez Jubindo, T. Sierra, *Liq. Cryst.*, 23, 275 (1997).
19. L.M. Blinov, V.G. Chigrinov, *Electro-optic Effects in Liquid Crystal Materials*, Springer, New York (1996).
20. A. Buka, L. Kramer, *Pattern Formation in Liquid Crystals*, Springer-Verlag, New York (1996).
21. W. Helfrich, *Phys. Rev. Lett.*, 24, 201 (1970).
22. D.A. Dunmur, K. Szumilin, T.F. Waterworth, *Mol. Cryst. Liq. Cryst.*, 149, 385 (1987).

23. K.H. Yang, *J. Appl. Phys.*, 54, 4711 (1983); L.M. Blinov, *Electro-Optical and Magneto-Optical Properties of Liquid Crystals*, Ch. 5, John Wiley and Sons Ltd., New York (1983).

24. See for example, T.C. Chieu, K.H. Yang, *Jap. J. Appl. Phys.*, 28, 2240 (1989).

25. W. Nernst, *Z. Physik. Chem.*, 2, 613 (1988).

26. A. Einstein, *Ann. Phys.*, 19, 371 (1906).

27. P. Debye, E. Hückel, *Phys. Z.*, 24, 49 (1923); and 25, 97 (1924).

28. W.B. Russel, D.A. Saville, W.R. Schowalter, *Colloidal Dispersions*, Cambridge University Press, (2001).

29. A.C. Becquerel, *Bull. Soc. Philomath, Paris*, Ser. 37, 149 (1820).

30. W.G. Cady, *Piezoelectricity*, Mc Graw-Hill Book Company, Inc., New York (1946).

31. J. Curie, P. Curie, Bulletin no. 4 de la Societee Mineralogique de France, 3, 90 (1880).

32. G. Lippmann, *C.R. Cad. Sc. Paris*, 92, 1049 (1881).

33. W. Voight, General theory Piezo and Pyroelectric properties of Crystals, *Abh. Gott.*, 36, 1 (1890).

34. Langevin, French Patent, 505, 703, 17 Sept. 1918.

35. W.G. Cady, *Phys. Rev.*, 27, 419 (1915).

36. B.M. Wul, I.M. Goldman, *Akad. Nauk.*, 49, 179 (1945). A. Von Hippel, R.G. Breckenridge, F.G. Chesley, L. Tisza, *Ind. Eng.*, 38, 1097 (1946); S. Roberts, *Phys. Rev.*, 71, 890 (1947).

37. Y. Wada, in *Electronic Properties of Polymers*, Ch. 4., Ed. J. Most, G. Pfister, Wiley-Interscience, New York (1982).

38. E. Fukada, *Ultrasonics*, 229 (1968), and *Adv. Biophys.*, 6, 121 (1974).

39. H. Kaway, *Jpn. J. Appl. Phys.*, 8, 975 (1970).

40. E. Fukada, I. Yasuda, *J. Appl. Physiol.*, 3, 117 (1964).

41. A recent overview of the Curie principle is written in the book: S.T. Lagerwall, *Ferroelectric and Antiferroelectric Liquid Crystals*, Wiley-VCH, Weinheim (1999).

42. J.F. Nye, *Physical Properties of Crystals*, Oxford University Press, Oxford (1957).

43. S. Garoff, R.B. Meyer, *Phys. Rev. Lett.*, 38, 848 (1977).

44. C.H. Bahr, G. Heppke, *Liq. Cryst.*, 2, 825 (1987); G. Andersson, I. Dahl, P. Keller, W. Kuczhynski, S.T. Lagerwall, K. Skarp, B. Stebler, *Appl. Phys. Lett.*, 51 (9), 640 (1987).

45. J. Prost, *J. Phys. (Paris)*, 39, 639 (1978).

46. H. Brand, *Makromol. Chem. Rapid Commun.*, 10, 441 (1989); W. Meier, H. Finkelmann, *Makromol. Chem. Rapid Commun.*, 11, 599 (1990).

47. N.V. Madhusudana, R. Pratibha, *Mol. Cryst. Liq. Cryst.*, 5, 43 (1987).

48. P. Pieranski, E. Guyon, P. Keller, *J. Phys.*, 36, 1005 (1975).

49. A. Jákli, L. Bata, *Mol. Cryst. Liq. Cryst.*, 201, 115 (1991).

50. A. Jákli, L. Bata, Á. Buka, N. Éber, I. Jánossy, *J. Phys. Lett.*, 46, L-759 (1985); A. Jákli, L. Bata, Á. Buka, N. Éber, *Ferroelectrics*, 69, 153 (1986).

51. A.P. Fedoryako et al., *Functional Materials*, 4(3), 375 (1997).

52. B. Bonev, V.G.K.M. Pisipati, A.G. Petrov, *Liq. Cryst.*, 6(1), 133 (1989).

53. S.U. Vallerien, F. Kremer, E.W. Fischer, *Makromol. Chem. Rapid Commun.*, 11, 593 (1990).

54. C.-C. Chang, L.-C. Chien, R.B. Meyer, *Phys. Rev. B*, 55, 534 (1997).

55. A. Jákli, T. Toth-Katona, T. Scharf, M. Schadt, A. Saupe, *Phys. Rev. E*, 66, 011701 (2002).

56. H. Kaway, *Jpn. J. Appl. Phys.* 8, 975 (1970).

57. A. Jákli A. Saupe, *J. Appl. Phys.*, 82(6), 2877 (1997).

58. L. Bata, N. Éber, A. Jákli, Hungarian Patent, 20052B (10.28. 1988); K. Yuasa, K. Hashimoto (Idemitsu Kosan Co., Ltd.) Jpn. Kokai Tokyo Koho JP 01,175,400 (11 Jul 1989); Seiko Epson (62–203132).
59. A. Jákli, A. Saupe, *Mol. Cryst. Liq. Cryst.*, 237, 389 (1993).
60. A. Jákli, M. Müller, D. Krüerke, G. Heppke, *Liq. Cryst.*, 24, 467 (1998).
61. H. Bock, W. Helfrich, *Liq. Cryst.*, 12, 697 (1992).
62. W.S. Williams, *Ferroelectrics*, 41, 2251 (1982) and references therein.
63. L.M. Blinov, S.A. Davidyan, A.G. Petrov, S.V. Yablonsky, *Zh. Eksp. Theor. Fiz. Lett.*, 48, 259 (1988).
64. R.B. Meyer, *Phys. Rev. Lett.*, 22, 918 (1969).
65. W. Helfrich, *Phys. Lett.*, 35A, 393 (1971); Z. *Naturforsch.*, 26A, 833 (1971).
66. A. Derzhanski, A.G. Petrov, *Phys. Lett.*, 36A, 483 (1971).
67. D. Schmidt, M. Schadt, W. Helfrich, Z. *Naturforsch.*, 27A, 277 (1972).
68. J.P. Marcerou, J. Prost, *Ann. Phys. (Paris)*, 38, 315 (1977).
69. D. Brewster, *Edinbg. J. Sci.*, 1, 208 (1824).
70. J. et P. Curie, *Compt. Rend.*, 91, 294, 383 (1880).
71. F. Pockels, *Ehrbuch der Kristalloptik*, B.G. Teubner, Leipzig, Berlin (1906).
72. F. Ratz, *Zeitschr. f. phys. Chemie*, 19, 305 (1896); K. Badecker, ibid., 36, 305 (1901).
73. P. Debye, *Physik. Zeitschr.*, XIII, 97 (1912).
74. E. Scrödinger, *Mathem.-Naturw. Klasse*; Bd.CXXI, Abt. IIa (1912).
75. J. Valasek, *Phys. Rev.*, 15, 537; 19, 478; ibid., 20, 639 (1920).
76. V.S. Bystrov, H.R. Leuchtag, *Ferroelectrics*, 155, 19 (1994); ibid., 22, 157 (1999).
77. M.H. Shamos, L.S. Lavine, *Nature*, 213, 267 (1967); H. Athenstaedt, *Nature*, 228, 830 (1970), and *Ferroelectrics*, 11, 365 (1976); E. Fukada, *Quart. Rev. Biophys.*, 16, 59 (1983).
78. H. Fröclich, *Riv. Nuovo Cimento.*, 7, 399 (1977).
79. A.R. von Hippel, *J. Phys. Soc. Japan*, 28, 1 (1969).
80. P. Fong, *Bull. Amer. Phys. Soc.*, 13, 613 (1968).
81. L.A. Beresnev, L.M. Blinov, E.I. Kovshev, *Dokl. Biophys.*, 265, 11 (1972); L.A. Beresnev, S.A. Pikin, W. Haase, *Cond. Mat. News*, 1 (8), 13 (1992).
82. A. Muller, *Phys. Lett.*, 96A, 319 (1983); A.G. Petrov, A.T. Todorov, B. Bonev, L.M. Blinov, S.V. Yablonsky, D.B. Fulachyus, N. Tsetkova, *Ferroelectrics*, 114, 415 (1991).
83. M. Born, *Ann. Phys.*, 55, 221 (1918).
84. D. Vorlander, M.E. Huth, Z. *Phys. Chem.*, A75, 641 (1911).
85. A. Saupe, *Mol. Cryst. Liq. Cryst.*, 7, 59 (1969).
86. R.B. Meyer, L. Liebert, I. Strelecki, P. Keller, *J. Phys. Lett. (Paris)*, 36, L69 (1975).
87. J.W. Goodby, R. Blinc, N.A. Clark, S.T. Lagerwall, M.A. Osipov, S.A. Pikin, T. Sakurai, K. Yoshino, B. Zeks (Eds.), *Ferroelectric Liquid Crystals: Principles, Properties and Applications*, Gordon and Breach, Philadelphia, PA (1991).
88. I. Musevic, R. Blinc, B. Zeks, *The Physics of Ferroelectric and Antiferroelectric Liquid Crystals*, World Scientific, Singapore (2000).
89. S. Chandrasekhar, *Liquid Crystals*, 2nd ed., Cambridge University Press, Cambridge (1992).
90. P.G. de Gennes, *The Physics of Liquid Crystals*, Clarendon Press, Oxford (1993).
91. S.T. Lagerwall, *Ferroelectric and Antiferroelectric Liquid Crystals*, Wiley-VCH, Weinheim (1999).
92. L.A. Beresnev, L.M. Blinov, M.A. Osipov, S. Pikin, *Mol. Cryst. Liq. Cryst.*, 158A, 1 (1988).
93. K. Schmitt, R.P. Herr, M. Schadt, J. Füfschilling, R. Buchecker, X.H. Chen, C. Benecke, *Liq. Cryst.*, 14, 1735 (1993).

94. N.A. Clark, S.T. Lagerwall, *Appl. Phys. Lett.*, 36, 899 (1980).

95. N.A. Clark, S.T. Lagerwall, Introduction to ferroelectric liquid crystals, in *Ferroelectric Liquid Crystals: Principles, Properties and Applications*, Ch. 1, Ed. J.W. Goodby, R. Blinc, N.A. Clark, S.T. Lagerwall, M.A. Osipov, S.A. Pikin, T. Sakurai, K. Yoshino, B. Zeks, Gordon and Breach, Philadelphia (1991).

96. L.A. Beresnev, L.M. Blinov, V.A. Bailkalov, E.P. Pozhidaev, G.V. Purvanetkas, A.I. Pavluchenko, *Mol. Cryst. Liq. Cryst.*, 89, 327 (1982).

97. Y. Galerne, L. Liebert, *Phys. Rev. Lett.*, 64, 906 (1990).

98. A.D.L. Chandani, E. Gorecka, Y. Ouchhi, H. Takezoe, A. Fukuda, *Jpn. J. Appl. Phys.*, 28, L1265 (1989).

99. M. Fukui, H. Orihara, Y. Yamada, Y. Yamamoto, Y. Ichibashi, *Jpn. J. Appl. Phys.*, 28, L849 (1989).

100. A.D.L. Chandani, Y. Ouchi, H. Takezoe, A. Fukuda, *Jpn. J. Appl. Phys. Lett.*, 28, L1261 (1989).

101. P. Mach, R. Pindak, A.M. Levelut, P. Barois, H.T. Nguyen, C.C. Huang, L. Furenlid, *Phys. Rev. Lett.*, 81, 1015 (1998); P.M. Johnson, D.A. Olson, S. Pankratz, H.T. Nguyen, J.W. Goodby, M. Hird, C.C. Huang, *Phys. Rev. Lett.*, 84, 4870 (2000).

102. Jan P.F. Lagerwall, Structures and properties of the chiral smectic C liquid crystal phases, Ph.D. dissertation, Chalmers University, Goteborg, Sweden (2002).

103. Y. Takanishi, A. Ikeda, H. Takezoe, A. Fukuda, *Phys. Rev. E.*, 51, 400 (1995).

104. J. Prost, Comptes Rendus du Colloque Pierre Curie, in *Symmetries and Broken Symmetries*, Ed. N. Boccara (1981).

105. H. Bock, W. Helfrich, *Liq. Cryst.*, 12, 697 (1992).

106. G. Scherowsky, X.H. Chen, *Liq. Cryst.*, 17, 803 (1994).

107. G. Heppke, D. Krüerke, C. Löhring, D. Lötcsch, D. Moro, M. Müller, H. Sawade, *J Mater Chem*, 10, 2657 (2000).

108. H. Bock, W. Helfrich, *Liq. Cryst.*, 18, 387 (1995).

109. H. Zimmermann, R. Poupko, Z. Luz, J. Billard, Pyramidic mesophases, *Z. Naturforsch.*, 40A, 149 (1985).

110. L. Lei, *Mol. Cryst. Liq. Cryst.*, 91, 77 (1983); J. Malthete, A. Collet, *J. Am. Chem. Soc.*, 109, 7544 (1987).

111. B. Xu, T.M. Swager, *J. Am. Chem. Soc.*, 115, 1159 (1993); ibid., 117, 5011 (1995); A. Serrette, T.M. Swager, *Angew. Chem. Int. Ed. Engl.*, 33, 2342 (1994).

112. M. Sawamura, K. Kawai, Y. Matsuo, K. Kanie, T. Kato, E. Nakamura, *Nature*, 419, 702 (2002); Y. Matsuo, A. Muramatsu, R. Hamasaki, N.Mizoshita, T. Kato, E. Nakamura, *J. Am. Chem. Soc.*, 126, 432 (2004).

113. A. Jákli, A. Saupe, G. Scherowsky, X.H. Chien, *Liq. Cryst.*, 22, 309 (1997).

114. M. Yamamoto, personal communication.

115. A. Roy, N.V. Madhusudana, P. Toledano, A.M. Figueiredo- Neto, *Phys. Rev. Lett.*, 82, 1466 (1999).

116. H.R. Brand, P.E. Cladis, H. Pleiner, *Eur. Phys. J.B.*, 6, 347 (1998).

117. T., Niori, T., Sekine, J., Watanabe, T., Furukawa, H., Takezoe, Distinct ferroelectric smectic liquid crystals consisting of banana-shaped Achiral molecules, *J. Mater. Chem.*, 6(7), 1231 (1996); T., Sekine, T., Niori, M., Sone, J., Watanabe, S.W., Choi, Y., Takanishi, H., Takezoe, Spontaneous helix formationin smectic liquid crystals comprising achiral molecules, *Jpn. J. Appl. Phys.*, 36, 6455 (1997).

118. A. Eremin, S. Diele, G. Pelzl, H. Nadasi, W. Weissflog, J. Salfetnikova, H. Kresse, *Phys. Rev. E,* 64, 051707-1-6 (2001).

119. D.R. Link, G. Natale, R. Shao, J.E. Maclennan, N.A. Clark, E. Körblova, D.M. Walba, *Science*, 278, 1924 (1997).
120. P.G. de Gennes, *The Physics of Liquid Crystals*, Clarendon Press, Oxford (1975).
121. D.R. Link, N. Chattham, N.A. Clark, E. Körblova, D.M. Walba, p. 322, Abstract Booklet FLC99, Darmstadt (1999).
122. A. Jákli, D. Krüerke, H. Sawade, G. Heppke, *Phys. Rev. Lett.*, 86 (25), 5715 (2001).
123. D.M. Walba, E. Körblova, R. Shao, J.E. Maclennan, D.R. Link, M.A. Glaser, N.A. Clark, *Science*, 288, 2181 (2000).
124. E. Gorecka, D. Pociecha, F. Araoka, D.R. Link, M. Nakata, J. Thisayukta, Y. Takanishi, K. Ishikawa, J. Watanabe, H. Takezoe, *Phys. Rev. E*, 62, R4524 (2000).
125. S. Rauch, P.Bault, H. Sawade, G. Heppke, G.G. Nair, A. Jákli, *Phys. Rev. E*, 66, 021706 (2002).
126. A. Jákli, S. Rauch, D. Lötzsch, G. Heppke, *Phys. Rev. E*, 57, 6737 (1998).
127. G. Heppke, A. Jákli, S. Rauch, H. Sawade, *Phys. Rev. E.*, 60, 5575 (1999).
128. A. Jákli, C.H. Lischka, W. Weissflog, G. Pelzl, S. Rauch, G. Heppke, *Ferroelectrics*, 243, 239 (2000).
129. A. Jákli, G.G. Nair, C.K. Lee, L.C. Chien, *Phys. Rev. E*, 63, 061710-1-5 (2001).
130. S. Rauch, P. Bault, H. Sawade, G. Heppke, A. Jákli, *Phys. Rev. E*, 66, 021706, (2002).
131. A. Jákli, G.G. Nair, H. Sawade, G. Heppke, *Liq. Cryst.*, 30 (3), 265 (2003).
132. M. Fukui, H. Orihara, Y. Yamada, Y. Yamamoto, Y. Ichibashi, *Jpn. J. Appl. Phys.*, 28, L849 (1989).

9

Applications

The first demonstrations of liquid crystals applications were the thermometers utilizing the temperature dependence of the pitch of cholesteric liquid crystals.[1] Then the alphanumerical displays based on dynamic light scattering effects were developed in the middle of the 1960s.[2] This latter effect has been forgotten by now due to its short lifetime and large power consumptions. In spite of this first failure, presently the main application for liquid crystals is in the flat panel display technologies (LCDs). One finds liquid crystals in PC monitors, notebooks, car navigation displays, PDAs, E-books, mobile phones, projectors and in flat panel high-definition TVs. The success of liquid crystals in displays is due to the anisotropic shape of the constituent molecules and the anisotropy of their response to the electric field, as we have seen in the previous chapters.

9.1 Liquid Crystal Displays

9.1.1 Display Structures

Although there are a number of different display modes, they are all common in the sense that they are parallel plate capacitors with polarizers bonded to the external surfaces. Although flexible displays, where the liquid crystal is bounded between plastic substrates, are already on the horizon, the boundary plates so far are mainly rigid glass plates separated by a few-micrometer spacing. The conductance of the inner substrates is mainly achieved by transparent indium tin oxide coating, which is patterned to ensure addressing pixels individually. In simple displays like in wrist watches or other alphanumerical displays, the addressing is done by the so-called multiplexing, and they are called passive-matrix LCDs. The number of electrodes is equal to the sum of the number of rows and of the columns. Each pixel is at the crossroad of a row and column electrodes and the pixel is fired when the potential difference between the column and row electrodes is larger than a threshold. The addressing of the individual pixels by this way is explained in Figure 9.1b. The rows or columns are connected to integrated circuits that control when a charge is sent down a particular column or row. To turn on a pixel, the integrated circuit sends a charge down the correct column of one substrate and a ground activated on the correct row of the other. The row and column intersect at the designated pixel, and that delivers the voltage to untwist the liquid crystals at that pixel. The gray level is controlled by the exact potential difference above the threshold. The simplicity of the passive-matrix system is beautiful, but it has significant drawbacks, notably slow response time and imprecise voltage control. Response time refers to the LCD's ability to refresh the image displayed. Imprecise control hinders the passive matrix's ability to influence only one pixel at a time. When voltage is applied to untwist one pixel, the pixels around it also partially untwist, which makes images appear fuzzy and lacking in contrast (crosstalk). A simplified scheme of time multiplexing with passive-matrix displays is shown in Figure 9.1c.

The pixels are addressed by gated AC voltages (only the envelopes are shown) with a complex temporal structure. A short pulse is applied periodically to the rows as a strobe signal, whereas the columns carry the information signals. A pixel is only selected if a difference in potential and hence an electrical

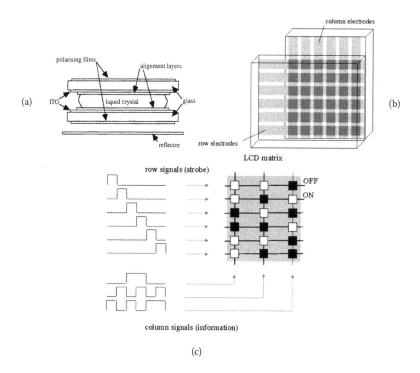

FIGURE 9.1
Brief explanation of the geometry of the liquid crystal displays (upper row), and their addressing method by time multiplexing (lower row).

field is present, i.e., only if row and column are not on a low or high level at the same time. More precisely, the pixel is selected if the RMS voltage is above the threshold for reorientation. An important consequence of passive time multiplexing is that the selection ratio U_{ON}/U_{OFF} approaches unity for large pixel numbers as, e.g., required with standard VGA computer displays or better. Therefore, liquid crystal modulators with rather steep electro-optical characteristics are required to achieve sufficient optical contrast with weak selection ratios.

The problems with the passive-matrix addressing can be solved by the active-matrix addressing method, where each pixel is addressed via a transistor integrated to the pixels.

In Thin-Film-Transistor (TFT) or active-matrix displays, shown in Figure 9.2, rows and columns, cross-over insulators, thin film transistors (amorphous silicon) and transparent pixel electrodes are placed on one of the glass plates by photo-lithographical techniques. The other plate contains the common counter electrode. Since the electro-optically active part of the display is reduced by the TFTs to typically between 30 and 60%, additional masks are required to block light leaking through the nonactive parts and, hence, to improve the contrast of the display. In total, six to nine lithographical steps are necessary, which makes the fabrication process of those displays more expensive. Such thin-film-transistor (TFT-LCD) methods were actually invented already 30 years ago[3] but became economically viable only in the 1990s. For color displays, each pixel is divided into three or four sub-pixels, which are covered with color filters to allow additive mixing of three basic colors. In four-sub-pixel arrangements, an additional neutral gray scale pixel is incorporated to improve brightness control.

Nowadays all laptop displays and TVs use this driving method, which actually can vary a lot and can be quite complicated, which we will not cover at all. The details of the addressing and driving methods are reviewed in a number of specialized papers and books.[4,5,6] Here we only review the most important display modes, and explain only their principles.

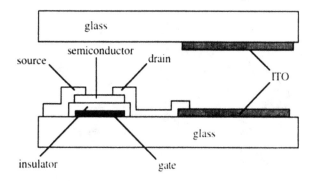

FIGURE 9.2
Schematics of the active-matrix method.

9.1.2 Nematic Liquid Crystal Displays

LCDs are optically passive in that they do not generate light to produce contrast. However, they modify the incoming light intensity or its polarization direction. For example, the polarization direction of linearly polarized light traveling along the helix of a chiral nematic or chiral smectic C structure follows the polarization of the director, if the pitch is much larger than the wavelength of the light, and if the birefringence is not too high. This waveguide regime is used in the so far most successful twisted nematic liquid crystal displays (TN-LCDs). The twisted nematic effect was patented independently by Schadt and Helfrich[7] in Europe and by Fergason in the United States[8] in 1971. The principle of this display mode is shown in Figure 9.3.

A nematic liquid crystal is filled between two glass plates, which are separated by thin spacers, coated with transparent electrodes and orientation layers inside. The orientation layer usually consists of a polymer (e.g., polyimide), which has been unidirectionally rubbed, e.g., with a soft tissue. As a result, the liquid crystal molecules are fixed with their alignment more or less parallel to the plates, pointing along the rubbing direction, which includes an angle of 90° between the upper and the lower plate. Consequently, a homogeneous twist deformation is achieved. The polarization of a linearly polarized light wave is then guided by the resulting quarter of a birefringent helix, if the orientation is not disturbed by an electrical field. The transmitted wave may pass, therefore, a crossed exit polarizer, and the modulator appears

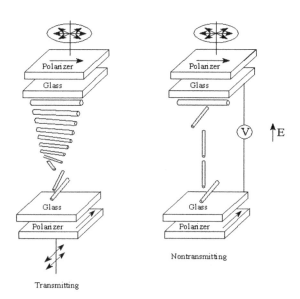

FIGURE 9.3
Principles of twisted nematic display mode.

bright. If, however, an AC voltage of a few volts is applied, the resulting electrical field forces the molecules to align themselves along the field direction, and the twist deformation is unwound. Now, the polarization of a light wave is not affected and cannot pass the crossed exit polarizer. The modulator appears dark. Obviously, the inverse switching behavior can be obtained with parallel polarizers. It must be noted further that gray scale modulation is achieved easily by varying the voltage between the threshold for reorientation (which is a result of elastic properties of LCs) and the saturation field.

The main advantage of this method is the large contrast and brightness, which made it extremely successful up to now. However, it is very difficult to achieve by this method fast enough displays that would be good enough for television purposes. Liquid crystal televisions use the so-called in-plane switching (IPS) method, where the electric field is not applied across the film, but along it by patterned electrodes in one plate. One of the switching methods by this method is shown in Figure 9.4. This display offers 12 ms response time that accommodates true moving image reproduction. Accordingly, it is suitable for high definition TV (HDTV) systems. Samsung's proprietary Plus Viewing Angle (PVA) technology ensures viewing angles of 170° in all directions. With the contrast ratio of 800:1, a 60-inch-diameter TV weights only 180 N and consumes 230 W.

In addition to the TN and IPS modes in nematic displays, hundreds of other displays methods were invented and tested over the last 30 years. Out of them the bistable nematic development based on surface flexoelectric interactions,[9] the bistable cholesterics displays based on switching between planar and focal conic textures[10] and the polymer dispersed liquid crystal displays[11] found some applications. None of them, however, offer better than a few milliseconds switching time.

Sub-milliseconds (and even sub-microseconds!) switching times so far are offered only by ferroelectric and antiferroelectric liquid crystals. Without aiming to review the majority of the possible display modes in ferroelectric smectic materials,[12,13] we just show the most known modes tested in SmC* materials.

9.1.3 Ferroelectric Smectic Displays

The first and best known ferroelectric liquid crystal display mode is the Surface Stabilized Ferroelectric Liquid Crystal Displays (SSFLCD), invented by Clark and Lagerwall in 1980.[14] The principle of this display mode is illustrated in Figure 9.5. The smectic layers are arranged in so-called uniform bookshelf geometry, i.e., they are standing in parallel planes normal to the substrates. The film thickness is typically smaller than of the helical pitch; thus the natural tendency of the director structure to form a helix is suppressed by the surface anchoring. One polarity of the electric field applied across the film rotates the ferroelectric polarization along the base of the tilt cone without distorting the layer structure. When the polarization is fully

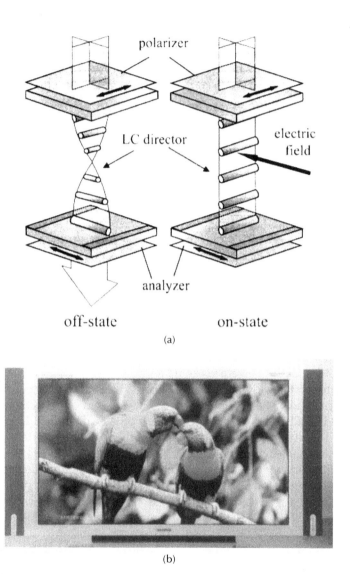

polarizer

LC director

electric
field

analyzer

off-state   on-state

(a)

(b)

FIGURE 9.4
In-plane switching method. (a) The schematics of the switching principle; (b) A 46-inch TFT-IPS-
LCD from Samsung Electronics Co., Ltd.

switched along the field, the director everywhere makes an angle θ (ideally $22.5°$)
with the smectic layers normal. If one of the crossed polarizers is set parallel
to this direction, the picture seen though the sample becomes dark. Now if
we reverse the polarity of the electric field, the direction of the spontaneous
polarization needs to be reversed, too. Consequently the director will rotate
around the cone by $180°$, which corresponds to a change of the optical axis

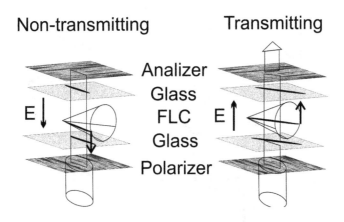

FIGURE 9.5
Illustration of the Surface Stabilized Ferroelectric Liquid Crystal Displays (SSFLCD).

by 2θ (ideally 45°). With the given directions of the crossed polarizers, the texture is bright now. As we have described in Section 8.8 (see Eg. 8.53), the switching time is inversely proportional to the electric field and the magnitude of the spontaneous polarization, and takes typically less than 100 µs. For ideally weak surface anchoring, both field-induced director structures would remain after field removal (bistability). This means that the last picture of the display would remain after turning off the power. This situation, however, in practice holds only for less than a few hours, and eventually the information fades away.

The advantages of SSFLC modes include the possible bistability (that requires weak surface anchoring), fast switching (<100 µs) and large viewing angle. The disadvantage is the difficulties to get gray scale, and the problems with the stability of uniform alignments. The last problems are related to the piezoelectricity (see Section 8.6) of the ferroelectric liquid crystal materials, which lead to all kinds of mechanical stability problems,[15] such as long-term pumping out of the material, or switching-induced misalignments.

For short pitch materials (where p < 0.7 µm), and at voltages below the helix unwinding threshold, the net polarization is proportional to the electric field, just like in electroclinic mode described in Chapter 8. In this case, the switching of the optical axis is due to the deformation of the helix, and it is called Deformed Helix FLC Mode (DHFLC). The principle of this mode is schematically illustrated in Figure 9.6.[16]

As the field is applied, the originally ideal sinusoidal polarization structure becomes distorted so that a larger part of the polarization will point parallel to the electric field. Whereas at zero electric field the sinusoidal variation results in an average optical axis parallel to the layer normal, the optical axis will be rotated when a field is applied. The rotation of the optical axis will be approximately proportional to the electric field and will change sign under

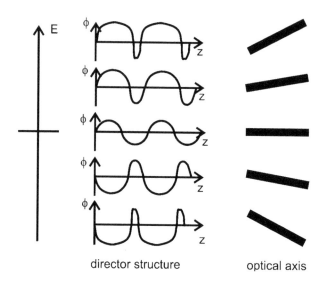

director structure   optical axis

FIGURE 9.6
Illustration of the Deformed Helix Ferroelectric Liquid Crystal Display (DHFLCD) mode.

field reversal. This allows an easy control of the gray level. Another advantage is that the switching time, $\tau \approx \frac{\gamma_1 p^2}{4\pi^2 K_{22}}$, is basically independent of the applied voltage and depends only weakly on the temperature. A disadvantage is the low brightness, because the tilt angle usually is much smaller than 45° if the material has an SmA phase above the SmC*, whereas if it appears directly from the nematic phase one has difficulty getting good alignment.

Interestingly, very similar V-shaped transmission–voltage characteristics, but with much better bright state, can be achieved even in surface-stabilized geometries for materials with high polarizations when strong anchoring favors the dipoles pointing out at the boundaries. When the polarization is high enough, the charge density, $\rho_P = \vec{\nabla} \cdot \vec{P}$, of splay deformations would result in a high electrostatic energy. To avoid that, the polarization rather becomes parallel to the substrates and splay occurs only in a thin region $\xi = (K_1\varepsilon/P_s^2)^{1/2}$ at the surfaces.[17] If the polarization is large enough, the torque of this kink $\tau_{PSK} \sim K/\xi = (KP^2/\varepsilon_{LC})^{1/2}$ becomes larger than that which the surface can hold ($\tau_{surf} \sim 10^{-4}\,J/m^2$), and the polarization becomes uniformly parallel to the surfaces. Because in this case the majority of the polarization is perpendicular to the electric field, the magnitude of the torque at small fields is proportional to PE, resulting in a rotation of the polarization without any threshold, and the optical transmission will be proportional to the electric field.[18] Actually the voltage dependence of the azimuth angle ϕ (see Figure 9.7a) can be calculated by the condition that $\vec{P}$ reorients to screen the field completely to zero in the liquid crystal.[19] The free charge and polarization charge on opposite surfaces of the insulating layer must then

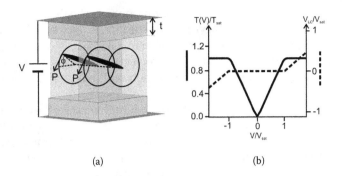

(a)   (b)

FIGURE 9.7
(a) Illustration of the polarization and director structure when external voltage is applied through insulating layers of thickness t. (b) When electric field is applied across the plates, a finite torque acts on the polarization, resulting in a switching of the director without a threshold and leading to the observed voltage dependence of the transmission.

be equal in magnitude, which means $\varepsilon_{IL} V = 2tP \cos\phi$, where ε_{IL} is the dielectric constant of the insulating layer of thickness t, and V is the applied voltage, as illustrated in Figure 9.7. This condition can be satisfied for $-2tP/\varepsilon_{IL} < V < 2tP/\varepsilon_{IL}$, or

$$V > 2tP/\varepsilon_{IL} : \quad \phi(V) = 0, \quad V_{LC} = V - 2tP/\varepsilon_{IL} \tag{9.1}$$

$$-2tP/\varepsilon_{IL} < V < 2tP/\varepsilon_{IL} : \quad \phi(V) = \cos^{-1}(\varepsilon_{IL} V/2tP), \quad V_{LC} = 0 \tag{9.2}$$

$$V < -2tP/\varepsilon_{IL} : \phi(V) = \pi, \quad V_{LC} = V + 2tP/\varepsilon_{IL} \tag{9.3}$$

The transmittance corresponding to $\phi(V)$ of Eq. (9.2) and the voltage drop inside the liquid crystals V_{LC} are shown in Figure 9.7b, where $V_{sat} = |2tP/\varepsilon_{LC}|$ is the voltage limit until $V_{LC} = 0$. It is interesting to note that the width of the "V-shaped" switching curve is proportional to the polarization, which is counter to the conventional intuition that switching should occur at lower voltage in higher polarizations. It is directly the result of the polarization screening.

Today's ferroelectric liquid crystals are made on small sizes on silicon chips (liquid crystals on silicon) and presently are serving only a relatively small market. This is partially because of the problems with field-induced mechanical stresses related to their piezoelectricity. However, we believe that sooner or later the ferroelectric displays should be good enough to replace the nematic displays. They are inherently bistable, i.e., only those pixels should be readdressed that are showing change, and they offer switching with 1–100 microsecond ranges. Their main drawback is that they have two-dimensional fluid structure, which does not allow spontaneous rehealing of damages of the uniform alignments. However, these problems can be overcome; for example, regulating the linear electromechanical effect is one clever way.[20] Maybe smectic ferroelectric and antiferroelectric displays will find themselves more useful in flexible displays, which will definitely be the displays of the near future.

Unfortunately, when people speak about liquid crystals they think about LCDs and believe that all the problems are solved concerning basic science. They may admit that there are some details to be solved, but those can be handled by the display industry alone. This situation is very similar to what happened in physics at the end of the nineteenth century, when Max Planck was advised not to study physics, since basically everything was solved there. Not only are we completely sure that basics research is needed to solve the challenges that liquid crystals have to face to play a dominant role in the future display technologies, but we are also confident that revolutionary new phenomena will arise, just like quantum mechanics evolved from the physics of the end of the nineteenth century, by further fundamental investigations of liquid crystals.

We should also keep in mind that the liquid crystals, these wonderful materials, are much more than displays. Even now there are a great number of nondisplay applications of liquid crystals.

9.2 Nondisplay Liquid Crystal Applications

9.2.1 Image and Signal Processing

Liquid crystals are used in display systems to spatially and temporally modulate light with information. With optical processing, computing systems represent and process information in the form of light, and they also need temporal and spatial modulation of light with information. For this reason commercially available liquid crystal devices often were used by researchers to study optical processing systems. The market for optical processing is presently small compared to displays, but it may have a potential for tremendous future growth.

9.2.2 Spatial Light Modulators (SLM)

Any device that controls the amount of light that passes through the device is called a light valve. A device that allows the transmittance to be independently controlled at different locations is called a spatial light modulator (SLM). An important example of such devices is an image projection system. If these projection systems are based on liquid crystals, we call them liquid crystal spatial light modulators (LCSLM). LCSLMs may be electrically controlled by a matrix of pixels or by the intensity of the light incident on the device.

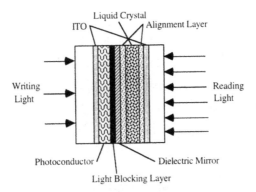

FIGURE 9.8
Schematics of an optically addressed spatial light modulator.

An electrically addressed LCSLM is identical in construction to an active matrix LCD. The construction of an optically addressed LCSLM is shown in Figure 9.8.

A two-dimensional picture is contained in the light striking the device from the left. This could be an image formed by a lens system or an image constructed by a scanning laser beam. The light lowers the resistance of a photo-conducting film, such as cadmium sulfide, in proportion to the light intensity, and is absorbed by the light-blocking layer. The decrease of the photoconductor resistance in a region where the incident light intensity is large causes the voltage appearing across the liquid crystal to increase above a threshold that switches the liquid crystal cell "ON" in this region. Accordingly, the light coming from right will be transmitted by the liquid crystal film and reflected by the dielectric mirror. In this way, the light intensity reflected back to the right will be proportional to the intensity of the light coming from the left. Note that the reading image is available as fast as the photoconductor and liquid crystal films response, thus allowing the device to capture, transfer and clear images at high speed. One application of LCSLM devices is in optical computing. Optical computing may have advantages over electronic computing in the sense that it allows parallel processing. If the intensity of the light striking the pixels of the LCSLM represents quantities, they can be manipulated independently, thus speeding up the processing tremendously. Summary of the issues, problems, challenges and possibilities of LCSLM devices is given by Owechko.[21]

9.2.3   Optical Communication Devices

Optical networks require a variety of active and passive devices to accomplish connectivity. The purpose of these devices is to allow generation, routing and detection of photons. The light can be affected by the device in several ways, e.g., by light absorption, light scattering, light diffraction, wave interference and waveguiding. The way liquid crystals affect light usually depends on the input polarization of the light. As we have learned in Chapter 5, the optical properties of the liquid crystals can perform the following elementary functions:

- Impose a shift on plane-polarized light.
- Change the ellipticity of the light by imposing a relative phase shift between the TE and TM polarized waves.
- Rotate the plane of the polarization of the light.

The ability to select a desired wavelength channel from a range of available wavelength channels is of great interest for advanced light-wave systems. This is especially important for the so-called high-density wavelength division-multiplexed (HD WDM) networks. We consider liquid crystal devices, constructed in the form of Fabry-Perot etalons.[22] To illustrate how liquid crystals may solve present telecommunication problems, we discuss briefly how to use them as tunable filters. In these devices the liquid crystal is sandwiched between two dielectric mirrors that reflect light by more than 99%. During the process of the light bouncing back and forth, only those wavelengths will eventually leak out the etalon, which constructively interfere. This is determined by the condition $n \cdot L = k \cdot \lambda \cdot \pi$, where n is the refractive index of the liquid crystal, L is the thickness of the etalon, and k is an integer number. Obviously, the change of the refractive index can be easily achieved by the field-induced realignment of the director. In this way the wavelength can be shifted by over 100 nm under the application of a few voltages. This effect and others, like liquid crystal Fabry-Perot devices in making tunable lasers, or for cross-connect filters, are discussed in detail by Patel.[23]

9.2.4   Photonic Applications

The search for structures exhibiting photonic band gaps, where the propagation of electromagnetic radiation is disallowed in a finite wavelength range, is a new branch of materials science. Photonic crystals are microstructures, in which the dielectric constant is modulated with a period comparable to the wavelength of light. Multiple interference between waves scattered from each unit cell may open a photonic band gap.[24,25,26,27] For many applications, it is advantageous to control stop band. LC materials are well suited for such applications because of their dielectric anisotropy and ready response to fields, as demonstrated, for example, by self-assembled opals infiltrated with a LC.

As discussed in Chapter 5, chiral liquid crystals exhibit modulated ground states and are therefore self-assembled PBG materials. The simplest of them is the cholesteric liquid crystal, which is a one-dimensional photonic band-gap material if the helical pitch is in the visible wavelength range. As illustrated in Chapter 5, such structure leads to selective reflection rendering a color to the material. The helical—and therefore the color—is sensitive to

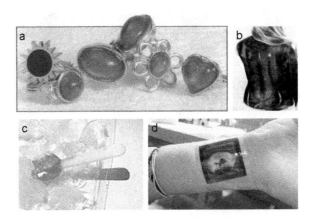

FIGURE 9.9
A few examples for the use of cholesteric liquid crystal thermometers.

the temperature, which makes them useful for temperature sensors, as first realized by J.L. Fergason in 1985.[28] A few examples of their applications are shown in Figure 9.9.

3D photonic crystals can be made by holographic lithography.[29] Switchable photonic crystals[30] can be made by electrically switchable Bragg gratings or holographic polymer dispersed LCs (H-PDLC).[31]

A holographic exposure can be used for fabrication.[32] The maxima and minima of light intensity in interference pattern dictate the modulated pattern of phase separating LC and polymer.[33] The LC can be reoriented by an electric field, thereby changing the refractive index in that location to create a truly switchable or tunable grating.[34] Applications include displays, optical switches, remote sensing, switchable lenses, agile beam steering, optical strain gauges, color filters, tunable waveguides, etc. Note that LC elastomers in cholesteric and blue phases are 1D and 3D band-gap materials, respectively, which can be tuned by mechanical stresses. These materials can be used for tunable lasers.[35]

9.3 Lyotropic Liquid Crystals and Life

Just as liquid crystals stand between the completely unorganized isotropic liquid and the strongly organized solids, life stands between the complete disorder and the complete rigidity. Indeed, lyotropic liquid crystals are very important in living organisms.

Lyotropic liquid crystals were discovered much earlier than thermotropics, yet their importance so far has been overshadowed. However, this situation may change soon, since the display applications are getting mature, whereas the fast development in modern biology requires deeper understanding of the physical properties of biological materials. This inevitably requires the understanding of the lyotropic liquid crystals, since life itself is based on them. In the following we will summarize the most important biological applications of the lyotropic liquid crystals.

Living organisms are composed of cells, which constitute the various organs and perform many and varied functions of life. In order to operate, biological organs must have some form and structure, whilst at the same time allowing movement in and out. Simple fluids do not have the structure to organize movement, while solids do have form and structure, but they cannot provide efficient transportation of materials. Accordingly, biological objects must be composed of structured fluids, just like lyotropic liquid crystals and related materials summarized in this book. Biological objects are all water-based lyotropic liquid crystals.

9.3.1   Biological Membranes

The plasma membranes of cells are constructed of lipids. Lipids have amphiphilic structure just as soaps and detergent surfactant discussed in Chapter 1.

A membrane in a cell wall fulfills a number of functions.[37,38] It acts as a barrier to prevent the contents of a cell from dispersing and also to exclude external agents such as viruses. The membrane, however, does not have a purely passive role. It also enables the transport of ions and chemicals such as proteins, sugars, nucleic acids in and out via the membrane proteins. Membranes appear not only in the external cell walls but also within the cell of eukaryotes (plants and animals, but not bacteria), where they subdivide the cell into compartments with different functions. A part of cell membrane is illustrated in Figure 9.10. It is built from a bilayer of lipids (usually phospholipids, except for the membranes of the brains, which is

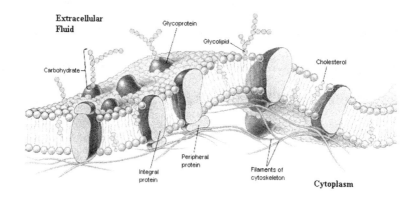

FIGURE 9.10
Schematic of a cell membrane.

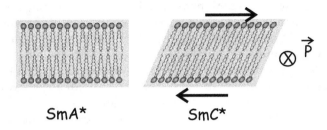

FIGURE 9.11
Illustration of the piezoelectric response of bilayers (such as cell membrane) to mechanical shear.

sphingomyelins that are much better electrical insulators than phospholipids) associated with membrane proteins and polysaccharides. The lipid bilayer is the structural foundation, and proteins and polysaccharides provide chemical functionality. The protein to lipid ratio is about 1:1 in most cells (except again the brain cells that contain only 18% proteins). Cells are usually sugar coated (glycosylated) with attached polysaccharides. The coating protects the cell and plays an important role in intercell binding, but at the same time, it makes the cell more prone to infectious bacteria. Cells are usually quite flexible due to the rapid rate of lateral diffusion of lipids within layers, which contrasts with the negligible rate of transverse diffusion. The rigidity of cells can be enhanced by inclusion of cholesterol in the membrane. However, too much cholesterol is not a good thing, because it can accumulate in the cell membrane of arteries, potentially leading to blockages and cardiovascular conditions. Integral proteins are very tightly bound within the membrane. On the other hand, transmembrane proteins are associated with a specific surface within the bilayer, for example the hydrophobic surface within the bilayer. These are important in the transport of ions of molecules across the cell membrane. Integral proteins are amphiphilic; the two ends extending into the aqueous medium contain hydrophilic groups, whereas the region within the bilayer is predominantly hydrophobic. Because of the apolar nature of the interior of lipid bilayers, they are impermeable to most ionic and polar molecules, which is the basis of the barrier activity of lipid membranes. Integral proteins are bound in the lipid bilayer and often act as channels for the transport of ions and molecules. These channels have to be highly selective to prevent undesirable material entering the cell and so are opened and closed as necessary. Membrane transport is also carried out by proteins that are not integral to the membrane. They are required to move ions, amino acids, sugars and nucleotides across the cell wall. They can either ferry these species across or away from channels.

The self-organization of lipids to form plasma membrane was a crucial step in the evolution of the life. Cell membranes demonstrate organic chemistry beyond covalent bonds. It is the interplay between molecular self-organization and molecular recognition of its individual constituents, which leads to the construction of this natural system. This fascinating phenomenon can be understood only if a wide range of modern scientific disciplines work together. One interesting question is whether it is possible to develop new materials which simulate natural supramolecular systems in that their functioning is determined by their organization. Since the structure of the cell membranes is too complex to study specific individual processes, model membranes are used to investigate biomembrane processes.[39]

Nerve and muscle membranes exhibit behavior interpretable as ferroelectric. Specifically, the Na[+] channel molecules have been proposed to supply the ferroelectricity of the membrane, which undergoes a ferroelectric para-electric transition analogous to the SmC*–SmA* transition. The surface charge density of the Na+ channels was measured to be about 2×10^{-2} C/m^2, which is in the order of known ferroelectrics. It was shown[40] that some membranes swell in response to voltage changes, and it was interpreted as evidence of a piezoelectric effect.[41] Ion channels sensitive to membrane stretch have been also observed in muscle cells.[42] Voltage-induced birefringence, similar to that known in ferroelectrics,[43] has also been reported.[44] Ferroelectricity was postulated to exist in cholesterol-containing lipid water systems and membranes with tilted chains. Since the molecules are chiral, the tilted director structure has the same symmetry as of SmC* materials.[45] From experiments on cholesterol–smectic C* mixtures, the polarization of myelin sheets that cover the nerve fibers and that contain 40% cholesterol, the spontaneous polarization was estimated as $P_s \sim 10^{-4}$ C/m^2. Bacteriorodpsin (bR), which is an integral membrane protein in purple membrane of bacterium (*Halobacterium salinarium*), recently was found to be ferroelectric. This was concluded by dielectric spectroscopy that indicates relaxation processes resembling the behavior of thermotropic liquid crystals near the SmC*-SmA* phase transition (so-called "soft," or electroclinic mode, see Chapter 8).[46]

In view of the discussions of the previous chapters we can easily realize that the cell membranes are piezoelectric and flexoelectric.[47] A mechanical shear would locally cause a tilt, which, due to the chirality of the molecules, leads to the appearance of macroscopic dipole moment, just as was described for SmA* materials in the previous chapter.

If we bend a membrane, we will induce a splay of the lipid molecules. The number of dipoles per unit area then will be smaller in the outer range than in the inner part. The gradient of the number of dipoles per unit area gives an electric polarization P_f. For simplicity, let us assume a one-dimensional curvature with curvature radius R. In this case the splay vector reads as $\vec{S} = \vec{n} \cdot div\vec{n} = \vec{n}\frac{1}{R}$, where $\vec{n}$ is the director (the average direction connecting lipids facing to each other). In a flat bilayer of thickness L the dipole density in the opposite layers average out, whereas in the splayed membrane the net density is:

$$\mu \frac{(R+L)^2 \pi - R^2 \pi}{R^2 \pi} \approx \mu \frac{2L}{R} \qquad (9.4)$$

The net polarization is the number density of the net dipole, i.e.,

$$P_f = \mu \frac{2L}{R} N_n = \mu \frac{2L}{R} \frac{N_{AV}\rho}{M_{mol}} \tag{9.5}$$

where N_n is the number of molecules in unit volume, $N_{AV} = 6 \times 10^{23}$ is the Avogadro number, M_{mol} is the molecular weight of the constituent molecules, and ρ is the mass density of the membrane. Comparing with the definition of the splay, the flexoelectric polarization becomes $\vec{P}_f = e_1 \vec{n} \frac{1}{R}$, and we get that the flexoelectric constant is:

$$e_1 = \mu \rho \frac{N_{AV}}{M_{mol}} 2L \tag{9.6}$$

Taking $\mu \sim 10^{-29}$ C·m (~3 *Debye*), $M_{mol} \sim 0.5$ kg, $\rho \sim 10^3$ kg/m³ and $L \sim 5$ nm, we get that $e_1 \sim 10^{-10}$ C/m. This value has a same order of magnitude as that of thermotropic nematic liquid crystals. It means that a curvature with radius comparable to that of the thickness of the membrane can result a polarization of about 10^{-2} C/m², i.e., comparable with the measured surface charge density of the Na+ channels. This indicates that the experimentally observed ferroelectric-type behaviors may actually be due to flexoelectricity. We note that, in reality, the lipids also carry negative charges, and there are free ions in the cells, which complicate the above simple picture.[48] The details of the flexoelectricity of membranes are discussed by Petrov.[38]

The cell membrane has a phase transition temperature, just like a thermotropic liquid crystalline material. In membranes the phase transition is called a gel point and corresponds to the freezing of the hydrocarbon chains and to the ordering of the head groups to a hexagonal pattern. The temperature of the liquid crystal to gel phase transition depends on the environment of the organism concerned. Homeotherapic animals (like human beings) control their own body temperature, so they normally are not exposed to large temperature changes. In these living organs the phospholipid membranes are composed of high proportion of saturated fatty acids, which give rise to a relatively high gel phase transition. On the other hand, poikilothermic organisms (e.g., fish) are subject to relatively large temperature ranges, and their phospholipid membranes include a high proportion of unsaturated fatty acids, which provide low gel phase transition temperature. It is important to note that, if the gelation temperature is much lower than the ambient temperature, the membranes are more prone to rupture, which can also cause death. Accordingly the gelation temperature has to be only moderately lower than the ambient temperature. Cells use eutectic mixtures to maintain wide operating temperatures, such as materials used in liquid crystal displays.

Interacting living amoeboid cells themselves form nematic liquid crystals phases.[36] It was observed that (1) a cluster of a polar nematic liquid crystal is formed by cells, which emit molecules for attracting other cells, and

(2) an apolar nematic liquid crystal is formed by elongated cells, which have an anisotropic steric repulsion.

9.3.2 Lyotropic Liquid Crystalline State of Biopolymers

Biopolymers such as proteins have a sequence of amino acid residues in the polymer chain. Due to bonding between –CO and –NH groups they may form α and β structures or double helix.

The organization of DNA in biological structures is largely unknown, but bears some resemblance to liquid crystalline phases observed *in vitro*.[50] It has been found that DNA forms at least three distinct liquid crystalline phases at concentrations comparable to those *in vivo*, with phase transitions occurring over relatively narrow ranges of DNA concentration.[51] A weakly birefringent, dynamic, "precholesteric" mesophase with microscopic textures intermediate between those of a nematic and a true cholesteric phase forms at the lowest concentrations required for phase separation. At slightly higher DNA concentrations, a second mesophase forms which is a strongly birefringent, well-ordered cholesteric phase with a concentration-dependent pitch varying from 2 to 10 μm.[52] At the highest DNA concentrations, a phase forms, which is two-dimensionally ordered and resembles smectic phases of thermotropic liquid crystals observed with small molecules. The formation of the anisotropic phase results from the competition between the orientation entropy and the volume excluded by a DNA molecule to another molecule.[53] In case of supercoiled DNA, the topology is anticipated to be the key factor in controlling the phase behavior. Indeed, supercoiling has been reported to be a major factor for plasmic DNA in the cytoplasm of bacteria.[54,55]

In contrast to chemical synthesis, nature uses DNA to produce viruses that are identical to each other, which results in highly monodisperse viruses. This high monodispersity of virus suspensions is the property that makes them an appealing system to study the phase behavior of hard rods experimentally. A virus consists of a core of nucleic acid (DNA or RNA, but never both) surrounded by a coat of antigenic protein. The virus provides the genetic code for replication, and the host cell (e.g., bacteria for bacteriophages) provides the necessary energy and raw material for the virus to grow.

Linear viruses are known to form liquid crystalline structures since tobacco mosaic virus (TMV) was found to form nematic liquid crystalline phase. This inspired Onsager to write his seminal paper about the isotropic–nematic transition in hard rods.[53] According to his theory (see Chapter 3), the isotropic phase becomes unstable with respect to the nematic arrangement when,

$$c\pi L^2 D = 16 \tag{9.7}$$

where c is the rod number density, and L (D) is the length (diameter) of the rod.

Viruses are charged; therefore, in addition to the steric repulsion of the hard rods, a long-range soft repulsion, due to electrostatic interactions, is present.

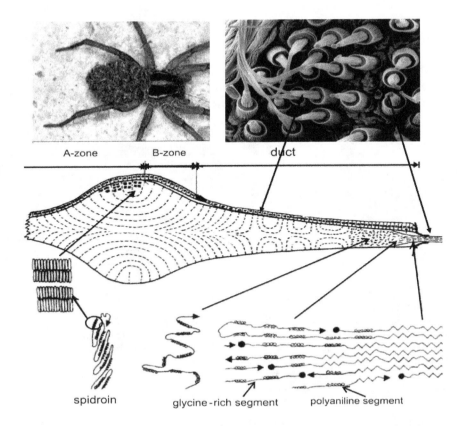

A-zone   B-zone   duct

spidroin   glycine-rich segment   polyaniline segment

FIGURE 9.12
Upper part: Pictures of a spider and the main organs producing silk threads consisting of cuticles of the funnel (diameter: 350mm), the valve (300mm); spigot (190mm), and duct (40mm). Bar represents 0.3mm length. Lower part: sketch of the structure of the silk material inside the duct. The main component of the silk is an oriented proteins, spidroin (block co-polymers) composed of transverse lamellae that show a regular alternation of axially oriented b crystallites and helices derived from glycine-rich segments. During the extruding process the spidroin becomes aligned in the water solution and forms oriented lamellar liquid crystalline structures. (Lower part of the figure is based on Figure 2 of Ref.[72].)

In this case the condition to the isotropic nematic transition will be changed so that the diameter D will be replaced by D_{eff}, which can be rigorously calculated, and turns out to be roughly equal to the distance between two rods, where the intermolecular potential is equal to the thermal energy $k_B T$. At high ionic strength, D_{eff} approaches the bare diameter, while at low ionic strength, it is of several Debye screening length, i.e., much larger than the bare diameter.

TMVs are rigid hollow cylinders with lengths of 300 nm and diameter of 18 nm. They are formed by 2130 protein subunits arranged in a single helix around the central hole of radius about 2 nm. In solutions of near neutral pH, each protein subunit has a fixed charge of about $-6e$, which is practically offset by strongly bound counterions.[56]

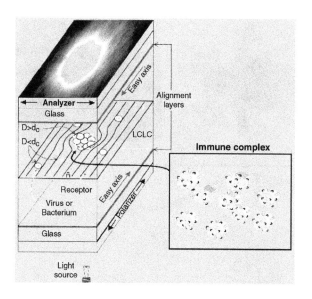

FIGURE 9.13

A scheme detailing the mechanism of the LCLC biosensor for the detection and amplification of immune complexes.[73] The inset shows bacterial immune complex formation whereby a single antibody cross-links two microbial antigens. The transmitted optical signal can be measured by an array of photo sensors on the plastic film placed at the opposite side of the cassette. The cassette is laminated with two polymer polarizing films with mutually perpendicular directions of polarization. (Figure courtesy of O. Laurentovich.)

The first definitive experimental observation of SmA ordering in TMV solutions was given in 1989.[57] Since then other viruses were studied as model systems for hard rod liquid crystals. Because of the outer protein structures of the viruses, it is impossible to decrease their surface charges by dissolving them in apolar or weakly polar solvents; however, it was shown that the effective diameter of the virus can be altered by covalently attaching polymer into the surface.[58]

Bacteriophage *fd* is a good model system to study liquid crystalline behavior because it is easily grown in large quantities, it is very monodisperse, and it is very stable in solution, because it is highly charged.[59] Bacteriophage *fd* is a chiral, monodisperse rod-like colloid with a long semiflexible rod with contour length of 880 nm, diameter of 7 nm and persistence length of 2.2 μm. The virus consists of a single-stranded circular DNA molecule coated with a protein layer of 2700 identical protein subunits. These subunits with a molecular mass of 5240 form about 99% of the total protein mass. In addition, because *fd* infects *E. coli,* it is easy to use standard molecular cloning techniques to alter the basic properties, such as contour length, of the virus. A review on the phase behavior with many general references can be found in Fraden.[60]

At concentrations of around 20 mg/ml coexistence between isotropic and nematic/cholesteric phases was observed.[61] The isotropic phase has short-range positional and orientational order and appears dark between crossed polarizers. The anisotropic phase has long-range orientational order and is birefringent, with "fingerprint" texture characteristic to cholesterics. At very high concentrations the formation of the smectic phase was observed. The periodic pattern is due to the smectic layers, which are composed of a two-dimensional liquid of essentially aligned rods. Due to large contour length of the virus, it is possible to directly observe smectic layers with optical microscope.

Very recently even chiral smectic C structures of M13 virus particles (L = 880 nm,[62] D = 6.8 nm) were observed.[63]

A famous and technologically significant liquid crystal is a lyotropic liquid crystalline polymer called Kevlar. It is a synthetic polyamide of rather simple structure. However, when Kevlar is dissolved in high concentrations in sulfuric acid, a liquid crystalline phase is formed. The strength of the polymer in general is derived from the orientation of the polymer chains when the polymer is extruded. The polymer chains of a liquid crystal polymer are inherently ordered, and so when extruded in the liquid crystal phase, they acquire extremely high strength, because the uniform alignment eliminates defects. For example, compared to nylon, which is extruded in the liquid phase, Kevlar is thirty times stronger, although it is only slightly denser. Compared to steel, which is five times stronger than Kevlar but eight times denser, we see that the specific strength of the Kevlar is about twice as large as steel.

Interestingly, the very same method is used by the spiders during spinning their fiber net, as has been demonstrated by Vollrath and Knight.[64] The spider silks have typically even better mechanical properties than that of Kevlar. Its strength of 1.1 GPa approaches that of typical high tensile engineering steel (1.3 GPa), but silks have a significantly lower density (1.3 g/cm³ instead of 7.8 kg/cm³ of the steel). This high strength is also due to the liquid crystalline phase of the silk during the extruding. During the extruding process, the silk contains about 50% water and 50% protein,[65] which are mainly spidroin I and II and the main proteins making up spider dragline silk.[66] As it is extruded, the diameter of the gland becomes smaller and smaller, thus the flow rate increases. This induces a flow alignment of the nematic liquid crystal, just as we learned in Chapter 4. During this realignment the viscosity decreases (shear thinning), thus making the extruding easier for the spider (see Figure 9.14). Rod-like molecules with length L in solutions of concentration $c \ll 1/L^3$ have an effective viscosity $\eta = \eta_0(1 + cL^3)$, which is basically the same as of the host material. However, when $c \gg 1/L^3$, $\eta = \eta_0(1 + cL^3)^3$, which is much larger than the host viscosity. This is true for isotropic orientation of the rods; however, if the rods become oriented, the effective viscosity will be $\eta = \eta_0(1 + cD^2L)^3$, where D is the diameter of the rods. If $D \ll L$, $c \cdot D^2L < 1$, and $\eta \sim \eta_0(1 + 3cD^2L)$. Thus, for very slender rods, the effective viscosity can be kept low by adjusting the alignment, which makes extruding equally easy for spider. Basically this allows the spider to spin fiber with the same energy both in the morning cold and in the heat of sunlight. We note that the molecular weight of the silk protein is about an order of magnitude larger than that of Kevlar, yet the extrusion forces used by spiders are smaller than those used by machines making the Kevlar fibers. Once the silk leaves the spigot, it eventually loses water and becomes solid, and its strength is determined by the defects, which are eliminated by the extruding process described above.

9.3.3 Lyotropic Chromonic Liquid Crystals (LCLCs)

The lyotropic chromonic liquid crystals (LCLCs) summarized in Chapter 1.3 may soon find applications as optical elements and biological sensors. For example, dried LCLC film can be used as both orienting and light-polarizing layer in flexible liquid crystal displays.[67] The advantage is that the LCLC polarizer can be put *inside* the cell, between the electrode and the regular electrically controlled thermotropic LC layer. The internal polarizer allows one to use cheap plastic substrates with significant birefringence in the development of flexible displays. The LCLC materials can be cast in the form of monomolecular nanoscale layer and multilayered stacks.[68]

Another exciting field of possible application of LCLC is real-time microbial detection.[69] The idea is that the LCLC matrix serves as an environment in which the microbes are exposed to the antibodies capable of binding to the antigens at the microbe's surface. The antibody–antigen reaction is very specific.[70] Since each antibody has two identical binding sites, it can bind together multiple microbes into an aggregate called an immune complex. In contrast to currently available tests that are lengthy, a single immune complex can distort the LC director and thus drastically increase the intensity of light passing through the sample.[71] This effect is substantial only when the particle size D is

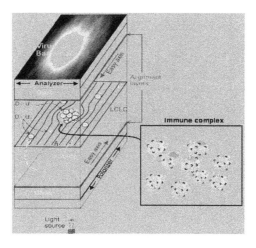

FIGURE 9.14

Stages of the fiber extrusion. (a) The thick cuticle of the funnel (350 μm); (b) the thinner inside duct, that allows surfactants and lubricants into the lumen (40 μm); (c) valve (300 μm); spigot (190 μm).

larger than the critical size d_c determined by elastic and surface properties of the liquid crystal: $d_c \approx K/W$ where K is the elastic constant and W is the anchoring strength at the surface of the complex. As we have explained in Chapter 4, this argument is true for all liquid crystals; however, the water-based but surfactant-free LCLCs are unique in the sense that they are not toxic. The principle of the real-time microbe detection[73] is illustrated in Figure 9.13.

We emphasize that these examples represent only the tip of the iceberg. One does not have to be a Nostradamus to foresee that in the near future significant new observations will find analogies between liquid crystal effects and biological phenomena. What thousands of researchers worked out in the area of thermotropic liquid crystals during the search for the ultimate displays, or discoveries just motivated by purely academic curiosity, soon will help us to understand biological processes.

References

1. J.L. Fergason, *Sci. Am.*, 211, 77 (1964).
2. R. Williams, *J. Chem. Phys.*, 39, 384 (1963); G.H. Heilmeier, L.A. Barton, L.A. Zanoni, *Appl. Phys. Lett.*, 13, 46 (1968).
3. B.J. Lechner, *Proc. IEEE*, 59, 1566 (1971); T.P. Brody, J.A. Assars, G.D. Dixon, *IEEE Trans. Elec. Dep.*, ED-20, 995 (1973).
4. S. Kobayashi, H. Hori, Y. Tanaka, Ch. 10; T. Scheffer, Ch. 11, in *Handbook of Liquid Crystal Research*, Ed. P.J. Collings, J.S. Patel, Oxford University Press, New York (1997), and references therein.
5. T. Scheffer, J. Nehring, Twisted nematic and supertwisted nematic mode LCDs, in *Liquid Crystals—Applications and Uses*, Vol. 1, Ch. 10, Ed. B. Bahadur, World Scientific, Singapore (1990), and references therein.
6. P. Yeh, C. Gu, *Optics of Liquid Crystal Displays*, Wiley, New York (1999).
7. M. Schadt, W. Helfrich, *Appl. Phys. Lett.*, 18, 127 (1971).
8. J.L. Fergason, US Patent 3, 731, 986 (1971).
9. R. Barberi, G. Durand, Controlled textural bistability in nematic liquid crystals, in *Handbook of Liquid Crystal Research*, Ch. 15, Ed. P.J. Collings, J.S. Patel, Oxford University Press, New York (1997), and references therein.
10. S.T. Wu, D.K. Yang, *Reflective Liquid Crystal Displays*, Wiley, New York (2001), and references therein.
11. P.S. Drzaic, *Liquid Crystal Dispersions*, World Scientific, Singapore (1995); G.P. Crawford, S. Zumer (Eds), *Liquid Crystals in Complex Geometries Formed by Polymer and Porous Networks*, Taylor and Francis, London (1996); G.P. Crawford, J.W. Doane, S. Zumer, Polymer dispersed liquid crystals: Nematic droplets and related systems, in *Handbook of Liquid Crystal Research*, Ch. 9, Ed. P.J. Collings, J.S. Patel, Oxford University Press, New York (1997), and references therein.
12. S.T. Lagerwall, *Ferroelectric and Antiferroelectric Liquid Crystals*, Wiley-VCH, Weinheim (1999).
13. J. Fünfschilling, M. Schadt, *Ferroelectrics*, 213, 195 (1998); P. Maltese, *Mol. Cryst. Liq. Cryst.*, 215, 57 (1992).
14. N.A. Clark, S.T. Lagerwall, *Appl. Phys. Lett.*, 36, 899 (1980), U.S. Patent, 4,367,924 (1983).

15. A. Jákli, *Mol. Cryst. Liq. Cryst.*, 292, 293 (1997).
16. L.A. Beresnev, L.M. Blinov, D.I. Dergachev, S.B. Kondratjev, *JETP Lett.*, 413 (1987); A. Jákli, L. Bata, L.A. Beresnev, *Mol. Cryst. Liq. Cryst.*, 177, 43 (1989); L.A. Beresnev, V.G. Chigrinov, D.I. Dergachev, E.P. Poshidaev, J. Fünfschilling, M. Schadt, *Liq. Cryst.*, 5(4), 1171 (1989).
17. Z. Zhuang, J.E. Maclennan, N.A. Clark, *Proc. SPIE*, 1080, 110 (1989).
18. P. Rudquist, J. Lagerwall, M. Buivydas, F. Gouda, S.T. Lagerwall, N.A. Clark, J. Maclennan, R. Shao, D. Coleman, S. Bardon, T. Bellini, D.R. Link, G. Natale, M. Glaser, D. Walba, M. Wand, X. Chen, *J. Mat. Chem.*, 9, 1257 (1999).
19. N.A. Clark, D. Coleman, J.E. Maclennan, *Liq. Cryst.*, 27 (7), 985 (2000).
20. A. Jákli, A. Saupe, *J. Appl. Phys.*, 82 (6), 2877 (1997).
21. Y. Owechko, Applications of liquid crystals in image and signal processing, in *Handbook of Liquid Crystal Research*, Ch. 13, Ed. P.J. Collings, J.S. Patel, Oxford University Press, New York (1997), and references therein.
22. L. Levi, *Applied Optics*, Vol. 2, Wiley, New York (1980).
23. J.S. Patel, Liquid crystals for optical communication devices, in *Handbook of Liquid Crystal Research*, Ch. 12., Ed. P.J. Collings, J.S. Patel, Oxford University Press, New York (1997), and references therein.
24. D. Kang, J.E. Maclennan, N.A. Clark, A.A. Zakhidov, R.H. Baughman, Electro-optic behavior of liquid crystal filled silica opal photonic crystals: Effect of liquid crystal alignment, *Phys. Rev. Lett.*, 86, 4052 (2001).
25. K. Busch, S. John, Liquid crystal photonic band gap materials: The tunable electromagnetic vacuum, *Phys. Rev. Lett.*, 83, 967 (1999).
26. K. Yoshino, Y. Shimoda, Y. Kawagishi, K. Nakayama, M. Ozaki, Temperature tuning of the stop band in transmission spectra of liquid crystal infiltration synthetic opal as tunable photonic crystal, *Appl. Phys. Lett.*, 75, 93 (1999).
27. S.W. Leonard, J.P. Mondia, H.M. van Driel, O. Toader, S. John, K. Busch, A. Birner, U. Gosele, V. Lehmann, Tunable 2D photonic crystals using liquid crystal infiltration, *Phys. Rev. B*, 61, R2389 (2000).
28. J.L. Fergason, Polymer encapsulated nematic liquid crystals for display and light control applications, *SID Int. Symp. Dig. Technol.*, 16, 68 (1985).
29. M. Campbell, D.N. Sharp, M.T. Harrison, R.G. Denning, A.J. Turberfield, Photonic crystals made by holographic lithography, *Nature*, 404, 53 (2000).
30. V.P. Tondiglia, Natarajan, R.L. Sutherland, D. Tomlin, T.J. Buning, Holographic formation of electro-optical polymer-liquid crystal photonic crystals, *Adv. Mater.*, 14, 187 (2002).
31. T.J. Bunning, V. Natarajan, V.P. Tondiglia, R.L. Sutherland, D.L. Vezie, W.W. Adams, The morphology of holographic transmission gratings recorded in polymer dispersed liquid crystals, *Polymer*, 36, 2699 (1995).
32. S. Shoji, S. Kawata, Photofabrication of three dimensional photonic crystals by multibeam laser interference into photopolymerization resin, *Appl. Phys. Lett.*, 76, 2668 (2000).
33. C.C. Bowley, G.P. Crawford, Diffusion kinetics of formation of holographic polymer dispersed liquid crystal display materials, *Appl. Phys. Lett.*, 76, 2235 (2000).
34. C.C. Bowley, A.K. Fontecchio, G.P. Crawford, J.J. Lin, L. Li, S. Faris, Multiple gratings simultaneously formed in holographic polymer dispersed liquid crystal displays, *Appl. Phys. Lett.*, 76, 523 (2000).
35. H. Finkelmann, S.T. Kim, A. Munoz, P. Palffy-Muhoray, B. Taheri, Tunable mirror-less lasing in cholesteric liquid crystalline elastomers, *Adv. Mater.*, 13, 1069 (2001).

36. H. Gruler, U. Dewald, M. Eberhardt, *Eur. Phys. J. B,* 11, 187 (1999).
37. I.W. Hamley, *Introduction to Soft Matter,* Wiley, Chichester (2000).
38. A.G. Petrov, *The Lyotropic State of Matter,* Gordon and Breach, Singapore (1999).
39. A. Reichert, H. Ringsdorf, A. Wagenknecht, Attempts to mimic biomembrane processes: Function of phospholipase A2 at lipid monolayers, in *Supramolecular Chemistry,* p. 325, Ed, V. Balzani, L. de Cola, Kluwer Academic Publishers, Netherlands (1992).
40. K. Iwasa, I. Tasaki, R.C. Gibbons, *Science,* 210, 338 (1980).
41. H.R. Leuchtag, *J. Theor. Biol.,* 127, 321 (1987).
42. B. Hille, *Ionic Channels of Excitable Membranes,* Sinauer, Sunderland (1992).
43. M.E. Lines, A.M. Glass, *Principles and Applications of Ferroelectrics and Related Materials,* Clarendon Press, Oxford (1977).
44. L.B. Cohen, B. Hille, R.D. Keynes, *J. Physiol.,* 211, 495 (1971).
45. L. Beresnev, L.M. Blinov, *Mendellev J. All-Union Chem. Soc.,* 28, 149 (1982).
46. I. Ermolina, A. Strinskovski, A. Lewis, Y. Feldman, *J. Phys. Chem. B,* 105(14), 2673 (2001).
47. A.G. Petrov, *Physical and Chemical Bases of Biological Information Transfer,* p. 167, Ed. J. Vassileva, Plenum Press, New York (1975).
48. A.T. Todorov, A.G. Petrov, J.H. Fendler, *J. Phys. Chem.,* 98, 3076 (1994).
49. D. Voet, J.G. Voet, *Biochemistry,* 2nd ed., Wiley, New York (1995).
50. F. Livolant, A. Leforestier, *Prog. Polym. Sci.,* 21, 1115 (1996).
51. J. Jizuka, J.T. Yang, in *Liquid Crystals and Ordered Fluids,* Ed. J.F. Johnson, Plenum Press, New York (1969).
52. E. Schenecal, G. Maret, K. Darnschfeld, *Int. J. Biol. Macromol.,* 2, 256 (1980).
53. L. Onsager, *Ann. N.Y. Acad. Sci.,* 51, 627 (1949).
54. Z.E. Reich, J. Wachtel, A. Minsky, *Science,* 264, 1460 (1994).
55. J.R.C. van der Maarel, S.S. Zkaharova, W. Jesse, C. Backendorf, S.U. Egelhaaf, A. Lapp, *J. Phys. Cond. Mat.,* 15, S183 (2003).
56. R.B. Scheele, M.A. Lauffer, *Biochemestry,* 6, 3076 (1967).
57. X. Wen, R.B. Meyer, D.L.D. Caspar, *Phys. Rev. Lett.,* 63, 2760 (1989).
58. Z. Dogic, S. Fraden, *Phil. Trans. R. Soc. Lond. A,* 359, 997 (2001).
59. D.A. Marvin, H. Hoffmann Berling, *Nature,* 197, 517 (1963).
60. S. Fraden, Phase transitions in colloidal suspensions of virus particles, in *Observation, Prediction and Simulation of Phase Transitions in Complex Fluids,* p. 113, Ed. M. Baus, L.F. Rull, J.P. Ryckaert, NATOAS Series C, vol. 460, Kluwer Academic Publishers, Dordrecht (1995).
61. Z. Dogic, S. Fraden, *Langmuir,* 16, 7820 (2000).
62. M. Baus, L.F. Rull, J. Ryckaert, *Observation, Prediction and Simulation of Phase Transition in Complex Fluids,* p. 113, Kluwer Academic, Boston (1995).
63. S.-W. Lee, B.M. Wood, A.M. Belcher, *Langmuir,* 19, 1592 (2003).
64. F. Vollrath, D.P. Knight, *Nature,* 410, 541 (2001); D.P. Knight, F. Vollrath, *Proc. R. Soc. Lond. B,* 266, 519 (1999).
65. D.H. Hijirada, K.G. Do, C. Michal, S. Wong, D. Zax, L.W. Jelinski, *Bipohis J.,* 71, 3442 (1996).
66. J.P. O'Brien, S.R. Fachnestock, Y. Termonia, K.C. Gardner, *Adv. Mater.,* 10, 1185 (1998).
67. T. Sergan, T. Schneider, J. Kelly, O.D. Lavrentovich, *Liq. Cryst.,* 27, 567 (2000); R. Penterman, S.L. Klink, H. de Koning, G. Nisato, D.J. Broer, *Nature,* 417(6884), 55 (2002).
68. T. Schneider, O. D. Lavrentovich, *Langmuir,* 16, 5227 (2000).

69. C.J. Woolverton, G.D. Niehaus, K.J. Doane, O. Lavrentovich, S.P. Schmidt, S.A. Signs, Detection of ligands with signal amplification, U.S. Patent # 6,171,802 (February 2001); O.D. Lavrentovich and T. Ishikawa, Bulk alignment of lyotropic chromonic liquid crystals, U.S. Patent # 6,411,354 (June 25, 2002).

70. A.D. Strosberg, J.E. Leyseng, Receptor-based assays, *Curr. Opin. Biotechnol.*, 2, 30 (1991).

71. V.K. Gupta, J.J. Skaife, T.B. Dubrovsky, N.L. Abbott, *Science*, 279, 2077 (1998).

72. D.P. Knight, F. Vollrath, *Phil. Trans. R. Soc. Lond. B*, 357, 155 (2002).

73. S.V. Shiyanovskii, T. Schneider, I.I. Smalyukh, T. Ishikawa, G.D. Niehaus, K.J. Doane, C.J. Woolverton, O.D. Lavrentovich, *Phys. Rev. E*, 71, 020702 (R) (2005).

Index

A

Abbe refractometer, 171
Acceptor–donor type impurities, 236
Achiral materials, 155
Active-matrix display, 269
Adhesion, work of, 47
AFLC materials, *see* Antiferroelectric liquid
        crystal materials
Aliphatic hydrocarbons, 294
L-amino acids, 14
Amoeboid cells, nematic liquid crystals
        phases of, 283
Amphiphiles
    mixed, 28
    polar–apolar duality of, 22
Amphiphilic block copolymers, 35
Amphiphilic polyhydroxy amphiphiles, 38
Amphiphilics, flexible, 38
Amphiphilic solvents, 37
Amphiphilic–water aggregation, 23
    concentration, 26–29
    optimal surface to tail volume ratio,
        23–26
Amphoteric surfactant molecule, 296
Amphotropic materials, best-investigated,
        38
Anionic surfactants, 295, 296
Anisotropic films, transmittance of, 165
Anisotropic materials, deformations of, 309
Anisotropic medium, propagation of light in,
        155
Anisotropic systems, piezoelectric
        components in, 313
Antibody–antigen reaction, 288
Antiferroelectric liquid crystal (AFLC)
        materials, 255, 271
Antiferroelectric state, chiral domains,
        259
Antistatic agent, 23
Applications, 267–292
    liquid crystal displays, 267–276
        display structures, 267–269
        ferroelectric smectic displays, 271–276
        nematic liquid crystal displays, 270–271
    lyotropic liquid crystals and life, 279–289
        biological membranes, 280–284
        lyotropic chromonic liquid crystals,
            288–289
        lyotropic liquid crystalline state of
            biopolymers, 284–288
    nondisplay liquid crystal applications,
        276–279
        image and signal processing, 276
        optical communication devices, 277–278
        photonic applications, 278–279
        spatial light modulators, 276–277
Aromatic hydrocarbons, 294
Arrhenius behavior, 229–230, 236
Asphaltenes, 297
Atomic Force Microscope, 79
p-Azoxyanisole (PAA), 2

B

Bactericide, 23
Bacteriophage, 286
Bacteriorodpsin, 282
Banana smectics, 20, 78, 79, 80
Batonnets, 131
Bend-twist coupling, SmC, 129
Bent-core liquid crystals, 18
Bent-core molecules, 259
    different phases of, 19
    liquid crystals of, 78
    packing of, 20
Benzene derivatives, 294
Berreman's model for planar anchoring, 146
Bertrand Lens, 175
Biaxial materials, melatopes of, 176
Biaxial nematics, 129
Bicontinuous structures form, amphiphile
        aggregate structure, 29
Bingham fluids, 140
Bingham model, 140
Bingham-plastics, 302
Bioferroelectricity, 249
Biological membranes, 280
Biological organs, fluids of, 280

327

Appendix A

Chemicals

A.1 Basics of Organic Chemistry

The study of organic chemistry is greatly simplified by considering hydrocarbons as parent compounds and describing other compounds as derived from them.

A summary of the structures and nomenclatures of hydrocarbons is shown in Figure A.1.

In general, an organic molecule consists of a skeleton of carbon atoms with special groups of atoms within or attached to the skeleton. These special groups of atoms are often called functional groups because they represent the most common sites of chemical reactivity (function). Possible functional groups in hydrocarbons are double or triple (i.e., π) bonds. Atoms other than C and H are called heteroatoms, the most common being O, N, S, P and the halogens. Most functional groups contain one or more heteroatoms.

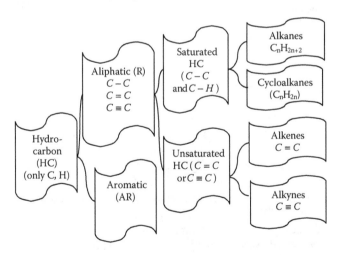

FIGURE A.1
Classification of hydrocarbons. (Based on table from K.W. Whitten, K.D. Gailey, R.E. Davis, *General Chemistry*, 4th ed., Saunders College Publishing, 1992.)

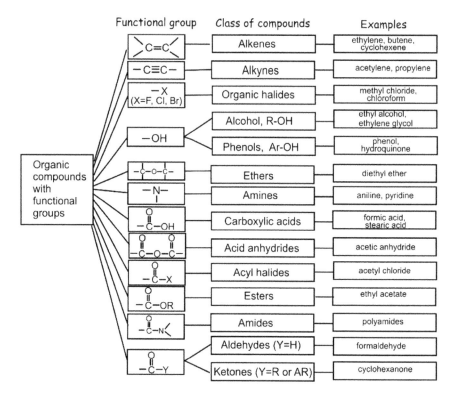

FIGURE A.2
Summary of functional groups. (Based on the table by K.W. Whitten, K.D. Gailey, R.E. Davis, *General Chemistry*, 4th ed., Saunders College Publishing, 1992.)

The summary of some functional groups and the classes of organic compounds are given in Figure A.2.

It is common to represent the aliphatic hydrocarbons (e.g. alkyl) with R- and the aromatic group (e.g., an aryl such as phenyl) with Ar -.

In the area of thermotropic liquid crystals, especially important aromatic hydrocarbons are the benzene (C_6H_6) derivatives (see Figure A.3).

In the field of biological materials the fats (solids) and oils (liquids), which are esters of glycerol and aliphatic acids of high molecular weight, have special importance. "Fatty acids" are organic acids that occur in fats and oils (as esters.) Saturated fatty acids are linear and tend to pack like sticks of wood, to form solid masses in blood vessels. The *trans* unsaturated fatty acids have a slight Z-shape kink in the chain, but are also linear. By contrast, *cis* unsaturated fatty acids are bent, so do not pack as well as linear structures and do not collect in blood vessels. Many natural vegetable fats contain esters of *cis* unsaturated fatty acids.

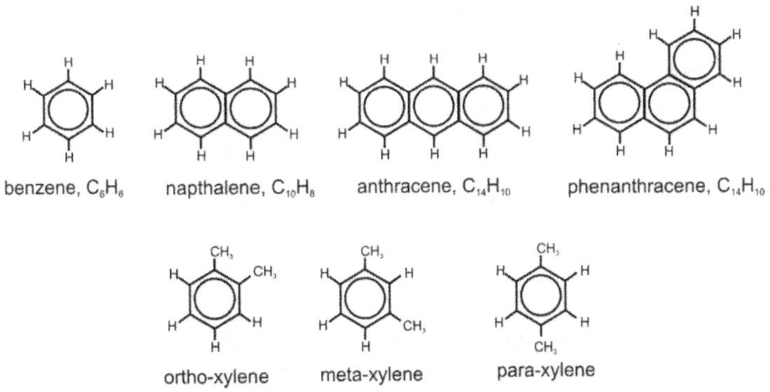

benzene, C_6H_6 napthalene, $C_{10}H_8$ anthracene, $C_{14}H_{10}$ phenanthracene, $C_{14}H_{10}$

ortho-xylene   meta-xylene   para-xylene

FIGURE A.3

Structures of various aromatic hydrocarbons. We note that it is customary to omit the hydrogen in the drawing of structures, and show only the functional groups. The bonds in the benzene rings are numbered from 1 to 6 starting from the top bond and going clockwise.

A.2 Surfactants

Anionic surfactants are dissociated in water in an amphiphilic anion, and a cation,[†] which is in general an alkaline metal (Na^+, K^+) or a quaternary ammonium. They are the most commonly used surfactants. They include alkylbenzene sulfonates (detergents), (fatty acid) soaps, lauryl sulfate (foaming agent), di-alkyl sulfosuccinate (wetting agent), lignosulfonates (dispersants), etc. (see Figure A.4).

Nonionic surfactants do not ionize in aqueous solution because their hydrophilic group is of a nondissociable type, such as alcohol, phenol, ether, ester, or amide. In the past decade glucoside (sugar-based) head groups have become popular because of their low toxicity. In glycolipids, the lipophilic group is often alkyl or alkylbenzene type, the former coming from fatty acids of natural origin (see Figure A.5).

Cationic surfactants are dissociated in water into an amphiphilic cation and an anion, most often of the halogen type. A very large proportion of this class corresponds to nitrogen compounds such as fatty amine salts and quaternary ammoniums, with one or several long chains of the alkyl type, often coming from natural fatty acids. These surfactants are in general more expensive than anionics, because of the high-pressure hydrogenation reaction to be carried out during their synthesis. They are mainly used in bactericides, i.e., to clean and aseptize surgery hardware, to formulate heavy-duty

---

[†] Anion: negatively (–) charged ion which moves toward anode during electrolysis; Cation: positively (+) charged ion which moves toward cathode.

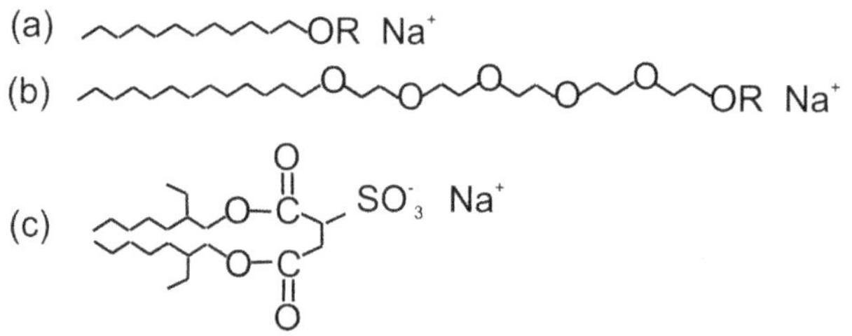

FIGURE A.4
Typical anionic surfactants. (a) Sodium alkyl sulfate ($R = SO_3^-$)/phosphate ($R = PO_3^{2-}$); (b) sodium alkyl ether sulphate($R = SO_3^-$)/phosphate ($R = PO_3^{2-}$); sodium bis (2-ethylhexyl)sulfo-sucinate (Aerosol OT).

disinfectants for domestic and hospital use, and to sterilize food bottles or containers, particularly in the dairy and beverage industries. In addition they are also used as positively charged substances which are able to adsorb on negatively charged substrates to produce antistatic and hydrophobic effect, often of great commercial importance such as in corrosion inhibition (see Figure A.6).

When a single surfactant molecule exhibits both anionic and cationic dissociations it is called *amphoteric* or *zwitterionic*. This is the case of natural substances such as amino acids and phospholipids. Amphoteric surfactants are generally quite expensive, and consequently their use is limited to very special applications such as cosmetics, where their high biological compatibility and low toxicity is of primary importance.

The past two decades have seen the introduction of a new class of surface-active substance, so-called *polymeric surfactants* or *surface active polymers*, which result from the association of one or several macromolecular structures exhibiting hydrophilic and lipophilic characters, either as separated blocks or as grafts. They are now very commonly used in formulating products as diverse as cosmetics, paints, foodstuffs and petroleum production additives. A macromolecule can obviously exhibit an amphiphilic structure.

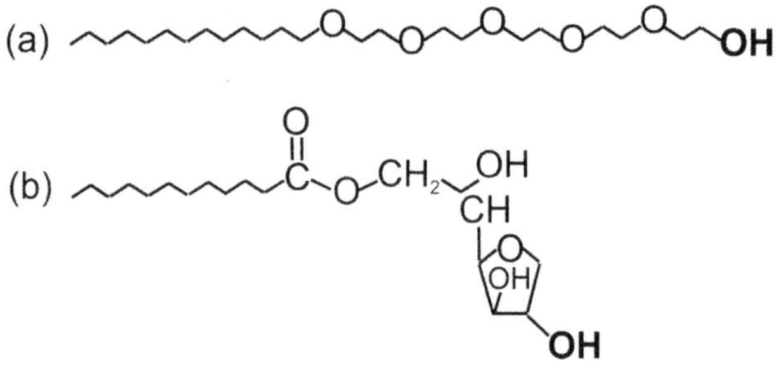

FIGURE A.5
Typical nonionic surfactants. (a) Fatty alcohol ethoxylate (C_mE_n); (b) sorbitan alkanota (sorbitan ester, "Span").

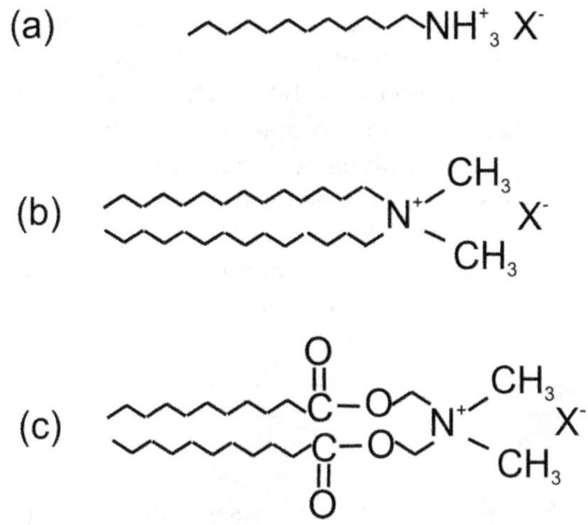

FIGURE A.6
Typical cationic surfactants. X = Br, Cl. (a) Fatty amine halide; (b) dialkyl trimethylammonium halide; (c) dialkyl ester trimethylammonium halide.

Asphaltenes, which are natural compounds found in crude oils, have polar and nonpolar groups. However, the location and segregation of these groups is often ill-defined, or at least less defined than in smaller molecules. There are two main configurations: *"block"* and *"graft,"* which are illustrated in Figure A.7, where H and L represent hydrophilic and lipophilic monomer units, respectively.

In the first case hydrophilic monomer units H are linked together to form a hydrophilic group, and lipophilic units L do the same to form a lipophilic group. The result is a macromolecular surfactant with well-defined and separated hydrophilic and lipophilic parts, which is much bigger than a conventional surfactant molecule. The most-used block polymer is the so-called copolymer of ethylene-oxide and propylene-oxide, either with two or three blocks. Although the hydrophilic and lipophilic parts are quite separated, the polymer polarity segregation is not that obvious, since both groups are slightly polar, one (PolyEO) just barely more polar than the other (polyPO). H-(OCH2-CH2)a-(O-CH[CH3]-CH2)b-O-(CH2CH2O)c-H.

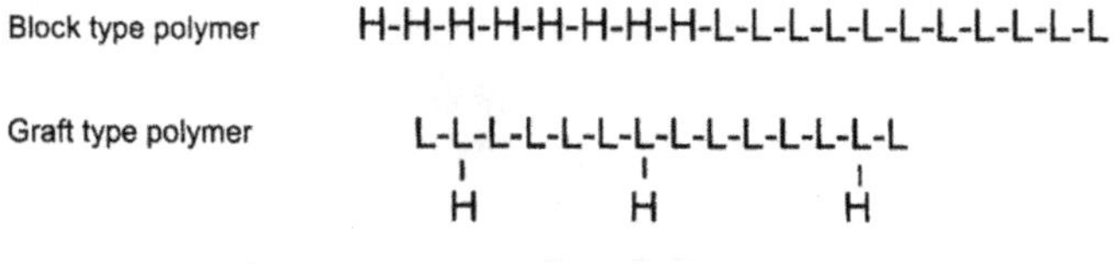

Block type polymer   H-H-H-H-H-H-H-H-L-L-L-L-L-L-L-L-L-L

Graft type polymer   L-L-L-L-L-L-L-L-L-L-L-L-L-L
                  H   H   H

FIGURE A.7
Schematic structures of block- and graft-type polymers.

These surfactants have many uses, in particular as colloid and nanoemulsion dispersants, wetting agents, detergents and even additive to dehydrate crude oils. However, most polymeric surfactants are graft-type, particularly synthetic products such as polyelectrolytes, which are not strictly surfactants or are not used for their surfactant properties. It is the case of hydrosoluble or hydrodispersible polyelectrolytes which are utilized for the antiredeposition, dispersant and viscosity-enhancing properties such as carboxymethyl cellulose, polyacrylic acid and derivatives.

For readers who would like to learn more about the chemistry and use of surfactant (amphiphilic) molecules, we recommend the following literature.[1,2,3,4]

References

1. J.L. Salager, Surfactants—Types and Uses (FIRP Booklet #300A), Merida-Venezuela, Version #2 (2002).
2. A.M. Schwartz, J.W. Perry, J. Berch, *Surface Active Agents and Detergents,* Vol. II, R. Krieger Pub. Co., New York (1977).
3. A. Davidson, B. Mildwidsky, *Synthetic Detergents,* Halsted Press, (1978); McCutcheon Detergents and Emulsifiers, McCutcheon Division Pub. Co., 175 Rock Road, Glen Rock, NJ 07452 (Annual, 3 volumes); M.J. Schick, Ed., *Noionic Surfactans,* Marcel Dekker, New York (1967).
4. Proceedings. World Conference on Soaps and Detergents, *JAOCS,* 55(1) (1978); A. O'Lenick, *J. Surfactants Detergents,* 3, 229 and 387 (2000); a wealth of information on surfactants is available at web site: www.surfactants.net/.

Appendix B

Rheology of Condensed Matters

Rheology ("rhei" means flow in Greek, so please do not let your spell-checker change it to "theology") is the interdisciplinary science of deformation and flow of matter. Physicists used to call it "continuum mechanics," but chemists and biologists do not think in terms of continua, and it is rather referred to as rheology when treating complex fluids, especially those with high biological importance.

The task of rheology is to find the relation between strain (ε), which is the relative change in length and stress, σ (in units of N/m^2 or Pa), which is the force acting on element of area A with its direction normal n_α.

$$\sigma_{\alpha\beta} = \frac{n_\alpha F_\beta}{A} \tag{B.1}$$

Diagonal elements of the stress tensor are pressure; off-diagonal elements are shear stresses.

In elementary level two kinds of condensed matters are distinguished: solids and fluids.

In *solids* the strain is a function of the applied stress, providing that the elastic limit is not exceeded. For small strains, there is a linear relation between strain and stress. Since in general both the strain and stress are second-rank tensors linking material properties, the elastic constants c should be elements of a fourth-rank tensor, so that:

$$\sigma_{\alpha\beta} = \sum_{\gamma,\delta=x,y,z} c_{\alpha\beta,\gamma\delta} e_{\gamma\delta} \tag{B.2}$$

The strain is independent of the time over which the force is applied, and if the elastic limit is not exceeded, the deformation disappears when the force is removed.

For isotropic or one-dimensional solids (B.2) simplifies to the well-known Hooke's law:

$$\frac{F}{A} = k\frac{x}{l} \tag{B.3}$$

where F is the magnitude of the force, k is the elastic constant, x is the displacement, and l is the initial length. Elastic energy is obtained by integrating the force F with respect to strain $e = x/l$.

$$U = \int F dx = \frac{1}{2} V k e^2 \qquad \text{(B.4)}$$

with $V = A \cdot l$ being the volume.

In *fluids* the rate of strain is proportional to the applied stress. A fluid continues to flow as long as the force is applied, and will not recover its original form when the force is removed.

The viscosity (η) of a fluid measures its resistance to flow under an applied shear stress. Representative units for viscosity are kg/(m·sec) = Pas, g/(cm·sec) (also known as Poise, designated by P). The centiPoise (cP), one hundredth of a Poise, is also a convenient unit, since the viscosity of water at room temperature is approximately 1 centiPoise. $[\eta] = \text{Pa·s}$ ($\eta_{water} \sim 1\ cP = 10^{-3}\ Pas$). The *kinematic viscosity* (ν) is the ratio of the viscosity to the density, $\nu = \eta/\rho$, and will be found to be important in cases in which significant viscous and gravitational forces exist.

The lower limit of the viscosity of a fluid can be estimated from the viscosity of an ideal gas, which is:

$$\eta = \frac{1}{3} \rho \bar{c} l \qquad \text{(B.5)}$$

where l is the mean free path, $\bar{c}$ is the mean thermal velocity of the fluctuating particles, and ρ is the mass density of the material.

We note that viscosity of gases increases with increasing temperature as $\eta = \eta_o(T/T_o)^n$ (T is the absolute temperature, η_o is the viscosity at an absolute reference temperature T_o, and n is an empirical exponent that best fits the experimental data).

For liquids, the mean-free-path is the intermolecular separation ($l \sim 5\ \text{Å}$), $\bar{c} \sim 10^3\ \text{m/s}$, i.e., $\eta_{min} \sim 0.3$ cp. In addition, in real liquids there is an energy barrier E_a to "hop" from one position to the other. The probability of hopping can be expressed as $f \sim \exp(-E_a / k_B T)$, which scales the viscosity as $\eta = \eta_{min}/f$. This means that the viscosities of liquids decrease with increasing temperature and generally vary approximately with absolute temperature T according to the Arrhenius behavior:

$$\ln \eta = a - \frac{E_a}{k_B} \frac{1}{T} \qquad \text{(B.6)}$$

Viscosities of liquids are generally two orders of magnitude greater than of gases at atmospheric pressure. For example, at 25°C, $\eta_{water} = 1\ cP$ and $\eta_{air} = 1 \times 10^{-2}\ cP$.

In the continuum descriptions of fluids, any small element is considered to be a continuum, containing large number of molecules. The relevant parameters are the velocity field vector $\vec{v}(\vec{r},t)$, the mass density $\rho(\vec{r},t)$ and the pressure $p(\vec{r},t)$.

Equation of continuity expresses that the gradient of mass flows in unit time across unit area of a surface is equal to the density change:

$$\vec{\nabla}(\rho \cdot \vec{v}) = -\frac{\partial \rho}{\partial t} \quad or \quad \frac{\partial(\rho \vec{v})_\alpha}{\partial x_\alpha} = -\frac{\partial \rho}{\partial t} \tag{B.7}$$

where x_α ($\alpha = 1, 2, 3$) are the coordinate axes. As a usual convention, summation is understood when Greek suffix appears twice, i.e., $a_\alpha b_\alpha = a_x b_x + a_y b_y + a_z b_z$.

In case of incompressible fluids $\rho(\vec{r},t) = $ const., i.e., $\vec{\nabla}\vec{v} = 0$ or $\partial v_\alpha / \partial x_\alpha = 0$. This is usually true for fluids at slow motions (where the speed is less than about 0.3 Mach), which means that phenomena like propagation of sound are not included in the description.

The equation of motion:

$$\rho \frac{d\vec{v}}{dt} = \vec{f} \tag{B.8}$$

corresponds to Newton's second law, where $d\vec{v}/dt$ refers to a particular fluid element (material time derivative). To characterize the velocity at a fixed coordinate in the laboratory system, we rather use the partial derivative $\partial \vec{v}/\partial t$. The relation between the two derivatives can be given by considering that in dt the fluid element moves by $d\vec{s} = \vec{v} \cdot dt$. Since $\vec{v}$ depends both on the position ($\vec{s}$) and time t, the variation dv is written as:

$$d\vec{v}(\vec{s},t) = \frac{\partial \vec{v}}{\partial \vec{s}} d\vec{s} + \frac{\partial \vec{v}}{\partial t} dt \tag{B.9}$$

Dividing both sides by dt we get that:

$$\frac{d\vec{v}}{dt} = \frac{\partial \vec{v}}{\partial \vec{s}} \vec{v} + \frac{\partial \vec{v}}{\partial t} = (\vec{v} \cdot \vec{\nabla})\vec{v} + \frac{\partial \vec{v}}{\partial t} \tag{B.10}$$

In (B.8), $\vec{f}$ is the force per unit volume (unit is small), and has two contributions:

1. The net force on the volume element due to pressure gradient: $-\vec{\nabla}p$.
2. The viscous term: $f_{vis,}$ where

$$f_{visc} = 2\eta \frac{\partial A_{\alpha\beta}}{\partial x_\beta} \tag{B.11}$$

Here η is the viscosity coefficient, and $A_{\alpha\beta} = \frac{1}{2}\left(\frac{\partial v_\beta}{\partial x_\alpha} + \frac{\partial v_\alpha}{\partial x_\beta}\right)$ is the symmetric part of the velocity gradient tensor, that is $A_{\alpha\beta} = A_{\beta\alpha}$. The antisymmetric part of the velocity gradient tensor $W_{\alpha\beta} = \frac{1}{2}\left(\frac{\partial v_\beta}{\partial x_\alpha} - \frac{\partial v_\alpha}{\partial x_\beta}\right)$ is related to the vorticity of the

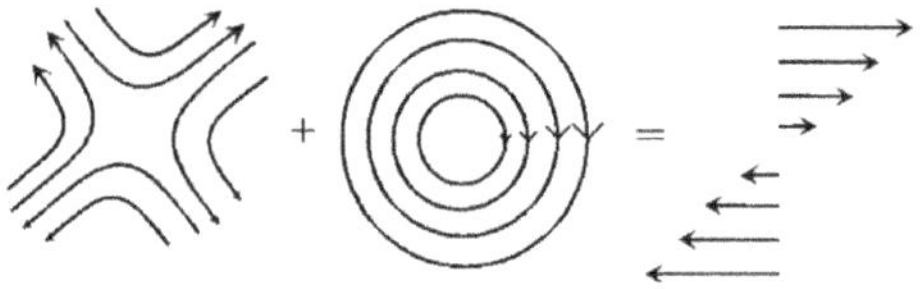

FIGURE B.1
Sketches of the irrotational (a), rotational (b) and simple shear flows (c).

flow: $\vec{\omega} = \frac{1}{2}\vec{\nabla}\times\vec{v} = \left(-W_{yz}, W_{zx}, W_{xy}\right)$, which describes the local angular velocity of the fluid. When $\hat{W} = 0$, we deal with irrotational flows (see Figure B.1a), whose streamlines never loop back on themselves. When $\hat{A} = 0$ we speak about a rotational flow that contains streamlines that loop back on themselves (see Figure B.1b). Typically, only inviscid (nonviscous) fluids can be irrotational. When both are present with the same amplitude the resulting flow corresponds to simple shear flow (see Figure B.1c).

Neglecting the external forces, we can write (B.8) as:

$$\rho\left(\frac{\partial\vec{v}}{\partial t} + \left(\vec{v}\cdot\vec{\nabla}\right)\vec{v}\right) = -\vec{\nabla}p + f_{visc} \tag{B.12}$$

Similar to solids, we can introduce the stress tensor $\hat{\sigma}$, which has the form $\sigma_{\alpha\beta} = -p\delta_{\alpha\beta} + 2\eta A_{\alpha\beta}$. This gives:

$$\rho\left(\frac{\partial\vec{v}}{\partial t} + (\vec{v}\cdot\vec{\nabla})\vec{v}\right) = \vec{\nabla}\left[-p\delta_{\alpha\beta} + \eta\left(\frac{\partial v_\alpha}{\partial x_\beta} + \frac{\partial v_\beta}{\partial x_\alpha}\right)\right] \equiv \nabla\hat{\sigma} \tag{B.13}$$

This is the Navier–Stokes equation of isotropic incompressible fluids.

Fluids with viscosity η independent of the shear rate are called Newtonian fluids. All gases and most liquids, which have simpler molecular formula and low molecular weight such as water, benzene, ethyl alcohol, CCl_4, hexane, and most solutions of simple molecules are Newtonian fluids.

Fluids with viscosity that depends on the shear rate are called non-Newtonian fluids (Figure B.2).

Generally non-Newtonian fluids are complex mixtures: slurries, pastes, gels, polymer solutions, etc.

Non-Newtonian fluids are usually divided into three categories.

- *Bingham-plastics* resist a small shear stress, but flow easily under larger shear stresses. Examples include toothpaste, jellies, some slurries, smectic liquid crystals, foams (shaving cream).
- *Pseudo-plastics* have viscosity, which decreases with increasing velocity gradient. They are also called *shear thinning fluids*. At low shear rates a shear thinning fluid is more viscous than a Newtonian fluid, and at high shear rates it is less viscous. Most non-Newtonian fluids

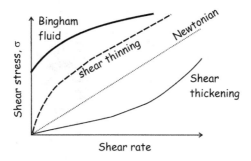

FIGURE B.2
Illustration of the non-Newtonian behaviors. For a Newtonian fluid the shear stress is proportional to the shear rate. In contrast, there are nonlinear dependencies if shear thickening or shear thinning occurs. A Bingham fluid is characterized by a finite stress at zero shear rates.

belong to this category. Examples include nematic liquid crystals, hard polymers in liquids.

- *Dilatant fluids* are characterized by viscosity that increases with increasing velocity gradient. They are uncommon, but suspensions of starch and sand behave in this way. They are also called as shear thickening fluids.

Although there are materials which possess rheological properties that depend on the duration of shear (e.g., in *thixotropic fluids* the dynamic viscosity decreases, and in *rheopectic fluids* the dynamic viscosity increases with the time for which shearing forces are applied), we will consider only those materials which do not change their properties with time.

Non-Newtonian and Newtonian fluids can easily be distinguished by simple physical experiments, such as those illustrated in Figure B.3. For example, a Newtonian fluid is depressed near a rotating rod (look at the shape of your cup of tea or coffee when you stir the sugar in it), whereas non-Newtonian liquids climb up near the rod (Figure B.3a). Spheres falling in Newtonian fluids get eventually closer, whereas in non-Newtonian fluids they separate during falling (Figure B.3b).

Non-Newtonian fluids have both viscous and elastic properties, and they are called viscoelastic fluids. An example is so-called "silly putty," which is made from poly-dimethyl-siloxane (silicone). It flows like a liquid out of the container, but when it forms a ball, it behaves as elastic, i.e., it bounces back. The crucial factor determining the viscous and elastic behavior is the time period of the force applied: short force pulse leads to elastic response, whereas long-lasting force causes flow. The viscoelasticity in polymers is due to shear-induced entanglements and nonlinear behavior of the chains, coils. A well-known natural viscoelastic material is for, example, the egg white, which springs back when a shear force is released. A polymer resembles both liquid and solids.

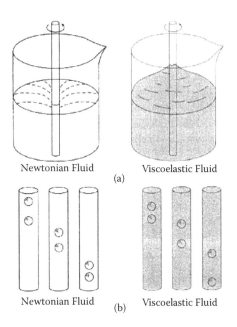

Newtonian Fluid   Viscoelastic Fluid
                    (a)

Newtonian Fluid   (b)   Viscoelastic Fluid

FIGURE B.3

Main differences between the Newtonian and non-Newtonian fluids. (a) Rotating rod experiment. (b) Falling spheres experiment.

The structured fluids we deal with here are in general viscoelastic, and their flow can be described by a complex viscosity:

$$\eta = \eta' + i\eta'' \tag{B.14}$$

The imaginary part is related to the real part of the elastic modulus G as $\eta'' = G'/\omega$, where the complex elastic modulus is defined as:

$$G = G' + iG'' \tag{B.15}$$

The imaginary part of the elastic constant G'' is related to the real part of the viscosity as $G'' = \eta_1\omega$.

In other words, the elastic component of the stress is in phase with the strain, and the viscous stress is out of phase of the strain. For periodic strain of form $\varepsilon = \varepsilon_0 \sin \omega t$, this means that $\sigma = \sigma' \sin \omega t + \sigma'' \cos \omega t$.

We now define an in-phase shear modulus as $G' = \sigma'/\varepsilon_0$ and an out-of-phase modulus as $G'' = \sigma''/\varepsilon_0$. The ratio $\tan \delta = G''/G'$ is a measure of energy loss per cycle. G' is called storage modulus, and G'' is the loss modulus, to reflect the energy transfer during the deformation.

Many important engineering problems cannot be solved completely by theoretical or mathematical methods. Problems of this type are especially common in fluid-flow, heat-flow, and diffusional operations. One method of attacking a problem for which no mathematical equation can be derived is empirical experimentation. For example, the pressure loss from friction in a long, round, straight, smooth pipe depends on the length, the diameter of the pipe, the flow rate of the liquid, the density and viscosity of the liquid. If any one of these variables is changed, the pressure drop also changes. The empirical method of obtaining an equation relating these factors to pressure drop requires that the effect of each separate variable be determined in turn by systematically varying that variable, while keeping all others constant. The procedure is laborious, and it is difficult to organize or correlate the results. However, there exists a method intermediate between formal mathematical development and a completely empirical study. It is based on dimensional analysis; even if we cannot give the theoretical equations among the variables affecting a physical process, we know that the equation must be dimensionally homogeneous. Because of this requirement, it is possible to group many factors into a smaller number of dimensionless groups of variables. The groups themselves, rather than the separate factors, appear in the final equation. Dimensional analysis does not yield a numerical equation, and experiment is required to complete the solution of the problem. The result of a dimensional analysis is valuable in pointing a way to correlations of experimental data suitable for engineering use.

Dimensional analysis drastically simplifies the task of fitting experimental data to design equations where a completely mathematical treatment is not possible; it is also useful in checking the consistency of the units in equations, in converting units, and in the scale-up of data obtained in physical models to predict the performance of full-scale model. Useful dimensionless parameters and their relevance are listed in Table B.1.

Probably the most important number in this list is the Reynolds number, which determines whether the flow is stationary ($Re < 10$), or turbulent ($Re > 10$). The most important stationary flow types are the following (see Figure B.4).

The simplest flow where the flow velocity is constant everywhere is called *plug flow*. This type of flow occurs when a liquid is pushed through porous media, or flow of smectics normal to the layers, or in cholesteric along the helical axis belong to this category. Also plug flow is a fair approximation to actual flow in the entrance range of a simple channel or pipe when the Reynolds number (in this case, based on the pipe diameter D) is small (see Figure B.4a).

Far from the entrance range of the pipe, the fully developed flow is called *Poiseuille flow*, which is characterized by a parabolic velocity profile. To illustrate this, we consider the flow in x-direction of a viscous fluid in a channel. The channel has a width in the y-direction of a, length l_z ($z \gg a$) in the z-direction, and a length l_x ($x \gg a$) in the x-direction. There is a pressure drop Δp along l_x, so that the pressure gradient is constant (such a pressure gradient could be supplied by gravity, for instance). We assume the flow is steady ($\partial v/\partial t$), and the speed depends only on the y-direction $\vec{v}(\vec{r}) = v_x(y)$. In this case, the nonlinear term in the Navier–Stokes equation (B.13) vanishes, and we are left with a simple equation for v_x:

$$\eta \frac{\partial^2 v_x}{\partial y^2} + \frac{\Delta p}{l_x} = 0 \tag{B.16}$$

TABLE B.1

Important Dimensionless Numbers in Fluid Mechanics

Dimensionless number	Symbol	Formula	Numerator	Denominator	Importance
Reynolds number	Re	$Dv\rho/\eta$	Inertial force	Viscous force	Fluid flow involving viscous and inertial forces
Froude number	Fr	v^2/gD	Inertial force	Gravitational force	Fluid flow with free surface
Weber number	We	$v^2\rho D/\sigma$	Inertial force	Surface force	Fluid flow with interfacial forces
Deborah number	De	τ/t_f	Relaxation time	Period of force $(t_f = 1/\omega)$	De >> 1: solid De << 1: liquid
Weissenberg number	Wi	$\dot{\gamma}\tau$	Shear rate	Frequency	Wi >> 1 (nonlinear flow effects)
Peclet number	Pe	$\dot{\gamma}\tau_D$	Shear rate	Diffusion rate	Colloids dynamics
Mach number	Ma	v/c	Local velocity	Sonic velocity	Gas flow at high velocity
Drag coefficient	C_D	$F_D/(\rho v^2/2)$	Total drag force	Inertial force	Flow around solid bodies
Friction factor	F	$\tau_w/(\rho v^2/2)$	Shear force	Inertial force	Flow though closed conduits
Pressure coefficient	C_P	$\Delta p/(\rho v^2/2)$	Pressure force	Inertial force	Pressure drop studies

Note: D is diameter in m, v and c are in units of ms^{-1}, r is the density in units of kgm^{-3}, h is the viscosity in units of kgm^{-1}s^{-1}, and p is the pressure in Nm^{-2}.

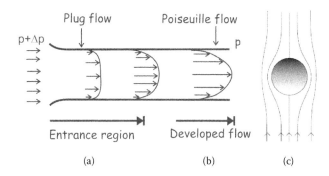

FIGURE B.4

Illustration of stationary flow types. (a) Plug flow at the entrance region of a pipe and (b) the Poiseuille flow in the fully developed flow region; (c) Stokes flow around a sphere.

Integrating twice, we obtain:

$$v_x(y) = -\frac{1}{2\eta}\frac{\Delta p}{l_x}y^2 + C_1 y + C_2 \qquad (B.17)$$

where C_1 and C_2 are integration constants. To determine these, we impose the no-slip boundary condition at the top and bottom edges of the channel: $v_x(y = \pm a/2) = 0$. We obtain:

$$v_x(y) = \frac{1}{2\eta}\frac{\Delta p}{l_x}[(a/2)^2 - y^2] \qquad (B.18)$$

This shows that the velocity profile is indeed a parabola, with the fluid in the center of the channel having the greatest speed. Once we know the velocity profile, we can determine the flow rate Q, defined as the volume of fluid which passes a cross section of the channel per unit time. This is obtained by integrating the velocity profile over the cross-sectional area of the channel:

$$Q = \int_0^{l_z} dz \int_{-a/2}^{a/2} dy v_x(y) = \frac{l_z a^3 \Delta p}{12\eta l_x} \qquad (B.19)$$

The analogous result for flow through a pipe of radius r and length l in the presence of a uniform pressure gradient $\Delta p / l$ would be:

$$Q = \frac{\pi r^4 \Delta p}{8\eta l} \qquad (B.20)$$

The important feature of both of these results is the sensitive dependence upon either the channel width a or the pipe radius r. For instance, for a pipe with a fixed pressure gradient, a 20% reduction in the pipe radius leads to a 60% reduction of the flow rate! This clearly has important physiological implications — small amounts of plaque accumulation in arteries can lead to very large reductions in the rate of blood flow.

A second important flow at small Reynolds numbers is *Stokes' flow* — the flow of a viscous fluid around a sphere of radius a, which is moving with a speed v. The derivation of the actual velocity field and the resulting drag force F_D are complicated, and we will just quote the result here, which reads:

$$F_D = 6\pi\eta av \qquad (B.21)$$

This is usually referred to as *Stokes' law*. Although the derivation is not trivial, apart from the numerical constant 6π, we may understand it by dimensional analysis. First we must ask: what can this drag force depend on? Obviously, it depends on the *size* of the ball: let's say the radius is a, and that has dimension meter (m). It also must depend on the *speed* v, which has dimension m/s. Finally, it depends on the *coefficient of viscosity* η which has dimensions $kgm^{-1}s^{-1}$. The drag force F_D has dimensions $[F_D] = mkgs^{-2}$. What combination of $[a] = m$, $[v] = ms^{-1}$ and $[\eta] = kgm^{-1}s^{-1}$ will give $[F_D] = kgms^{-2}$? It's easy to see immediately that F_D must depend linearly on η; that is the only way to balance the kg term. Now let's look at F_D/η, which can only depend on a and v. $[F_D/\eta] = m^2s^{-1}$. The only possible way to get a function of a and v having dimension m^2s^{-1} is to take the product av. So, the dimensional analysis establishes that the drag force is given by: $F_D = Ca\eta v$, where C is a constant that cannot be determined by dimensional considerations.

Stokes' law is important in determining things such as the settling of dust (from a volcanic explosion, say), or the sedimentation of small particles (pollutants) in a river.

Appendix C

Symmetry at Work

Here we demonstrate in two examples how symmetry considerations help eliminate coupling constants. The first example describes Frank's considerations in determining the relevant elastic constants of nematic liquid crystals,[1] and the second example will show us how many piezoelectric coupling constants are possible in systems with different symmetries. In both cases the nonvanishing coefficients are determined by the symmetry of the material and using the Curie principle, which states that tensor coefficients characterizing material properties should be invariant under the symmetry transformations of the substance.[2]

C.1 Elastic Constants of 3D Anisotropic Fluids

Let us consider the possible deformations (splay, twist and bend) of anisotropic materials. As illustrated in Figure C.1, without loosing generality we choose the coordinate system so that the z-axis be parallel to the undistorted director field ($\vec{n} \| z$).

In this coordinate system the director curvature can be separated into six components:

$$\text{"splays"} \quad s_1 = \frac{\partial n_x}{\partial x}; \quad s_2 = \frac{\partial n_y}{\partial y} \qquad (C.1)$$

$$\text{"twists"} \quad t_1 = \frac{\partial n_y}{\partial x}; \quad t_2 = +\frac{\partial n_x}{\partial y} \qquad (C.2)$$

$$\text{"bends"} \quad b_1 = \frac{\partial n_x}{\partial z}; \quad b_2 = \frac{\partial n_y}{\partial z} \qquad (C.3)$$

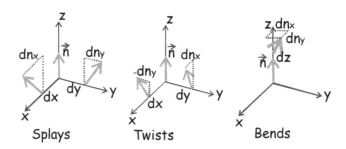

FIGURE C.1
Representation of the director deformation field for splays, twists and bends.

For small deformations we can write the director components as:

$$n_{x'} = -a_1 x + a_2 y + a_3 z + O(r^2)$$

$$n_{y'} = a_4 x - a_5 y - a_6 z + O(r^2) \tag{C.4}$$

$$n_z = 1 + O(r^2)$$

where $r^2 = x^2 + y^2 + z^2$.

Comparing (C.1), (C.2) and (C.3) with (C.4), we get that: $s_1 = a_1$, $t_2 = a_2$, $b_1 = a_3$, $-t_1 = a_4$, $s_2 = a_5$ and $b_2 = a_6$.

In the limit of small distortions $a\nabla \bar{n} \ll 1$ (a: molecular size), the distortion free energy density approaches zero, which means that F_d can be expressed in powers of $\vec{\nabla} \bar{n}$.

For small director deformations it can be written as:

$$F_d = \frac{1}{2} K_{ij} a_i a_j + K_i a_i, \quad (i, j = 1, \dots 6; \quad K_{ij} = K_{ij}) \tag{C.5}$$

Here, as usual, summation over repeated suffices is implied. So far, we did not have any symmetry considerations, so (C.5) would mean that we need to deal with $36 + 6$ elastic constants, which would make any analysis extremely complicated. Fortunately, the symmetry of the material reduces the number of nonzero elastic constants considerably.

Because there was arbitrariness in our choice of coordinate system, we require that when we replace this by other permissible one (x', y', z'), in which we have new curvature elastic components, F_d should be the same as before. As the Curie principle states, this implies that the same conditions will be valid for the elastic constants determined by a_i and $a_i a_j$ as for those determined by a_i' and $a_i' a_j'$.

- For uniaxial symmetry the choice of x and y are arbitrary, so the properties are invariant with respect to 90 degrees rotation around the uniaxis. This means that the $x' = y$, $y' = -x$, $z' = z$ and $n_x' = n_y$, $n_y' = -n_x$,

$n_z' = n_z$ transformations are permissible. Comparing these conditions with (C.4), we arrive at the equations:

$$n_{x'} = a_1 x' + a_2 y' + a_3 z' = a_1 y - a_2 x + a_3 z = n_y = a_4 x + a_5 y + a_6 z \qquad (C.6)$$

This means that:

$$-x(a_2 + a_4) + y(a_1 - a_5) + z(a_3 - a_6) = 0 \qquad (C.7)$$

which is equivalent to the statement that: $a_2 = -a_4, a_a = a_9, a_3 = a_6$.

The invariance of F_d with respect to this transformation means that $K_1 = K_5$, $K_2 = -K_4$, $K_3 = K_6$; furthermore, $K_{11} = K_{55}$, $K_{22} = K_{44}$, $K_{33} = K_{66}$, $K_{12} = -K_{45}$ and $K_{14} = -K_{25}$. Taking into consideration that $x' = y$ means, for example, that $K_{13} = K_{23}$, but $y' = -x$ means that $K_{23} = -K_{13}$, we get that $K_{13} = K_{23} = 0$. This means that all terms where the indices 1, 2, 4 or 5 appear only once will be zero. Accordingly, $K_{16} = K_{26} = K_{34} = K_{35} = K_{46} = K_{56} = 0$. A rotation of $45°$ gives a further equation $K_{11} = K_{15} - K_{22} = K_{24} = 0$, and a rotation by another arbitrary angle gives $K_{12} + K_{14} = 0$. These two latest constraints can be combined as $K_{12} = -K_{14} = K_{25} = -K_{45}$.

- For nonpolar nematic phase, the sign of $\bar{n}$ is arbitrary (head–tail symmetry). It is a significant convention in our definition of curvature components that z is positive in the positive direction of $\bar{n}$. In addition, to retain the right-handed coordinate system, either x or y should change sign when z changes sign. Hence the absence of polarity is equivalent to the following set of transformations: $\bar{n}' = -\bar{n}$, $x' = x$, $y' = y$ and $z' = -z$. These altogether give:

$$n_{x'}' = -a_1 x' + a_2 y' + a_3 z' + O(r^2)$$
$$n_{y'}' = -a_4 x' + a_5 y' + a_6 z' + O(r^2) \qquad (C.8)$$

The coefficients with indices 1, 5 and 6 has changed sign, so the required invariance provides that those constants should be zero, where indices 1, 5, or 6 appear once, i.e.,

$$K_{12} = K_{13} = K_{14} = K_{25} = K_{26} = K_{35} = K_{36} = K_{45} = K_{46} = 0, \text{ and } K_1 = K_5 = K_6 = 0.$$

Most of this information is already contained by the uniaxiality condition. The only additional effects of the nonpolar condition are that K_1 and $K_{36} = K_{12} = 0$.

- For nonchiral materials, where the left and right hands are the same, the $x' = x$, $y' = -y$, $z' = z$ transformations are permissible. This means that the terms where 2 or 4 appears only once, will be zero, i.e., K_{12} = K_{32} = K_{14} = K_{34} = K_{54} = K_{64} = K_{52} = K_{62} = 0 and K_2 = K_4 = 0. Compare to the constraints given by the uniaxiality and the head–tail invariance; these bring only two new constraints, namely: K_{12} = K_2 = 0.

Summarizing, for a uniaxial, nonpolar and nonchiral materials (just as normal nematic liquid crystals) $K_i = 0$ for all $i = 1, \ldots 6$, and the energy density can be written as:

$$F_d = \frac{1}{2} \sum_{i,j=1,6} K_{ij} a_k a_j \tag{C.9}$$

where the nonvanishing components of the second-rank tensor K_{ij} are:

$$K_{ij} = \begin{pmatrix} K_{11} & 0 & 0 & 0 & K_{11} - K_{22} - K_{24} & 0 \\ 0 & K_{22} & 0 & K_{24} & 0 & 0 \\ 0 & 0 & K_{33} & 0 & 0 & 0 \\ 0 & K_{24} & 0 & K_{22} & 0 & 0 \\ K_{11} - K_{22} - K_{24} & 0 & 0 & 0 & K_{11} & 0 \\ 0 & 0 & 0 & 0 & 0 & K_{13} \end{pmatrix} \tag{C.10}$$

Taking into account (C.5), we can write the distortion-free energy as:

$$F_d = \frac{1}{2} K_{11}(s_1 + s_2)^2 + \frac{1}{2} K_{22}(t_1 + t_2)^2 + \frac{1}{2} K_{33}\left(b_1^2 + b_2^2\right)$$
$$- (K_{22} + K_{24})(s_1 s_2 + t_1 t_2) \tag{C.11}$$

Keeping in mind that $\vec{n} \cong (0,0,1)$ and $\dfrac{\partial n_z}{\partial z} \approx 0$,

$$(t_1 + t_2) = \frac{\partial n_x}{\partial y} - \frac{\partial n_y}{\partial x} = n_z \cdot (\vec{\nabla} \times \vec{n})_z \cong \vec{n} \cdot \vec{\nabla} \times \vec{n} \tag{C.12}$$

$$s_1 + s_2 = \frac{\partial n_x}{\partial x} + \frac{\partial n_y}{\partial y} \cong \vec{\nabla} \cdot \vec{n} \tag{C.13}$$

$$b_1^2 + b_2^2 = \left(\frac{\partial n_x}{\partial z}\right)^2 + \left(\frac{\partial n_y}{\partial z}\right)^2 \cong [(\vec{n} \cdot \vec{\nabla})\vec{n}]^2 \tag{C.14}$$

With these, and by utilizing the well-known operator identity, $a \times (b \times c) = b(a \cdot c) - c(a \cdot b)$, we can rewrite the distortion-free energy density in coordinate independent tensorial form as:

$$F_d = \frac{1}{2}[K_{11}(\vec{\nabla} \cdot \vec{n})^2 + K_{22}(\vec{n} \cdot \vec{\nabla} \times \vec{n})^2 + K_{33}(\vec{n} \cdot \vec{\nabla} \times \vec{n})^2]$$

$$+ (K_{22} + K_{24}) \cdot (\vec{\nabla} \cdot \{\vec{n}\vec{\nabla} \cdot \vec{n} + \vec{n} \times \vec{\nabla} \times \vec{n}\})$$

(C.15)

C.2 Piezoelectric Components in Anisotropic Systems

From the definitions of the piezoelectricity (see 8.41 and 8.42), and from the Curie principle, it follows that $d_{i,jk}$ will transform as the product of x_i, x_j, x_k. Accordingly, in a system with inversion symmetry, transformations like $x \to -x$, $y \to -y$, and $z \to -z$, would require that $d_{-i,-j-k} = (-1)^3 d_{i,jk}$ for any i,j,k, which is equivalent to the statement that systems with inversion symmetry cannot be piezoelectric. At the other side of the symmetry range, the materials with C_1 symmetry allow the presence of all piezoelectric constant. Since $d_{i,jk} = d_{i,kj}$, it means $3 \times 6 = 18$ constants.

In general, the number of the possibly nonzero components can be determined for each crystallographic group. In listing these, we treat separately those that allow, and those that exclude polar order. We will follow the so-called Schoenfliess notation.[†]

Those symmetry groups which do not permit polar order are the following:

- D_2 that has three mutually perpendicular two-fold axes of symmetry, say along x, y and z. Rotations through 180° around these axes change the sign of two out of three coordinates. The only nonzero components are therefore those that have three different suffixes: $d_{x,yz}$, $d_{z,xy}$ and $d_{y,zx}$ (the other nonzero components are the same due to the symmetry in the *jk* indices).

- D_{2d} is the same as D_2, except that two planes of symmetry passing through one axis (say the z) and bisecting the angles between the other two, are added. Reflection in one of these planes gives the transformation: $x \to y$, $y \to x$, $z \to z$. Hence the components that differ by interchanging x and y are equal, so there are only two independent constants present: $d_{z,xy}$; and $d_{y,zx} = d_{x,yz}$.

- T is obtained from D_2 by adding four diagonal three-fold axes of symmetry, rotations about, which effect a cyclic permutation such as $x \to z$, $y \to x$, $z \to y$. Hence all three components in D_2 are equivalent: $d_{z,xy} = d_{y,zx} = d_{x,yz}$.

[†] For those who are not familiar with the Schoenfliess, or other notations, we recommend for example: E.A. Wood, *Crystals and Light*, 2nd ed., Dover Publications, New York, 1977.

- T_d is the same as T, except that two planes of symmetry passing through one axis (say the z) and bisecting the angles between the other two are added. The same result is obtained as for the class *T.*

- D_4 has one four-fold axis of symmetry (say, z axis) and four twofold axes lying in the xy plane. Here the symmetry transformations of the D_2 are supplemented by the rotation through 90° about the z-axis, i.e., the transformation x → y, y → −x, z → z. Consequently, one of the components of D_2 must be zero ($d_{z,xy} = -d_{z,yx} = 0$), and the other two are the same but opposite in sign: $d_{y,zx} = -d_{x,yz}$.

- D_6 is the same as D_4, except for the six-fold axis (say in the z direction) with six two-fold axes in the xy plane, with the same coefficients.

- S_4 includes the transformations x → y, y → −x, z → −z and x → −x, y → −y, z → z. The nonzero components are $d_{z,xy}$; $d_{x,yz} = d_{y,xz}$; $d_{z,xx} = -d_{x,zx} = -d_{y,zy}$.

- D_3 has one three-fold symmetry axis (say, the z axis) and three twofold axes lying in the xy plane. The determination of the nonzero components is quite complicated, but it can be shown[3] that two coefficients will be nonzero: $d_{y,zx} = -d_{x,zy}$ and $d_{y,xy} = -d_{x,xx} = d_{x,yy}$.

- D_{3h} is obtained from D_3 by adding a plane of symmetry (the xy plane) perpendicular to the three-fold axis, and only the second component of D_3 remains.

- C_{3h} has a three-fold axis and a plane of symmetry perpendicular to it. It has the same two components as of D_3, but either of them can be made to vanish by a suitable choice of the x and y indices.

The polar groups have ten elements. Four of them are the C_1, C_2, C_3, C_4 and C_6, corresponding to systems with two-fold, three-fold, four-fold and six-fold symmetry axes (say, in the z direction), respectively. In the classes C_{nv} (n = 1, 2, 3, 4, 6) the xz plane is the plane of symmetry, and in C_s z is the perpendicular to the plane of symmetry.

The nonzero components in these classes are the following:

- C_1: all $d_{i,jk}$
- C_s: all in which the suffix z appears twice
- C_{2v}: $d_{z,xx}$; $d_{z,yy}$; $d_{z,zz}$, $d_{x,xz}$; $d_{y,yz}$ (either x or y may appear only twice)
- C_2: where x or y appear twice (just as for C_{2v}) and where x and y appear together: $d_{x,yz}$; $d_{y,xz}$; $d_{z,xy}$
- C_{4v}: $d_{z,xx} = d_{z,yy}$; $d_{z,zz}$, $d_{x,xz} = d_{y,yz}$
- C_4: the same as C_{4v} together with $d_{x,yz} = -d_{y,xz}$
- C_{3v}: $d_{z,zz}$; $d_{x,xz} = d_{y,yz}$; $d_{x,xx} = -d_{x,yy} = -d_{y,xy}$; $d_{z,xx} = d_{z,yy}$
- C_3: the same as C_{3v} together with $d_{x,yz} = -d_{y,xz}$; $d_{y,xx} = -d_{y,yy} = d_{x,xy}$
- C_{6n}: $d_{z,zz}$, $d_{x,xz} = d_{y,yz}$; $d_{z,zx} = d_{z,yy}$
- C_6: the same as C_{6n} together with $d_{x,yz} = -d_{y,xz}$

In addition to the above crystallographic symmetry groups, in fluids one also needs to consider continuous point groups introduced by P. Curie, which are: C_∞, $C_{\infty v}$, $C_{\infty h}$, D_∞, $D_{\infty h}$, $SO(3)$, $O(3)$. Note that the last two terms represent spherical groups, where no Schoenfliess notation exists, and we used the mathematical notation. $O(3)$ is used to describe all rotations and reflections. $SO(3)$ contains only rotations. Three of these seven symmetries, C_∞, D_∞ and $SO(3)$, represent chiral symmetry; they can appear in both handedness (enantiomeric forms). $C_{\infty h}$ has a mirror plane, and it cannot represent any handedness. Note that the first two groups (C_∞ and $C_{\infty v}$) are polar, so they obviously allow piezoelectricity. In addition to this, the nonpolar D_∞ symmetry also allows piezoelectricity. The nonzero components of these three classes are:

- C_∞: It consist the $x \to -x$, $y \to -y$, and $x \to y$, $y \to -x$ transformations, therefore only those remain unchanged where x and y appear twice, or none, which mean: $d_{z,zz}$; $d_{z,xx}$; $d_{z,yy}$.
- $C_{\infty v}$ has two mirror planes with intersection along the infinite-fold symmetry axis, so x and y become equivalent, so there are only two independent constants: $d_{z,zz}$; and $d_{z,xx} = d_{z,yy}$.
- D_∞: Similar to the arguments given at describing the symmetry of D_2, the only nonzero components are those that have three different suffixes: $d_{x,yz}$; $d_{z,xy}$ and $d_{y,zx}$. In addition, due to the equivalence of x and y, we get that $d_{zxy} = -d_{z,yx} = 0$ and $d_{x,yz} = -d_{y,xz}$.

References

1. F.C. Frank, *Discussions Faraday Soc.*, 25, 19 (1958).
2. A recent overview of the Curie principle is written in the book: S.T. Lagerwall, *Ferroelectric and Antiferroelectric Liquid Crystals*, Wiley-VCH, Weinheim (1999).
3. L.D. Landau, Electrodynamics of continuous media, Pergamon Press, Oxford (1980).

Appendix D

Dielectric Spectroscopy

D.1 Dielectric Relaxation

The buildup of the electric field-induced polarization requires finite times. In quasi-DC fields all polarization mechanisms will be present, but at increasing frequencies they eventually drop out.

To show this, let us follow the time dependence of the induced polarization P after turning on a static electric field $\bar{E}$ across the material. Obviously, after a very long time P will reach the final value P_f given by:

$$P_f = \chi(0)\varepsilon_o E \tag{D.1}$$

Here $\chi(0)$ is the zero frequency susceptibility. It is reasonable to assume that its change rate will be proportional to the deviation from the equilibrium value. This is the basic assumption of the irreversible thermodynamics describing how a thermodynamic variable relaxes back toward equilibrium. Accordingly:

$$\dot{P} = \frac{P_f - P}{\tau} \tag{D.2}$$

where τ is the switching time. Integrating it, we get:

$$-\ln(P_f - P) = \frac{t}{\tau} + const. \tag{D.3}$$

or, if we require that $P(0) = 0$, we have:

$$\frac{P_f - P}{P_f} = e^{-t/\tau} \tag{D.4}$$

which gives:

$$P = P_f(1 - e^{-t/\tau}) \tag{D.5}$$

If we turn off the field from the value P_f, we get that the polarization decays to zero as:

$$P = P_f e^{-t/\tau} \tag{D.6}$$

It is important to note that the switching time and the relaxation time are the same for small fields.

We emphasize that τ^{-1} is not the angular velocity of a single rotating dipole, but it is a macroscopic relaxation frequency and is related to viscosity of the fluid. For quantitative estimates of the relaxation time τ_r, we can describe the motion of the particles as rotational Brownian motion in a mean-field potential, where the relaxation of a rotation with angle θ can be related to the rotational diffusion constant D_θ as:

$$\langle \theta^2 \rangle^{1/2} = (4 D_\theta t)^{1/2} \tag{D.7}$$

Defining the relaxation time τ as the time when the angle variation is unity, we get:

$$\tau_r = (2D_\theta)^{-1} \tag{D.8}$$

For an isotropic distribution of angular momentum, the rotational diffusion constant can be related to the effective molecular radius a and a microscopic viscosity η as:

$$D_\theta = k_B T / 8\pi \eta a^3 \tag{D.9}$$

Combining this with (D.8), we get that:

$$\tau = \frac{4\pi \eta a^3}{k_B T} \tag{D.10}$$

Taking, for example, a spherical particle with radius a = 1 nm, and the medium with viscosity of $\eta = 0.1$ Pas, we get that the relaxation time in room temperature would be in the order of 0.1 µs, corresponding to relaxation frequency ω_0 in the range of 100 MHz.

If we applied a periodic field, $E = E_0 e^{i\omega t}$ to a dielectric material with a given relaxation time τ, the response of the medium was described by the frequency dependent susceptibility $\chi(\omega)$. In this case, P can be written as:

$$P = \chi(\omega)\varepsilon_o E_o e^{i\omega t} \tag{D.11}$$

Assuming that the induced polarization varies with the same frequency as the external field ($P(t) = P \cdot e^{i\omega t}$), and that (D.2) is valid at any instant, we get:

$$i\omega P = i\omega \chi(\omega)\varepsilon_o E = \frac{\chi(0)\varepsilon_o E - \chi(\omega)\varepsilon_o E}{\tau} \tag{D.12}$$

hence,

$$i\omega\tau\chi(\omega) = \chi(0) - \chi(\omega) \tag{D.13}$$

or

$$\chi(\omega) = \frac{\chi(0)}{1+i\omega\tau} \tag{D.14}$$

Splitting it in real and imaginary parts, we get:

$$\chi'(\omega) = \frac{\chi(0)}{1+\omega^2\tau^2}$$
$$\chi''(\omega) = \frac{\omega\tau\chi(0)}{1+\omega^2\tau^2} \tag{D.15}$$

This result was first published by Peter Debye in 1927,[1] and is called a Debye-type relaxation mechanism.

As generally true for periodic phenomena, a convenient way to characterize the frequency-dependent dielectric spectrum is through complex notation: ε' (real) = in-phase response; ε'' (imaginary) = out-of-phase response.

$$\varepsilon^*(\omega) = \varepsilon'(\omega) - i\varepsilon''(\omega) \tag{D.16}$$

The phase angle between the electric field and the field-induced polarization is $\tan^{-1}(\varepsilon''/\varepsilon')$. The out-of-phase component appears at higher frequencies, when the director or some part of it cannot follow the field.

Generally, the frequency dependencies of the real and imaginary components of the dielectric constants can be given as:

$$\varepsilon'(\omega) - \varepsilon'(\infty) = \frac{\varepsilon'(0) - \varepsilon'(\infty)}{1+\omega^2\tau^2}$$
$$\varepsilon''(\omega) = \frac{\omega\tau[\varepsilon'(0) - \varepsilon'(\infty)]}{1+\omega^2\tau^2} \tag{D.17}$$

where $\varepsilon'(0) - \varepsilon'(\infty) = \frac{N\mu^2}{3\varepsilon_0 k_B T}$ is the susceptibility of the given dipole relaxation mode, with N as the number density of dipoles with dipole moment μ. Such a behavior is schematically plotted in Figure D.1.

So far we have considered only one dielectric mode, but in reality there can be several modes (for example, due to internal dipole motion in flexible molecules or collective dipole motion). If these modes are sufficiently separated, it is possible to apply (D.17) to each relaxation processes.

These curves can be interpreted as each microscopic polarization effect is counteracted by viscous forces and, hence, cannot be driven at infinite speed.

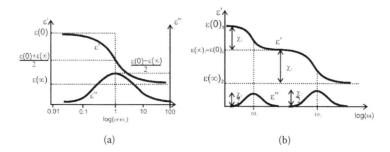

(a) (b)

FIGURE D.1
(a) Schematic plot of the real and imaginary parts of the complex permittivity as a function of relative frequency $\omega_o = 2\pi/\tau$. (b) Typical behavior of the real (upper graph) and imaginary (lower graph) components of the dielectric permittivity as a function of frequency, if two separated modes fall in the observation window. Note, for each mode a Debye analysis is performed.

If the field oscillates at too high frequencies, the polarization would not have time to build up before the driving field changes sign; consequently, these modes will not contribute to the dielectric constant. At a frequency (called relaxation frequency), the induced polarization decreases to half of its maximum, and the electric energy is absorbed and transformed to heat. For example, the dielectric relaxation time of water molecules is about 3×10^{-15} s, which corresponds to $\lambda \sim 10$ μm of the electromagnetic wave. This is the basis of the microwave heating of water-containing foods.

D.2 The Cole–Cole Plot

Eliminating the variable $\omega\tau$ from Eq. (D.17), we obtain:

$$\varepsilon''(\omega)^2 + \left\{\varepsilon'(\omega) - \frac{1}{2}[\varepsilon'(0) + \varepsilon'(\infty)]\right\}^2 = \frac{1}{4}[\varepsilon'(0) - \varepsilon'(\infty)]^2 \qquad (D.18)$$

With the transformation $\chi = \varepsilon'(0) - \varepsilon'(\infty)$, (D.18) reads:

$$\left(\varepsilon' - \left(\varepsilon'(\infty) + \frac{\chi}{2}\right)\right)^2 + \left(\varepsilon''\right)^2 = \left(\frac{\chi}{2}\right)^2 \qquad (D.19)$$

This is the equation of a circle of radius $\chi/2$ centered around the point $\varepsilon' + \chi/2, \varepsilon'' = 0$; therefore a Debye-type process should produce a semicircle (no negative ε'' values are possible).

The corresponding plot is called Cole–Cole plot after the scientists introduced it in 1941 (Figure D.2).[2]

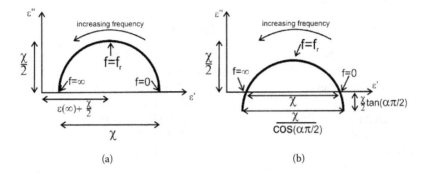

FIGURE D.2
The Cole–Cole plot for the case of one Debye-type process (a) and for one distributed (Cole–Cole) process.

Several pieces of quantitative information can be collected from this representation. Although it has no frequency axis, because ε' is monotonously decreasing with increasing frequency, we know that the higher is ε', the lower is the frequency. Since the maximum absorption occurs at the relaxation (or maximum absorption) frequency, we also know that the top of the semicircle corresponds to this frequency. The radius of the semicircle is $\chi/2$, and the graphs cuts the ε' axis at $\varepsilon'(\infty)$. The Cole–Cole plot is often used to characterize the nature of absorption and the susceptibility of the mode.

In real systems, when a symmetric distribution of relaxation times exists, the complex permittivity is described by the function:

$$\varepsilon^*(\omega) - \varepsilon'(\infty) = \frac{\varepsilon'(0) - \varepsilon'(\infty)}{1 + (i\omega\tau)^{1-\alpha}} \tag{D.20}$$

The parameter α is introduced by Cole and Cole. After some calculations one obtains:[3]

$$\left(\varepsilon' - \left(\varepsilon'(\infty) + \frac{\chi}{2}\right)\right)^2 + \left(\varepsilon'' + \frac{\chi}{2}\tan\left(\frac{\alpha\pi}{2}\right)\right)^2 = \left(\frac{\chi}{2\cos\left(\frac{\alpha\pi}{2}\right)}\right)^2 \tag{D.21}$$

The effect of α is to shift the center of semicircle below the abscissa by $\frac{\chi}{2}\tan(\alpha\pi/2)$.

For separable relaxations:

$$\varepsilon^*(\omega) - \varepsilon'(\infty) = \sum_j \frac{\chi_j}{1 + i\omega\tau} \tag{D.22}$$

where x_j are the weighting factors.

D.3 Dielectric Measurements Techniques

Nowadays, standard measurements are carried out on indium tin oxide (ITO)-coated cells, which are transparent to visible light, so allow simultaneous optical observations of the material alignment.[†] Thin ITO coating unfortunately has an absorption in the 1–10 MHz range, which decreases with decreasing cell thickness, so in thin films a low-resistive ITO is needed, which has reduced transparency. At higher than 10 MHz precise measurements are possible with gold or copper surfaces.

The measuring field strength should be in the linear regime, i.e., the field should not cause observable reorientation in the liquid crystal. In nematics, it means voltages below the Freedericks transition (see Chapter 4), which practically means U < 1 V. In every serious measurement, care should be taken to verify that the measurements are in the linear regime.

It is advised that the measurements be taken both in heating and in cooling, since in liquid crystals the behavior depends strongly on this condition (for example, in case of monotropic phases that appear only in cooling from the isotropic phase).

During dielectric measurements one basically measures the magnitude and the phase of the current flowing through the cell containing the dielectric material when a sinusoidal periodic voltage $V_0 e^{i\omega t}$ is applied.

The measured current determines the impedance Z of the sample as:

$$\frac{1}{Z} = \frac{I_0 e^{i(\omega t + \phi)}}{V_0 e^{i\omega t}} = \frac{I_0}{V_0} e^{i\phi} \tag{D.23}$$

Assuming that the material is a perfect insulator with a complex dielectric constant $\varepsilon^* = \varepsilon'(\omega) - i\varepsilon''(\omega)$, the inverse impedance $1/Z$ of a film with area A and thickness d is:

$$\frac{1}{Z} = iC\omega = \frac{A}{d}\varepsilon_0 \omega \varepsilon^* = \frac{A}{d}\varepsilon_0 \omega \cdot (i\varepsilon' + \varepsilon'') \tag{D.24}$$

The real part of $1/Z$ can be interpreted as $1/R$, where R is a resistance in parallel with their capacitance C. We note that, in reality, fluid materials are leaky dielectric with a finite *DC* resistance R_{DC}. This adds a contribution $\frac{1}{R_{DC}} = \sigma_{DC}\frac{A}{d}$ (σ_{DC} is the static Ohmic conductivity) to $1/R$, giving:

$$\mathrm{Re}(1/Z) = \frac{A}{d}[\sigma_{DC} + \varepsilon_0 \omega \varepsilon''] \tag{D.25}$$

[†] Metals become transparent at the plasma frequency ω_p, which depends on the carrier density of the conductor. Most metals are opaque for visible light, but ITO has a carrier density ~1%, and ω_p is in the near infrared range, thus is transparent in the visible range. [See J.D. Livingston, *Electronic Properties of Engineering Materials*, Wiley, MIT Series in Materials Science & Engineering (1999.)]

It is important to emphasize that the terms due to the Ohmic conductivity appear as a contribution to the dielectric loss, which decays at low frequency with $1/\omega$. Fitting the calculated ε'' curve with a $1/\omega$ function, therefore, one can calculate the Ohmic conductivity. Both the dielectric and Ohmic loss contribute to the dissipation, which is defined as

$$D \equiv \frac{\text{Re}(I)}{\text{Im}(I)} = \frac{\varepsilon''(\omega)}{\varepsilon'(\omega)}.$$

As we illustrate in Figure D.3, usually real measurement cells are quite complicated, especially those that contain polymer alignment layers, which are characterized with a resistance R_p and capacitance C_p in parallel to the material under study. In addition, the ITO electrodes and the wires going from the electrode to the instrument have a finite resistance r_e, which are in series with the electric circuit of the polymer layers and the studied material.

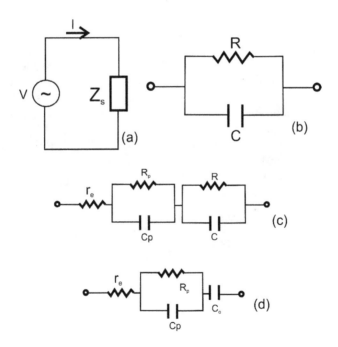

FIGURE D.3
Illustration of the principle of the impedance analyzer (a) and the equivalent electric circuits (b) of a leaky dielectric film (R is the resistance, and C is the capacitance of the material studied). The equivalent electric circuit of a real sample containing the electrodes with resistance r_e, polymer alignment layers with resistance R_p and capacitance C_p (c); the equivalent circuit of the empty cell with capacitance C_o (d).

In this case, the impedance of the sample becomes much more complicated and is described as:

$$Z_s = r_e + \frac{R_p}{1+(\omega C_p R_p)^2} + \frac{R}{1+(\omega CR)^2} - i\left\{ \frac{\omega C_p R_p^{\,2}}{1+(\omega C_p R_p)^2} + \frac{\omega CR^2}{1+(\omega CR)^2} \right\} \quad (D.26)$$

These unknown parameters can be calculated if one measures the empty cell, before filling it with the material to be studied. The imaginary and the real parts of impedance of the empty cell Z_e can be given as:

$$\mathrm{Im}(Z_e) = \frac{\omega C_p R_p^{\,2}}{1+(\omega C_p R_p)^2} + \frac{1}{\omega C_o} \quad (D.27)$$

and

$$\mathrm{Re}(Z_e) = r_e + \frac{R_p}{1+(\omega C_p R_p)^2} \quad (D.28)$$

We see from these expressions that the capacitance of the empty cell (more precisely filled with air, which has a dielectric constant very close to 1), $C_o = \varepsilon_o \frac{A}{d}$, can be determined from the value of $Im(Z_e)$ measured at low frequencies, when the contribution of the alignment layer is negligible. The electrode resistance is basically given by the $Re(Z_e)$ at high frequencies (~1 MHz), when the contribution of the alignment layer becomes negligible. The polymer resistance is determined from $Re(Z_e)$ at very low frequencies, and finally, C_p is determined from $Im(Z_e)$ measured at high frequencies with the known value of C_o.

After determining the above values, from (D.26), (D.27) and (D.28), we can calculate both R and C using the real and imaginary impedance values measured in the empty and filled cells, as[†]

$$\mathrm{Re}(Z_s) - \mathrm{Re}(Z_e) = \frac{R}{1+(\omega CR)^2} = \alpha \quad (D.29)$$

and

$$\mathrm{Im}(Z_s) - \mathrm{Im}(Z_e) + \frac{1}{\omega C_o} = \frac{\omega CR^2}{1+(\omega CR)^2} = \beta \quad (D.30)$$

From these equations, one can get:

$$\frac{1}{R} = \frac{\alpha}{\alpha^2 + \beta^2} \quad (D.31)$$

[†] We note that, in practice, more precise curves are obtained by using the fitted curves for the empty cells instead of the row data.

and

$$\omega C = \frac{\beta}{\alpha^2 + \beta^2} \tag{D.32}$$

which with (D.24) provide the frequency dependence of the complex dielectric constant of the material as:

$$\varepsilon'(\omega) = \frac{\beta}{\omega C_o (\alpha^2 + \beta^2)} \tag{D.33}$$

and

$$\varepsilon''(\omega) = \frac{\alpha}{\omega C_o (\alpha^2 + \beta^2)} \tag{D.34}$$

The commonly used LCR meters offer quick and easily computerized measurements in many frequencies, usually up to about 10 MHz. At higher frequencies, up to about 2 GHz, microwave technologies are needed. At even higher frequencies, which would test the displacement polarization, absorption spectroscopy (IR: intramolecular reorientation; UV-VIS: electronic polarization) techniques are used.

References

1. P. Debye, *Polar Molecules*, Chemical Catalogue Co., New York (1927), also available as a Dover Reprint, New York (1945).
2. K.S. Cole, R.H. Cole, *J. Chem. Phys.*, 9, 341 (1941).
3. K.S. Cole, R.H. Cole, *J. Chem. Phys.*, 10, 98 (1942).